Advances in
PARASITOLOGY

VOLUME 30

Advances in PARASITOLOGY

Edited by

J. R. BAKER
Cambridge, England

and

R. MULLER
International Institute of Parasitology
St Albans, England

VOLUME 30

ACADEMIC PRESS

Harcourt Brace Jovanovich, Publishers
London San Diego New York Boston
Sydney Tokyo Toronto

ACADEMIC PRESS LIMITED
24/28 Oval Road
LONDON NW1 7DX

United States Edition published by
ACADEMIC PRESS INC.
San Diego CA 92101

A catalogue record for this book is available from the British Library

ISBN 0–12–031730–3

Filmset by Bath Typesetting Ltd., London Road, Bath
Printed in Great Britain by Galliard (Printers) Ltd., Great Yarmouth

CONTRIBUTORS TO VOLUME 30

J. Barrett, *Department of Biological Sciences, University College of Wales, Aberystwyth, UK*

J. R. Barta, *Department of Pathology, Ontario Veterinary College, University of Guelph, Guelph, Ontario, Canada N1G 2W1*

B. Fried, *Department of Biology, Lafayette College, Easton, Pennsylvania 18042, USA.*

E. G. Garcia, *College of Public Health, University of the Philippines, Manila, Ermita 1000, Philippines*

R. A. Khan, *Department of Biology and Ocean Sciences Centre, Memorial University of Newfoundland, St John's, Newfoundland, Canada A1C 5S7*

G. F. Mitchell, *The Walter and Eliza Hall Institute of Medical Research, Melbourne, Victoria 3050, Australia*

L. T. Stableford, *Department of Biology, Lafayette College, Easton, Pennsylvania 18042, USA*

J. Thulin, *The National Environmental Protection Board, Marine Section, Box 584, S-74071, Öregrund, Sweden*

W. U. Tiu, *College of Public Health, University of the Philippines, Manila, Ermita 1000, Philippines*

PREFACE

This volume starts with a comprehensive review of an interesting but little-studied group of protists, the Dactylosomatidae, by Dr John Barta. Dr Barta's study, based largely on work done by himself and his colleagues at the University of Toronto and elsewhere, shows that the family should contain only the two genera *Babesiosoma* and *Dactylosoma*, and he reviews the validity of the species which have been ascribed to these two genera. The life-cycles of a representative species in each genus are also described as fully as possible, and illustrated by a series of excellent micrographs. Dr Barta concludes that these parasites are biologically intermediate between the more common genera of Adeleidae, such as *Haemogregarina* and *Karyolysus*, and the piroplasms.

While there have been numerous studies on the total amino acid composition of helminths, particularly cestodes, the metabolism of amino acids has been relatively neglected. Dr John Barrett provides a comprehensive review of the synthesis and catabolism of these major constituents of biological material. He points out important differences in the metabolism of helminths and mammals which may be exploitable in the search for new anthelmintic drugs.

The chick embryo is widely used in biological and biomedical studies and has proved to be the most fruitful ectopic site for the cultivation of helminths, particularly digenetic trematodes, outside the normal host. Drs Bernard Fried and Louis Stableford have reviewed extensively the biology, physiology and development of helminths on the chorioallantoic membrane and on other chick extra-embryonic sites, and have also related these to what is known of the structure, physiology and biochemistry of the chick embryo and its extra-embryonic membranes. Fertile hens' eggs provide a cheap, sterile, easily obtained substrate and studies require few facilities. Most studies have been with avian digeneans with progenetic metacercariae and in some cases fertile adults have been obtained of species for which the definitive hosts are not available.

In recent years, inbred mouse strains have proved to be useful laboratory models for studying the immunobiological reactions of schistosomes. Most studies have utilized *Schistosoma mansoni*, but the present review by Drs Mitchell, Tiu and Garcia considers infection with the much less worked-on species *S. japonicum*; the authors demonstrate that there are important differences between the two species and probably also between two strains of *S. japonicum*. They believe that analyses of immune responses of *S. japonicum* in the mouse model may well be more useful with regard to vaccines to prevent disease, such as by the prevention of embryonation, than in the primary identification of resistance to infection.

The volume then ends with a very topical contribution which cuts across the conventional boundary between "protozoology" and "helminthology", by Drs Rasul Khan and Jan Thulin, on the effects of marine pollution on the parasites of marine animals and the potential usefulness of these parasites as indicators of the extent of pollution. Most of the limited evidence available at present suggests that pollution may serve to increase the hosts' susceptibility to parasites and, perhaps, the severity of the pathological effects arising from parasitism, and also that parasitism may increase the susceptibility of the hosts to the pollutants to which they are increasingly likely to be exposed.

J. R. Baker
R. Muller

CONTENTS

Infection Characteristics of *Schistosoma japonicum* in Mice and Relevance to the Assessment of Schistosome Vaccines

G. F. MITCHELL, W. U. TIU AND E. G. GARCIA

Influence of Pollution on Parasites of Aquatic Animals

R. A. KHAN AND J. THULIN

The Dactylosomatidae

JOHN ROBERT BARTA

Department of Pathology, Ontario Veterinary College, University of Guelph, Guelph, Ontario, Canada N1G 2W1

I. Introduction

Members of the family Dactylosomatidae have remained until recently some of the most poorly understood members of the phylum Apicomplexa, despite the long period that has elapsed since their discovery. These intra-erythrocytic parasites of cold-blooded vertebrates were first described from amphibians in the last century. The observation of *Dactylosoma ranarum* (Lankester, 1882). Wenyon, 1926 by Lankester in 1871 is said to have generated considerable academic excitement (Levine, 1971).

Several related species were described from fishes and amphibians throughout the first half of this century. The family Dactylosomatidae Jakowska and Nigrelli, 1955 was erected to recognize this group of "Babesioidea in erythrocytes of cold-blooded vertebrates [where] four to 16 merozoites are produced" (Jakowska and Nigrelli, 1955). A year later, these authors created the genus *Babesiosoma* for the two described dactylosomatid parasites which produced only four merozoites; the five species producing more than four merozoites during intra-erythrocytic replication remained in the genus *Dactylosoma* Labbé, 1894 (see Jakowska and Nigrelli, 1956).

ADVANCES IN PARASITOLOGY VOL. 30
ISBN 0-12-031730-3

From 1956 to the present, three additional species of *Dactylosoma* and ten new species of *Babesiosoma* have been described. Despite the extended period over which species descriptions were made, virtually nothing was known concerning the biology of these parasites beyond light microscopic observations of the intra-erythrocytic stages in the vertebrate hosts. In this review of the family Dactylosomatidae, recent observations on the sporogonic development of *D. ranarum* in an experimental leech host and detailed observations on the life-history of *B. stableri* in the laboratory and in the field will be used to illustrate the biological relationships of the dactylosomatid parasites with their intermediate vertebrate hosts and definitive annelid hosts. These observations have clarified the relationships between the dactylosomatid genera *Dactylosoma* and *Babesiosoma*, and have more clearly defined the taxonomic status of species in the family Dactylosomatidae and their relationships with other Apicomplexa.

II. Taxonomic Considerations

The taxonomic affinities of the family Dactylosomatidae have been uncertain since the family was erected. Historically, these organisms have been included with the non-pigmented haemosporinids (Jakowska and Nigrelli, 1956; Manwell, 1964; Misra and Nigrelli, 1973) or piroplasms (Poisson, 1953; Kudo, 1966; Levine, 1971, 1985). The observations on the sporogonic development of *B. stableri* and *D. ranarum* indicate that the inclusion of the family Dactylosomatidae in the order Eucoccidiida by Boulard *et al.* (1982) was justified. The sporogony and subsequent merogonic replication of *B. stableri* in its definitive invertebrate host and vector are unique in the order (see Levine, 1985). This development exhibits characteristics intermediate between that of adeleid blood parasites of the genera *Haemogregarina*, *Cyrilia* and *Karyolysus*, and that of the piroplasms, *Babesia* and *Theileria* species (see Barta and Desser, 1989, for a discussion). The presence of a conoid and the general structural similarities of the gamonts to those of the haemogregarines (Desser and Weller, 1973; Paterson *et al.*, 1988) indicate that *B. stableri* is most closely related to the adeleids.

A phylogenetic analysis of the class Sporozoea (see Barta, 1989) demonstrated that the piroplasms and adeleids (as currently classified by Levine, 1985) were a monophyletic group excluding the eimeriids and haemosporinids. The blood-dwelling haemosporinids and piroplasms, despite some morphological and biological similarities (Vivier, 1979, 1982; Mehlhorn *et al.*, 1980), do not apparently form a monophyletic group within the phylum Apicomplexa. Therefore, the dactylosomatids represent a relatively advanced group of haemogregarine-like adeleid parasites.

A. GENERA IN THE FAMILY DACTYLOSOMATIDAE

The generic composition of the family Dactylosomatidae has been the subject of debate. The synonymy of the genus *Babesiosoma* with the genus *Haemohormidium* Henry, 1910, suggested by Mackerras and Mackerras (1961) and Laird and Bullock (1969) and supported by Levine (1971, 1984, 1988), has been strongly opposed (Becker, 1970; Misra and Nigrelli, 1973; Chaudhuri and Choudhury, 1983; Barta and Desser, 1989). An examination of fish blood infected with *H. cotti* and *H. beckeri* has convinced me that the genera *Haemohormidium* and *Babesiosoma* are not synonymous (see Figs 1–3, 5–10). *Haemohormidium* species are characterized by their small size and pleomorphism (Henry, 1910, 1913). The parasites undergo replication within erythrocytes and some tetranucleate meronts are formed. However, the structural consistency and number of progeny formed during each replicative cycle exhibited by species of the genus *Babesiosoma* is not evident in *Haemohormidium* species. The lack of close affinities between these genera was supported by Davies (1980), who examined *H. cotti* ultrastructurally. She was able to demonstrate that this organism is "piroplasm-like", but was unable to determine its taxonomic affinities more closely. Khan (1980) has demonstrated the biological transmission of a *Haemohormidium* species by a piscicolid leech.

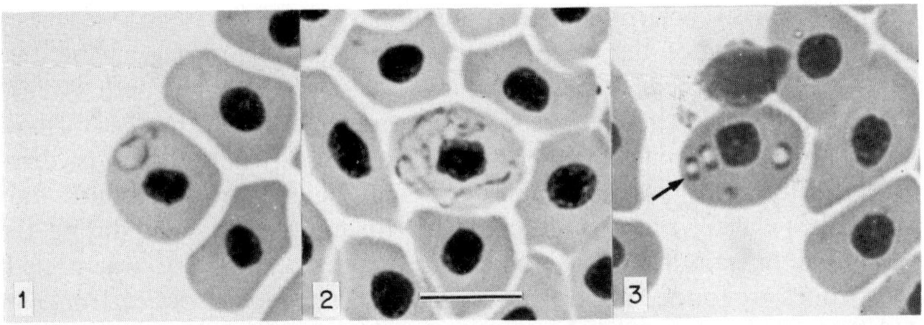

FIGS 1–3. Light micrographs of *Haemohormidium cotti* Henry, 1910 within the erythrocytes of *Cottus bubalis*. Note the pleomorphic outline of the organisms (Figs 1 and 2) and predominantly binucleate replicative stages (Fig. 3, arrowed). Bar = 10 μm.

In the absence of this synonymy, the family Dactylosomatidae contains only the general *Dactylosoma* Labbé, 1894 and *Babesiosoma* Jakowska and Nigrelli, 1956, as originally proposed by Jakowska and Nigrelli (1956). The genera *Haemohormidium* Henry, 1910 and *Sauroplasma* Du Toit, 1938, which had been included in the family Dactylosomatidae by Levine (1971), have been moved to the family Haemohormidiidae Levine, 1984. Levine (1988) erroneously included the genus *Schellackia* Reichenow, 1919 in the family Dactylosomatidae; this genus belongs in the eimeriorine family Lankesterellidae Nöller, 1920.

The criteria proposed by Jakowska and Nigrelli (1956) to separate *Babesiosoma* species from *Dactylosoma* species were: "(1) a less granular and more vacuolated cytoplasm in all stages; (2) a *Babesia*-like nucleus, without a definite karyosome; (3) reproduction by typical schizogony, binary fission, or budding; and (4) development of not more than four merozoites, usually arranged in a rosette or cross-shaped". Of these criteria, clearly the "less granular cytoplasm" which was "more vacuolated" in all stages when compared with *Dactylosoma* species at the light microscope level reflects the presence of considerably fewer amylopectin inclusions in the cytoplasm of the latter species than in *Babesiosoma* species. A "karyosome" in the nuclei of *Babesiosoma* species, although obscured at the magnification available using light microscopy, is readily apparent when the parasite is examined ultrastructurally (Barta and Desser, 1986). The interpretation of "binary fission" in *B. jahni* by Nigrelli (1929, 1930) was apparently erroneous. Later, Jakowska and Nigrelli (1956) stated that typical cruciform meronts, producing four merozoites without a residual mass, were most commonly encountered. The observation that merozoites of *Babesiosoma* species may remain closely associated in the host cell after being fully formed (Paperna, 1981; Barta and Desser, 1986) suggests that paired mature merozoites may have been the stages reported as undergoing "binary fission".

Despite the facts discussed above, continued separation of the two dactylosomatid genera, *Babesiosoma* and *Dactylosoma*, recognized by Jakowska and Nigrelli (1956), appears to be warranted. All of the *Dactylosoma* species examined produce higher numbers of progeny during intra-erythrocytic replication than do *Babesiosoma* species; the latter produce four merozoites during each merogonic cycle (in either host) and during the formation of gamonts. This number is doubled in oocysts, where eight sporozoites are formed during sporogony. In *Dactylosoma* species, from 6 to 16 merozoites are formed during merogonic replication in the vertebrate hosts (Jakowska and Nigrelli, 1956). Likewise, this number doubles in the oocysts of *D. ranarum*, where it is evident that the number of sporozoites formed in each oocyst (ca. 30) is about double the number of merozoites formed in each meront (6–16). This characteristic justifies the separation of the two dactylosomatid genera.

1. Babesiosoma *species*

Species of the genus *Babesiosoma* have been described from a wide variety of hosts including frogs, toads, newts and fish. All of the *Babesiosoma* species recognized herein have been found in hosts which have contact with water, indicating that the definitive hosts and vectors for all *Babesiosoma* species may be haematophagous leeches. The *Babesiosoma* species described have a cosmopolitan distribution in both marine and freshwater environments. (See Table 1 for a summary of the species described, their hosts and the localities from which they have been reported. Deviations from the nomenclature or taxonomic placements used by the describing authors are discussed below.)

Dactylosoma hannesi Paperna, 1981 was described from the blood of mullets (Mugilidae) from South Africa. Despite acknowledged similarities between this organism and *Babesiosoma* species, Paperna (1981) concluded that the bilobate nuclei located in the developing meronts and within the resulting four merozoites of this parasite were pairs of nuclei. Thus, eight merozoites were believed to be formed by a process of asymmetrical cytoplasmic fragmentation. Paperna's (1981) inability to find any infected cells containing the expected eight merozoites, and his illustrations and photomicrographs, clearly indicate that *D. hannesi* is a *Babesiosoma* species. The range of morphological variation exhibited by *D. hannesi* is within that demonstrated by other *Babesiosoma* species such as *B. stableri* (see Schmittner and McGhee, 1961) and *B. jahni* (see Nigrelli, 1930; Jakowska and Nigrelli, 1956). Thus, this parasite from grey mullets in South Africa should be named *Babesiosoma hannesi* (Paperna, 1981) n. comb.

Four species descriptions of doubtful validity from India are listed in Table 1. The first involves the description of *B. hareni* Haldar, Misra and Chakravarty, 1971 from fish, *Ophicephalus punctatus*, collected near Calcutta, India. These same authors (Misra *et al.*, 1969) had previously described *B. ophicephali* Misra, Haldar and Chakravarty, 1969 from the same host collected at the same locality. The second questionable report is that of Chaudhuri and Choudhury (1983), who described *B. batrachi* from *Clarias batrachus*, again from the same area. Two additional dactylosomatid species from fishes of India have evidently been assigned incorrectly to the genus *Dactylosoma*. *"D." striata* Sarkar and Haldar, 1979 and *"D." notopterae* Kundu and Haldar, 1984 were described from the erythrocytes of freshwater teleosts. In both descriptions, "tetra to octanucleate schizonts" are described which have nuclei arranged in two clusters of four each; the description of *D. notopterae* also includes a description of "nuclei arranged in four clusters of two each" (Kundu and Haldar, 1984). In neither description were infected erythrocytes containing eight merozoites observed and, in both cases, the authors provided illustrations that suggest the organisms more closely resemble *Babesiosoma* species than *Dactylosoma* species. Thus, as many as five *Babesiosoma* species have been described from freshwater teleosts obtained from West Bengal, India since 1969.

The morphometric and morphological criteria upon which these five species have been separated are not convincing (see Chaudhuri and

TABLE 1 Babesiosoma *species currently recognized and their host(s)*

Parasite	*Host(s)*	*Distribution*	*Reference(s)*
Babesiosoma aulopi	*Aulopus purpurissatus*, *Parma microlepis*	Sydney, Australia	Mackerras and Mackerras (1925), Laird and Bullock (1969)
B. jahni	*Triturus viridescens*	Pennsylvania, USA	Nigrelli (1929)
B. mariae	*Haplochromis nubilus*, *H. cinereus*, *H. serranus*, *H.* sp.	Lake Victoria, Uganda	Hoare (1930)
	Tilapia esculenta *T. variabilis*, *T. nilotica*, *Labeo victorianus*, *Astatoreochromis alluaudi*	Lake George, Uganda	Baker (1960)
B. rubrimarensis	*Lethrinus xanthochilus*, *L. variegatus*, *Cephalopholis miniatus*, *C. hemistictus*, *Scarus harid*, *Mugil troscheli*, *Epinephelus summana*, *Cheilinus diagrammus*	Red Sea	Saunders (1960)
B. stableri	*Rana pipiens pipiens*, *R. p. sphenocephala*, *R. catesbeiana*, *Bufo americanus*, *B. woodhousei*, *B. terrestris*	Wisconsin, USA	Schmittner and McGhee (1961)
	Rana septentrionalis, *R. catesbeiana*, *R. clamitans*	Ontario, Canada	Barta and Desser (1984)
B. tetragonis	*Catostomus* sp.	California, USA	Becker and Katz (1965)
B. hannesi n. comb. syn. *Dactylosoma hannesi*	*Mugil cephalus* *Liza richardsoni* *L. dumerili*	South Africa	Paperna (1981)
B. ophicephali[a]	*Ophicephalus punctatus*	West Bengal, India	Misra *et al.* (1969)
syn. *B. hareni*	*Ophicephalus punctatus*	West Bengal, India	Haldar *et al.* (1971)
syn. *B. batrachi*	*Clarias batrachus*	West Bengal, India	Chaudhuri and Choudhury (1983)
syn. *D. striata*	*Ophicephalus striatus*	West Bengal, India	Sarkar and Haldar (1979)
syn. *D. notopterae*	*Notopterus notopterus*	West Bengal, India	Kundu and Haldar (1984)
syn.(?) *Dactylosoma* sp.	*Mystus vittatus*	India	Mandal (1979)

TABLE 1 (*continued*)

Parasite	*Host(s)*	*Distribution*	*Reference(s)*
B. anseris[b]	*Cygnopsis cygnoides*	Egypt	Haiba and El-Shabrawy (1967)
B. gallinarum[b]	*Gallus gallus*	Egypt	Fahmy *et al.* (1979)
B. ptyodactyli[b]	*Ptyodactylus hasselquitii*	Egypt	El-Naffer *et al.* (1979)

[a] See discussion in text.
[b] Species probably related to the piroplasms (see discussion in text).

Choudhury, 1983 and Kundu and Haldar, 1984 for summaries of criteria). The morphometric and morphological differences noted between primary and secondary merogonic stages of *B. stableri* during experimental infections in a single animal (see Barta and Desser, 1986) are as great as the differences noted among the five dactylosomatid species reported from fish in India. The lack of feeding specificity (see Sawyer, 1986; Oosthuizen, 1989; Jones and Woo, 1990) exhibited by most rhynchobdellid leeches (which probably transmit all parasites of the genus *Babesiosoma*) would be expected to allow the exchange of parasites among these fish. Reliance on strict host specificity for the vertebrate host as a means of distinguishing *Babesiosoma* species does not appear warranted (see Table 1), and the morphometric descriptions of these species overlap sufficiently to make measurements unsuitable for diagnostic purposes; there remains no reliable criterion by which to separate these species. Therefore, *B. hareni* Haldar, Misra and Chakravarty, 1971, *D. striata* Sarkar and Haldar, 1979, *B. batrachi* Chaudhuri and Choudhury, 1983 and *D. notopterae* Kundu and Haldar, 1984 should all be considered junior synonyms of *B. ophicephali* Misra, Haldar and Chakravarty, 1969 (see Table 1) until definitive cross-transmission experiments have been conducted to justify any further specific designations.

Finally, there are three additional parasites described as *Babesiosoma* species which demonstrate few, if any, similarities with other dactylosomatid parasites. One example is *B. anseris* Haiba and El Shabrawy, 1967, found infecting the erythrocytes of the goose, *Cygnopsis cygnoides*, in Egypt. The description of this remarkably pleomorphic organism indicates that the parasite undergoes binary fission and "schizogony" forming two, four or six organisms. Clearly, this parasite is not a member of the genus *Babesiosoma* based on the description of the genus by Jakowska and Nigrelli (1956). The organism observed in the blood of the geese more closely resembles a piroplasm, possibly a species of *Babesia* (see Levine, 1971). For the same reasons, *B. gallinarum* and *B. ptyodactyli* are considered unlikely to belong to the genus *Babesiosoma* and are probably piroplasms, perhaps related to the genera *Sauroplasma* Du Toit, 1938 or *Haemohormidium* Henry, 1910, which are members of the family Haemohormidiidae Levine, 1984.

2. Dactylosoma *species*

Described species in the genus *Dactylosoma*, their intermediate vertebrate hosts and localities, are listed in Table 2. Included in this list are two species of doubtful affinity with the family Dactylosomatidae. *D. amaniae* (syn. *Lankesterella amania* Awerinzew, 1914) in *Chamaeleon fischeri* has never been reported a second time and the organisms described by Awerinzew (1914) are of ill-defined affinities. Awerinzew (1914) described tiny spherical bodies in which approximately six nuclei arise and from which six organisms are formed by fragmentation of the parent cell. Parasites that could be interpreted as gamonts were not described in this report. These observations suggest that this species may not be a member of the Dactylosomatidae. Fantham (1905) described a similar organism, *Dactylosoma tritonis* (syn. *Lankesterella tritonis*), from the blood of *Triton cristatus*, which does not have characteristics that correspond to those of the type species of the genus *Dactylosoma*. Levine (1988) has suggested that "*D.*" *amaniae* may actually be inclusions produced by rickettsial organisms, perhaps of the genus *Aegyptianella*; thus, both parasites may not belong to the Dactylosomatidae.

III. Biology of the Dactylosomatid Parasites

The biology of members of the family Dactylosomatidae will be examined using as an example from each of the two genera, *B. stableri* Schmittner and McGhee, 1961 and *D. ranarum* (Lankester, 1882) Wenyon, 1926.

A. *BABESIOSOMA STABLERI*

1. *Life-cycle*

The life-cycle of *B. stableri* Schmittner and McGhee, 1961 is known in detail; the remaining *Babesiosoma* species have been observed only within the erythrocytes of their intermediate vertebrate hosts. However, well-characterized *Babesiosoma* species share numerous morphological similarities. In addition, the vertebrate hosts of all dactylosomatid parasites are closely associated with aquatic environments, suggesting that leeches may act as vectors for all these parasites (Manwell, 1964; Paperna, 1981; Barta and Desser, 1989). Therefore, the life-cycles of all recognized *Babesiosoma* species may be similar to that depicted in Fig. 4.

TABLE 2 Dactylosoma *species currently recognized and their host(s)*

Parasite	*Host(s)*	*Distribution*	*Reference(s)*
Dactylosoma ranarum	*Rana esculenta*, *Rana temporia*,	Europe	Lankester (1871, 1882), Kruse (1890), Labbé (1894),
	Triton sp.		Laveran (1898),
	frogs		Hintze (1902)
	toads	Brazil	Durham (1902)
	frogs	Tunis, North Africa	Billet (1904)
	Rappia marmorata, *Rana galemensis*, *R. oxyrhynchus*, *R. macarensis*, *Bufo regularis*	The Gambia, Africa	Dutton *et al.* (1907)
	frogs	Caucasus	Finkelstein (1908)
		Europe	França (1908)
	Rana güntheri	Tonkin, Indochina	Mathis and Léger (1912)
	Rana esculenta	Europe	Nöller (1913)
	Bufo regularis	Transvaal, South Africa	Fantham *et al.* (1942)
	Bufo marinus	Central and South America	Walton (1946, 1947, 1948, 1949, 1950)
	Rana albilabris	Africa	
	R. labialis	Siam	
	R. ridibunda	North Africa	
	R. tigerina	Burma, Ceylon and India	
	B. marinus	Costa Rica	Ruiz (1959)
	R. nigromaculata	Japan	Tanabe (1931)
D. salvelini	*Salvelinus fontinalis*	Eastern Canada	Fantham *et al.* (1942)
D. sylvatica	*Rana sylvatica*	Eastern Canada	Fantham *et al.* (1942)
D. lethrinorum	*Lethrinus nebulosus*, *L. mahsenoides*	Red Sea	Saunders (1960)
D. taiwanensis	*Rana limnocharis*	Taiwan	Manwell (1964)
D. tritonis[a]	*Triton cristatus*	Europe	Fantham (1905)
D. amaniae[a]	*Chameleon fischeri*	West Africa	Awerinzew (1914)

[a] Possibly inclusions containing rickettsial organisms related to the genus *Aegyptianella* according to Levine (1988).

B. stableri infects erythrocytes of ranid frogs and toads of eastern North America. This parasite is transmitted between frogs by a glossiphoniid leech, *Desserobdella picta* (Verrill, 1872) Barta and Sawyer, 1990, which acts as the definitive host for the parasite. The leech feeds exclusively on amphibians throughout its eastern North American range but is otherwise not host-specific (Barta and Sawyer, 1990). *Dess. picta* is known to feed on

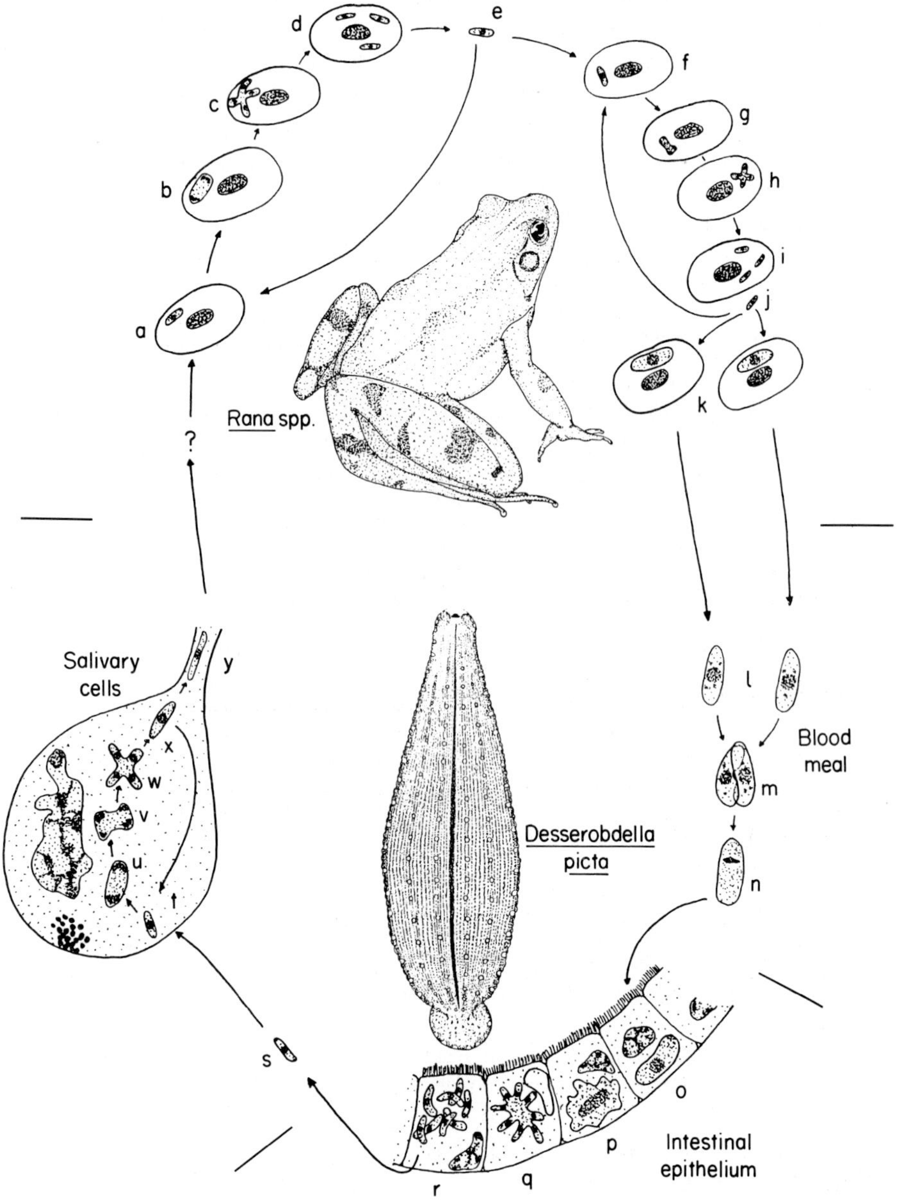
a
b
c
d
e
f
g
h
i
j
k
Rana spp.
?
Salivary cells
y
x
w
v
u
t
l
Blood meal
m
n
Desserobdella picta
o
p
q
r
s
Intestinal epithelium

common newts, *Trituris viridescens*, and therefore may also be involved in the transmission of *B. jahni*, which parasitizes these animals. The likelihood that *B. stableri* and *B. jahni* share definitive hosts and the reported lack of host specificity of *B. stableri* for its intermediate vertebrate hosts (Schmittner and McGhee, 1961; Barta and Desser, 1984) suggest that these parasites may be closely related. Attempts at transmitting *B. stableri* to newts by intraperitoneal injection of infected frog blood have been unsuccessful.

Other glossiphoniid leeches, such as *Dess. phalera* (Graf, 1899) Jones and Woo, 1990, which are known to feed on fishes, may be involved in the transmission of related dactylosomatid parasites infecting fishes of North America, such as *D. salvelini* of trout or *B. tetragonis* of suckers. *B. mariae*, which has been found in the blood of a wide variety of African freshwater fishes, is probably transmitted by another glossiphoniid leech, *Batracobdelloides tricarinata* (Blanchard, 1897) Oosthuizen, 1989, which is a common haematophagous ectoparasite of these fishes in southern Africa (Oosthuizen, 1989). Baker (1960) identified this leech as an ectoparasite on the fishes he examined from Lake Victoria. *Batracobdelloides* species are also found throughout the Indian subcontinent (Sawyer, 1986); therefore, leeches of this genus may be involved in the transmission of *B. ophicephali* in Indian fishes as well.

FIG. 4. Summary of the life-cycle of *Babesiosoma stableri* in frogs (*Rana* spp.) and the leech *Desserobdella picta*. Primary merogony (a–e) is initiated when a merozoite enters an erythrocyte of a frog (a). The merozoite undergoes a nuclear division forming a binucleate meront (b), which subsequently undergoes a second nuclear division to form a tetranucleate meront (c). This intra-erythrocytic, cruciform meront is typical of the genus *Babesiosoma*. Four merozoites bud simultaneously from the surface of the meront and separate (d) without the formation of a residual body. Merozoites resulting from primary merogony (e) penetrate other erythrocytes and may either repeat a cycle of primary merogonic replication (a–e) or initiate secondary merogony (f–j). A secondary merozoite (f) forms a binucleate meront (g) and cruciform, tetranucleate meront (h) in the same manner as occurs during primary merogony. Merozoites resulting from secondary merogony (i) leave the host cell (j) and penetrate other erythrocytes. These merozoites may either repeat a cycle of secondary merogonic replication (f–j) or mature into gamonts (k).

When ingested by a leech, gamonts are freed from the frog erythrocytes (l) and associates in syzygy (m) within the blood meal in the crop of the leech. The paired gamonts mature into gametes and fuse forming an ookinete (n), which penetrates intestinal epithelial cells to initiate sporogony (o–r). The young oocyst (o) enlarges (p), and then forms eight sporozoites by a process of peripheral budding (q). Mature sporozoites (r) leave the intestinal epithelium (s) and make their way to the salivary cells of the leech (t). Therein, they undergo merogonic replication similar to that which occurs in the frog erythrocytes (t–x). The resulting merozoites (x) may repeat a cycle of merogonic replication (t–x) or enter the ductules of the salivary cells (y). When the leech next feeds, the merozoites are injected into the frog. Pre-erythrocytic development (?) may or may not occur before merozoites re-enter erythrocytes (a).

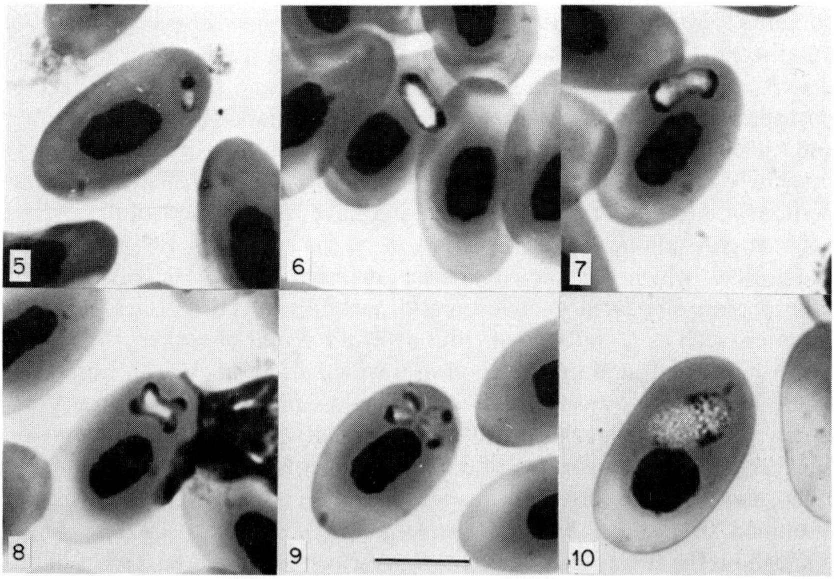

FIGS 5–10. Light micrographs of *Babesiosoma stableri*. Bar = 10 μm. Fig. 5. Merozoite. Fig. 6. Binucleate meront in which the pair of nuclei have migrated to opposite ends of the oblong organism. Fig. 7. Late binucleate meront in which the nuclei at each pole are initiating the second mitotic division. Fig. 8. Early development of the merozoites in a tetranucleate meront. Fig. 9. Advanced cruciform meront typical of the genus *Babesiosoma*. These four merozoites separate without the formation of a residual body. Fig. 10. Large mononucleate gamont that has abundant amylopectin granules visible as numerous unstained spots in the cytoplasm of the parasite. *B. stableri* is isogamous. (Figs 5, 6 and 8–10 from Barta and Desser, 1986. Reproduced with the permission of the Society of Protozoologists.)

2. *Intra-erythrocytic development*

Primary merogony of *B. stableri* is initiated when a merozoite enters a frog erythrocyte. The merozoite undergoes a nuclear division forming a binucleate meront that subsequently undergoes a second nuclear division to form a tetranucleate cruciform meront typical of the genus *Babesiosoma*. Four merozoites bud simultaneously from its surface and separate. The progeny leave the host cell without causing its lysis. Merozoites resulting from primary merogonic replication penetrate other erythrocytes and may either repeat a cycle of primary merogony or initiate secondary merogony (Figs 5–9). Unlike *D. ranarum*, which produces more numerous merozoites during primary merogony than secondary merogony, stages involved in the secondary merogonic development of *B. stableri* are significantly smaller and more densely stained than comparable stages in the primary merogonic cycle (Barta and Desser, 1986). Apart from size and staining characteristics, secondary merozoites form binucleate meronts and cruciform, tetranucleate meronts in the same manner as occurs during primary merogony. Progeny produced during secondary merogony leave their host cell and penetrate other erythrocytes, wherein they either begin another cycle of secondary merogonic replication or mature into gamonts (Fig. 10). Both merozoites and gamonts are highly motile, both within the cytoplasm of erythrocytes and free in the plasma. Gamonts of *B. stableri* can readily penetrate erythrocytes and exit again with no apparent deleterious effect on the host cell.

Ultrastructurally, intra-erythrocytic stages of *B. stableri* have typical apicomplexan features including about three rhoptries, numerous micronemes, a trilaminate pellicle, two pre-conoidal or apical rings and a short conoid (Fig. 11). Merozoites of *B. stableri* have about 40 subpellicular microtubules around their periphery, which is more than many apicomplexan taxa (Chobotar and Scholtyseck, 1982). Throughout intra-erythrocytic development, *B. stableri* has more numerous amylopectin granules than *D. ranarum*, which may suggest why *Babesiosoma* species appeared more "vacuolated" to Jakowska and Nigrelli (1956) using light microscopy, who therefore included this characteristic in their definition of the genus *Babesiosoma*.

All intra-erythrocytic development occurs within a parasitophorous vacuole (Fig. 11). The initial mitotic division in the parasite results in the formation of a binucleate meront that retains the remnants of the apical complex from the merozoite (Fig. 12). The nuclei migrate to opposite ends of this oblong meront. The second mitotic division occurs simultaneously with the budding of merozoites from the periphery of the meront; the nuclei of the binucleate meront divide as they are being incorporated into the merozoite anlagen (Fig. 13). The nuclei of the tetranucleate meront are fully separated by the time the meront assumes a cruciform shape. Fully formed merozoites separate without the formation of a residual body (Fig. 14) before entering other cells. Occasionally, pairs of fully formed merozoites remain in close contact within infected cells (Barta and Desser, 1986). This phenomenon has convinced some authors (e.g. Paperna, 1981) that binary fission occurred after initial separation of the merozoites. There is no ultrastructural evidence from either dactylosomatid genus to support this hypothesis. Penetration of host erythrocytes by merozoites is accomplished in much the same way as occurs with *Plasmodium* species (Aikawa *et al.*, 1978). The parasites enter red blood cells by forming an invagination in the host cell plasmalemma at the point of contact with the apical end of the parasite. The parasite enters this invagination by translocating posteriorly a circular region of close contact between the parasite outer pellicular membrane and the host cell plasmalemma (Barta and Desser, 1986).

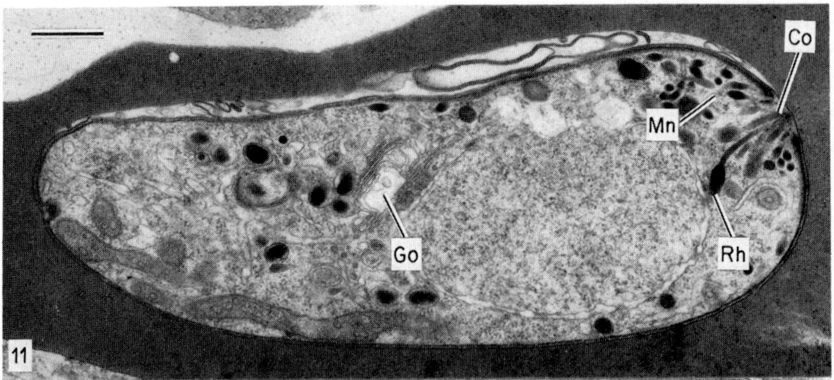

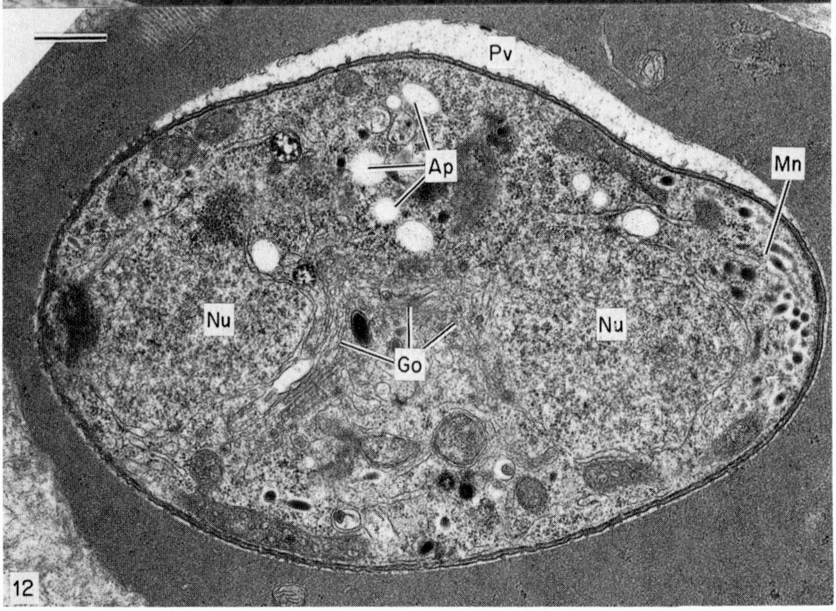

FIG. 11. Typical intra-erythrocytic merozoite of *Babesiosoma stableri* within a parasitophorous vacuole. A large nucleus and perinuclear Golgi body (Go) are posterior to a complete apical complex. Co, conoid; Mn, microneme; Rh, rhoptry. Bar = 0.5 μm. FIG. 12. Binucleate meront of *Babesiosoma stableri* shortly after first nuclear division. Note that the pair of nuclei (Nu) located at opposite ends of the meront share a single large Golgi body (Go). The remains of the apical complex of the merozoite are at one end of the meront. Ap, amylopectin; Mn, micronemes; Pv, parasitophorous vacuole. Bar = 0.5 μm.

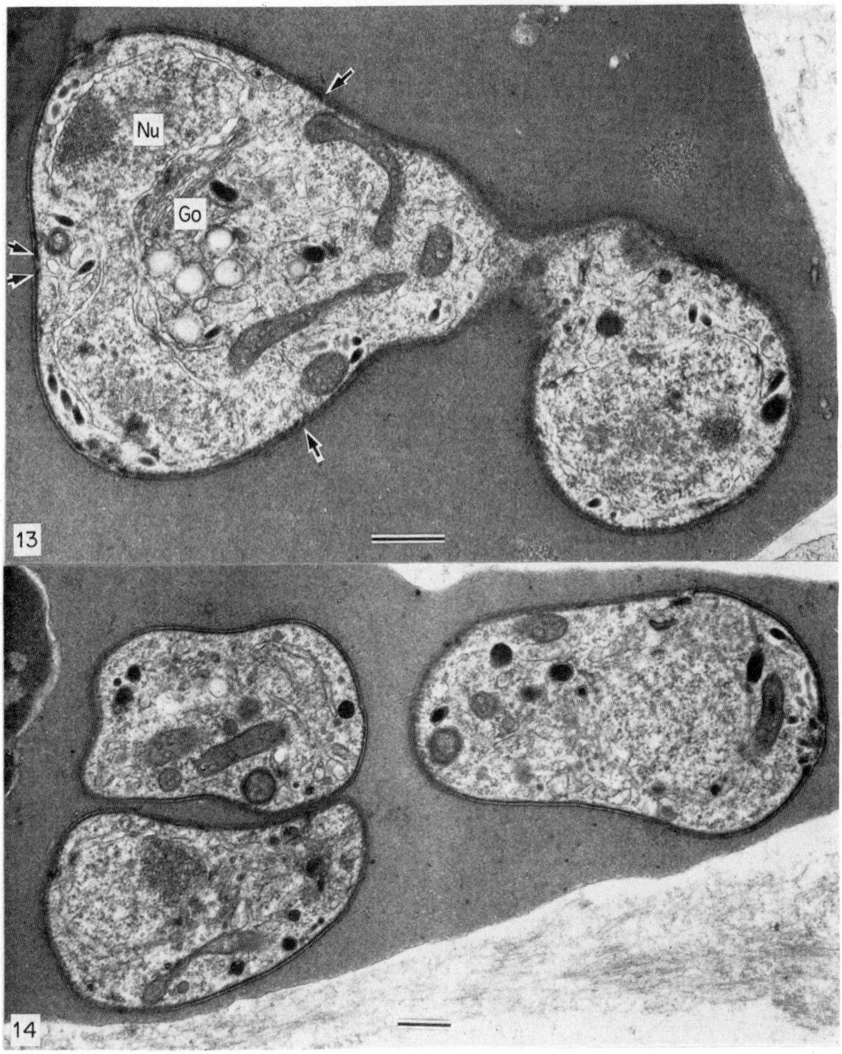

FIG. 13. Budding merozoite anlagen of *Babesiosoma stableri* at one pole of an early tetranucleate meront. A single bilobed nucleus (Nu) extends between the pair of developing merozoites; the limits of the developing pellicles are indicated by arrows. Meronts are tetrahedral during merozoite development, and therefore only three or four merozoites can usually be observed in thin sections (cf. Fig. 14). Go, Golgi body. Bar = 0.5 μm. FIG. 14. Late tetranucleate meront of *Babesiosoma stableri*, in which the resulting merozoites have completely formed. The apical complex of each merozoite has completed development. Bar = 0.5 μm.

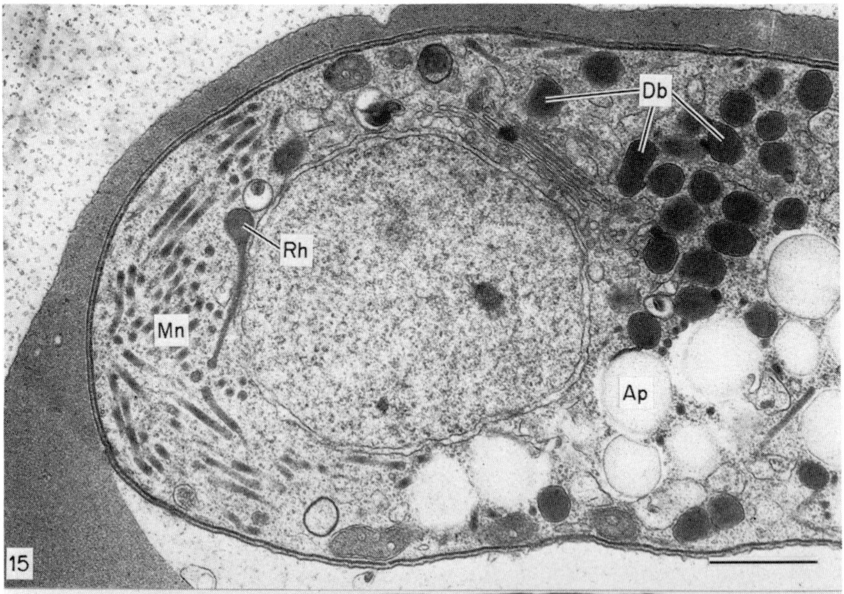

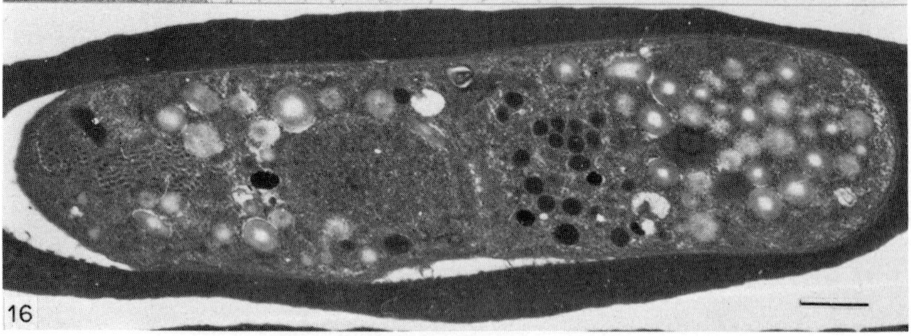

FIG. 15. Apical end of an intra-erythrocytic gamont of *Babesiosoma stableri*. Gamonts have a complete apical complex including micronemes (Mn) and rhoptries (Rh). Distinctive perinuclear dense bodies (Db) and numerous amylopectin inclusions (Ap) pack the remaining cytoplasmic areas. Bar = 0.5 μm. FIG. 16. Longitudinal section of a gamont of *Babesiosoma stableri*. Bar = 0.5 μm.

After repeated cycles of primary and secondary merogonic replication, the morphologically distinct gamonts appear (Fig. 10). Gamonts are characterized by a central nucleus, perinuclear dense bodies and large accumulations of amylopectin (Figs 15, 16). These granules stain with the periodic acid-Schiff reaction and can be resolved using light microscopy as small unstained regions throughout the cytoplasm of the organisms (Fig. 10). In addition to amylopectin, less numerous lipid inclusions are found in gamonts. Unlike the gamonts of *D. ranarum*, those of *Babesiosoma* are fully extended within host erythrocytes (Fig. 16). Like all other motile stages of *B. stableri*, gamonts have a complete apical complex including a conoid, numerous narrow micronemes, rhoptries and about 55 subpellicular microtubules (Fig. 16).

Gamonts persist in the blood of infected hosts for protracted periods (more than 6 months), which provides an extended period during which the parasite can be ingested by its definitive leech host (Schmittner and McGhee, 1961; Barta and Desser, 1986).

3. *Development in* Desserobdella picta

When ingested by the leech, gamonts are freed from the frog erythrocytes and associate in syzygy within the blood meal that is stored in the crop (Figs 17, 18). The paired gamonts mature into gametes (gametogenesis) and fuse (syngamy) forming a motile zygote (ookinete). Condensed chromatin drawn across the nuclei of newly formed zygotes (Fig. 19) suggests that a meiotic division may occur immediately after syngamy as occurs during ookinete formation of other apicomplexan parasites (see Sinden, 1985). The ookinetes are formed during the first 4 days after the leech feeds, but develop no further until they are passed with the blood meal from the crop to the intestine of the leech. There, the ookinetes penetrate epithelial cells to initiate sporogony. Unlike intra-erythrocytic stages of this parasite, intracellular development of *B. stableri* in its leech host occurs directly within the cytoplasm of infected cells: no parasitophorous vacuole is formed. The manner in which ookinetes penetrate intestinal epithelial cells without causing their lysis is unknown. Ookinetes are observed within the intestinal epithelium on the second day after feeding (Fig. 20) and continue to arrive for at least a week as the leech gradually digests the blood meal in its crop.

The intracellular ookinete becomes less organized and the components of the apical complex disaggregate (Fig. 21). The young oocyst enlarges, and then forms eight sporozoites by a process of simultaneous peripheral budding (Fig. 22). As in the intra-erythrocytic meronts, the sporozoite anlagen develop before completion of all sporogonic nuclear divisions. At least three developing sporozoites have been observed to incorporate portions of a single large nucleus during elongation. Mature sporozoites (Fig. 23) leave the intestinal epithelium and migrate to the salivary cells of the leech. Ultrastructurally, the sporozoites are similar to the merozoites but are more slender and contain fewer (ca. 32) subpellicular microtubules than do merozoites or gamonts (Fig. 24).

Once the sporozoites reach the salivary cells of the leech, merogonic replication similar to that which occurs in the frog erythrocytes is initiated

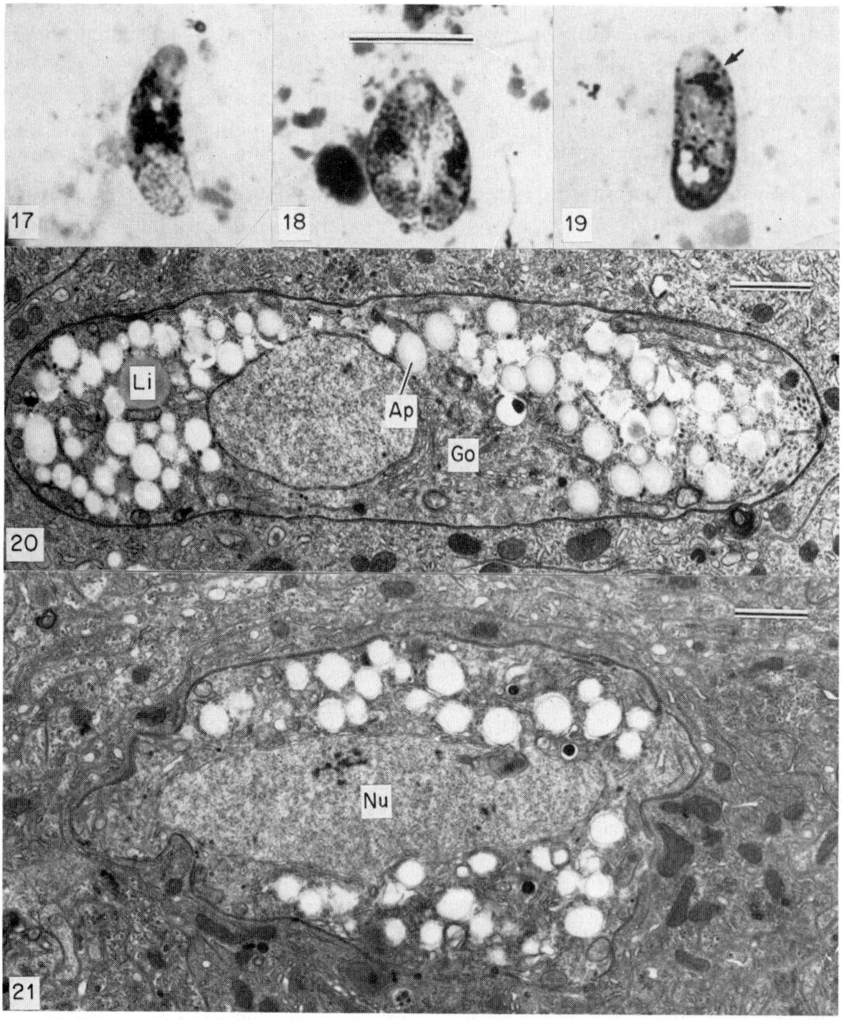

Figs 17–19. Light micrographs of stages of *Babesiosoma stableri* observed within the blood meal stored in the crop of *Desserobdella picta*. Bar = 5 μm. Fig. 17. Free gamont, note the diffuse nuclear material. Fig. 18. A pair of gamonts associated in syzygy. Fig. 19. Gamonts in syzygy have fused to form a zygote or ookinete; note the condensed chromatin (arrow) suggestive of a meiotic division. Fig. 20. Zygote of *Babesiosoma stableri* within an intestinal epithelial cell of *Desserobdella picta* 4 days after feeding on a frog infected with *B. stableri*. The prominent amylopectin granules (Ap) persist in the zygote. Note that the zygote is found directly within the cytoplasm of the infected cell; no parasitophorous vacuole is formed. Go, golgi body; Li, lipid. Bar = 1 μm. (Figs 17–20 from Barta and Desser, 1989. Reproduced with the permission of the Society of Protozoologists.) Fig. 21. During early sporogonic development, the outline of the early oocyst of *Babesiosoma stableri* becomes irregular and the nucleus (Nu) enlarges considerably. Bar = 1 μm.

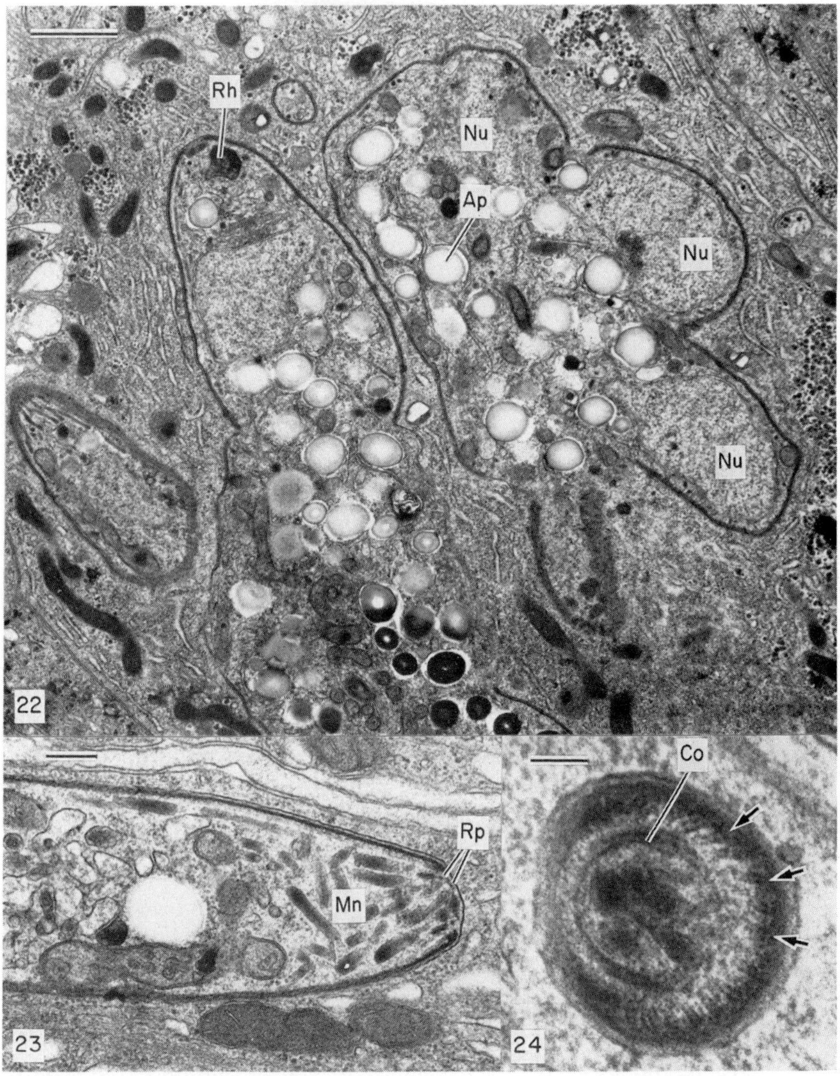

FIG. 22. *Babesiosoma stableri*: sporozoite formation is accomplished by simultaneous peripheral budding of sporozoite anlagen from the periphery of the oocyst. Ap, amylopectin: Nu, nucleus; Rh, immature rhoptry. Bar = 1 μm. FIG. 23. Apical complex of a mature sporozoite of *Babesiosoma stableri* during migration to the anterior of the leech. Mn, micronemes; Rp, preconoidal or accessory rings. Bar = 0.25 μm. FIG. 24. Apex of a mature sporozoite of *Babesiosoma stableri* demonstrating 32 subpellicular microtubules (arrows) and central conoid (Co). Bar = 0.1 μm.

(Fig. 25). The salivary cells are located in the anterior third of this leech and each cell possesses a long ductule that extends from the salivary cell body into the proboscis. The resulting merozoites (Figs 26, 27) may repeat a cycle of merogonic replication within the same large salivary cell body, infect other salivary cells, or move into the ductule that leads to the proboscis. Repeated merogonic replication within the salivary cell body can produce massively infected cells containing hundreds of merozoites which almost completely displace the salivary secretion granules (Fig. 28). Eventually, the salivary cell ductules are also filled with these merozoites which, like merozoites infecting erythrocytes of the vertebrate host, contain ca. 40 subpellicular microtubules (Fig. 27). The salivary cell contents are released during blood feeding through pits located in the leech cuticle near the proboscis tip (Barta and Sawyer, 1990). When the leech next takes a blood meal, merozoites are injected directly into the frog during the release of the salivary secretions from infected salivary cells. The sequence of developmental stages is strikingly similar to that exhibited by some *Babesia* species, where sporogony occurs in the intestinal epithelium, and merogony is initiated in salivary cells and in other tissues (Mehlhorn and Schein, 1984; Büscher *et al.*, 1988).

Once infected, leeches apparently remain so for the remainder of their lives. Merogonic replication within salivary cells restores the numbers of infective merozoites between blood meals. Leeches which had initiated intra-erythrocytic infections with *B. stableri* in frogs by feeding on them contained massively infected salivary cell bodies and ductules 4 months later. Pre-erythrocytic development has not been demonstrated for *B. stableri*; however, the lengthy pre-patent period observed in this species suggests that pre-erythrocytic asexual replication may be occurring (Barta and Desser, 1989).

4. *Epizootiology*

The natural prevalence of most *Babesiosoma* species in fish and amphibians has been reported to be about 10% (see Nigrelli, 1930; Jakowska and Nigrelli, 1956; Haldar *et al.*, 1971; Chaudhuri and Choudhury, 1983; Barta and Desser, 1984); one notable exception is *B. mariae* from African fresh-water fishes, for which a prevalence of approximately 50% was identified (Hoare, 1930; Baker, 1960).

The incidence of *B. stableri* in three sympatric ranid populations (*Rana catesbeiana*, *R. clamitans* and *R. septentrionalis*) in Ontario, Canada, was established using a mark–recapture procedure during the open water seasons of 1985, 1986 and 1987. Over 3000 animals were sampled during the 3-year period. Seasonal fluctuations and host-specific differences in incidence were

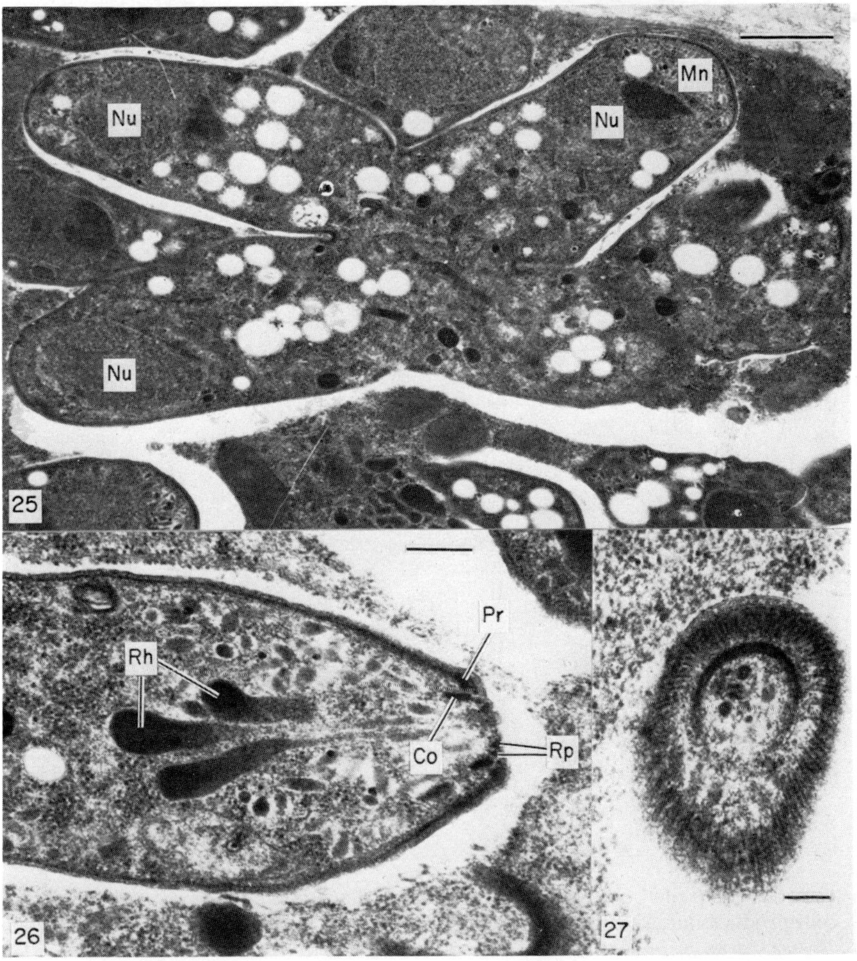

FIG. 25. Merogonic development of *Babesiosoma stableri* within the salivary cells of *Desserobdella picta*. A cruciform tetranucleate meront produces four merozoites by simultaneous peripheral budding in the same manner as occurs within anuran erythrocytes. Mn, micronemes; Nu, nucleus. Bar = 0.5 μm (from Barta and Desser, 1989. Reproduced with the permission of the Society of Protozoologists.) FIG. 26. Apical end of a merozoite of *Babesiosoma stableri* within a leech salivary cell. A conoid (Co), a pair of preconoidal rings (Rp) and a polar ring (Pr) are located anterior to three flask-shaped rhoptries (Rh). Bar = 0.25 μm. FIG. 27. Like the intra-erythrocytic merozoites, merozoites of *Babesiosoma stableri* infecting leech salivary gland cells have about 40 subpellicular microtubules. Bar = 0.1μm.

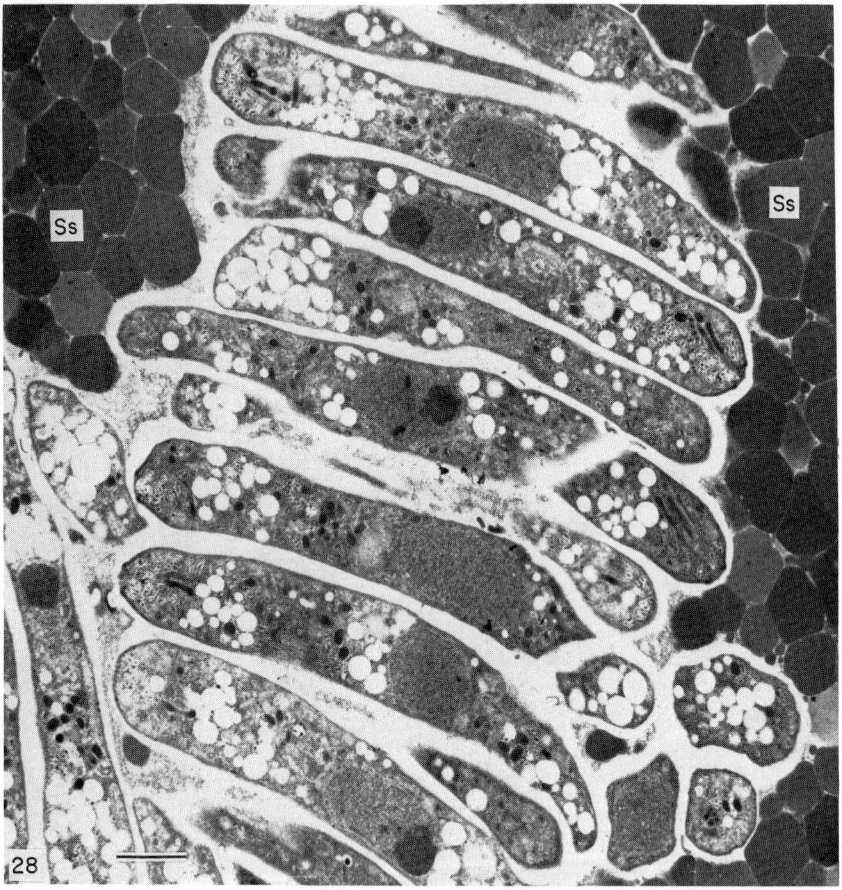

FIG. 28. Repeated merogonic replication of *Babesiosoma stableri* within salivary cells produces large numbers of elongate merozoites that are infective to frogs. When *Desserobdella picta* next feeds, the merozoites are injected into the bite wound when the leech secretes the electron-dense salivary secretions (Ss). Bar = 1 μm.

examined in the context of host and vector biology. The incidence of *B. stableri* in the three sympatric ranid hosts during the study is summarized in Fig. 29. A total of 216 infections (6.16% of animals examined) of *B. stableri* was detected.

R. septentrionalis and *R. catesbeiana* are the primary vertebrate hosts of *B. stableri* in the study area (Barta and Desser, 1984). *R. clamitans* were only rarely infected with this parasite. This difference in prevalence between species of vertebrate host is most probably not related to the susceptibility of the hosts to infection (see Schmittner and McGhee, 1961). Instead, the

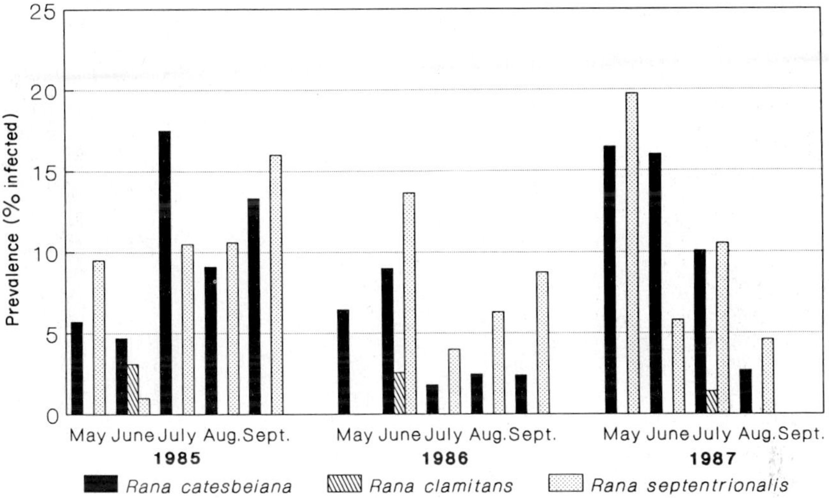

FIG. 29. Prevalence of *Babesiosoma stableri* in *Rana catesbeiana* (bullfrogs), *R. clamitans* (green frogs) and *R. septentrionalis* (mink frogs) during each month of a 3-year study at Algonquin Provincial Park, Ontario, Canada. Bullfrogs and mink frogs demonstrated comparable prevalences throughout the study; green frogs were only rarely infected.

observed low prevalence in *R. clamitans* is explained by the lack of contact between the vector and this host. Observations while collecting animals suggest that *R. catesbeiana* and *R. septentrionalis* were usually partially submerged in shallow water (and therefore ideally situated for attacks by *Dess. picta*), whereas *R. clamitans* were normally observed completely out of the water. In addition, *R. clamitans* are known to migrate to temporary ponds (where the leeches could not survive) for the summer months and return in late summer to permanent waters to prepare for overwintering and breeding during the following spring (Martof *et al.*, 1980). *R. clamitans* are ecologically poor hosts, despite their known natural and experimental susceptibility to infection with *B. stableri*. Conversely, the microhabitats occupied by the other two ranids place them in ideal ecological locations for effective transmission of the parasite. Infections with *Babesiosoma* species may not be greatly restricted by the susceptibility of the vertebrate hosts to the parasites (Schmittner and McGhee, 1961; see also Table 1). Rather, the ecological suitability of the potential vertebrate hosts as feeding targets for the local leech vector appears to be more important in determining which vertebrate hosts are infected at a particular locality.

No significant difference in incidence was related to the sex of the host. However, the prevalence of *B. stableri* in *R. catesbeiana* and *R. septentrionalis* generally increased with the size (age) of the hosts (Figs 30, 31). This may

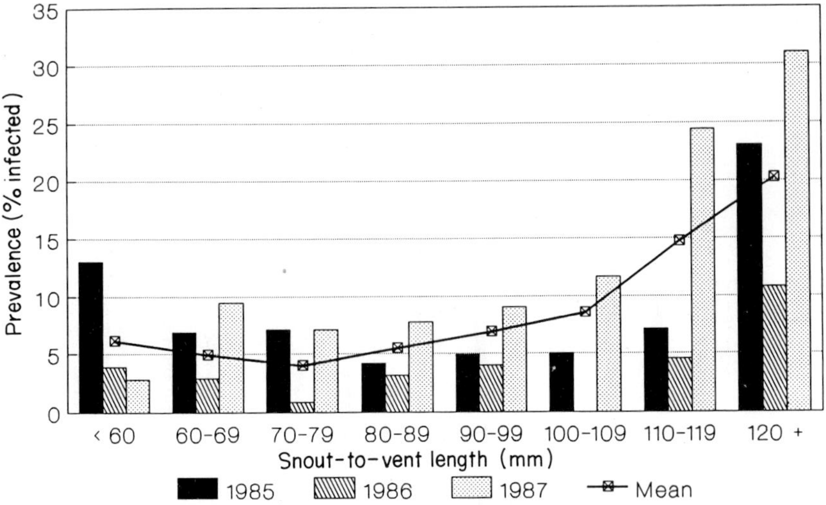

FIG. 30. Prevalence of *Babesiosoma stableri* infections in *Rana catesbeiana* correlated with the size (snout-to-vent length) of the host. Prevalence of *B. stableri* was observed generally to increase with increasing size (age) of the host. The clear association observed between the size of *R. septentrionalis* and the prevalence of *B. stableri* (Fig. 31) was not seen with the bullfrogs, despite an apparent positive correlation between host size and prevalence.

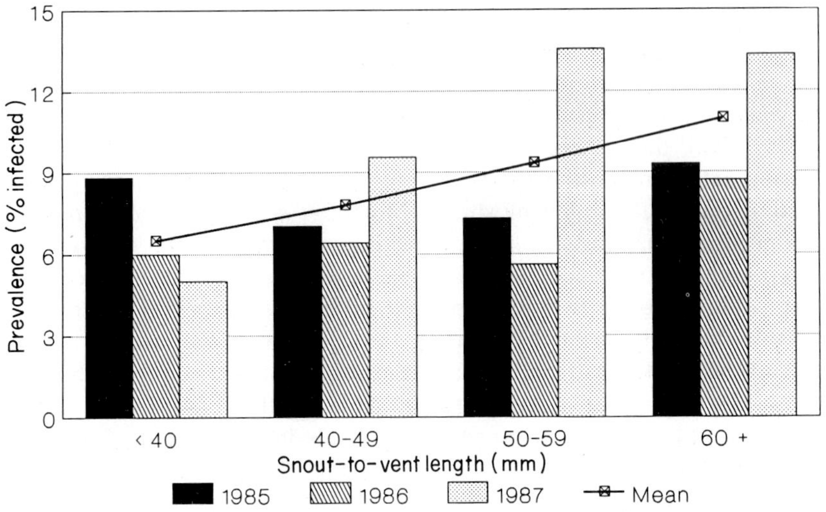

FIG. 31. Prevalence of *Babesiosoma stableri* infections in *Rana septentrionalis* correlated with the size (snout-to-vent length) of the host. The prevalence of *B. stableri* was observed to increase with increasing size (age) of the host.

reflect a low, but perhaps essentially constant, exposure to the parasite during each open water season. Once infected, the frogs remained infected for protracted periods (Schmittner and McGhee, 1961). *Dess. picta* completes one brooding cycle per year at the study location, which may limit transmission to open water periods; both frogs and leeches are inactive during the winter months. In tropical areas, transmission would be expected to continue more or less constantly throughout the year. Indeed, Oosthuizen (1989) recorded feeding and reproduction of *Batracobdelloides tricarinata*, a likely definitive host for *B. mariae*, throughout the year. This may explain the significantly higher prevalences recorded for *B. mariae* in African fishes by Baker (1960) and Hoare (1930) than those of other *Babesiosoma* species.

B. *DACTYLOSOMA RANARUM*

1. *Intra-erythrocytic development*

Only the type species of the family, *D. ranarum*, has been examined in any detail. This parasite undergoes two morphologically distinct merogonic cycles within the erythrocytes of its anuran hosts. During early merogonic development in red blood cells, large trophozoites and meronts that stain palely with Giemsa's stain are observed, which produce up to 16 merozoites by simultaneous peripheral budding (Fig. 32). Frequently, the budding occurs along one side of the meront giving rise to a hand-likc appearance for which the genus was named. After primary replicative cycles, secondary replication begins. Small, densely stained meronts produce six merozoites by peripheral budding (Fig. 33). These smaller merozoites are destined to produce gamonts (Fig. 34).

Ultrastructurally, the type species undergoes all intra-erythrocytic development within a parasitophorous vacuole (Boulard *et al.*, 1982). Membraneous extensions of the parasitophorous vacuole (Fig. 38) are located in the cytoplasm of infected cells and connect directly with the lumen of the parasitophorous vacuole (Barta *et al.*, 1987). Within the vacuole, merozoites of the primary cycle are characterized by a lower electron-density than merozoites of the secondary merogonic cycle. Both types of merozoites have a typical apicomplexan ultrastructure including a conoid. Rhoptries of *D. ranarum* have been observed with crystalline contents (Boulard *et al.*, 1982; Barta *et al.*, 1987). Merozoites of *D. ranarum* have ca. 30 subpellicular microtubules (Boulard *et al.*, 1982). Multiple nuclear divisions occur within the meront before the appearance of merozoite anlagen (Fig. 36). The final nuclear divisions occur simultaneously with the elongation of the developing merozoites. Merozoites (up to 16 during primary merogony) are formed at the periphery of the meront and separate without the formation of a residual body. Amylopectin granules are relatively rare in intra-erythrocytic stages of *D. ranarum* in comparison with *B. stableri*, and are only infrequently observed in secondary merogonic stages (Fig. 37).

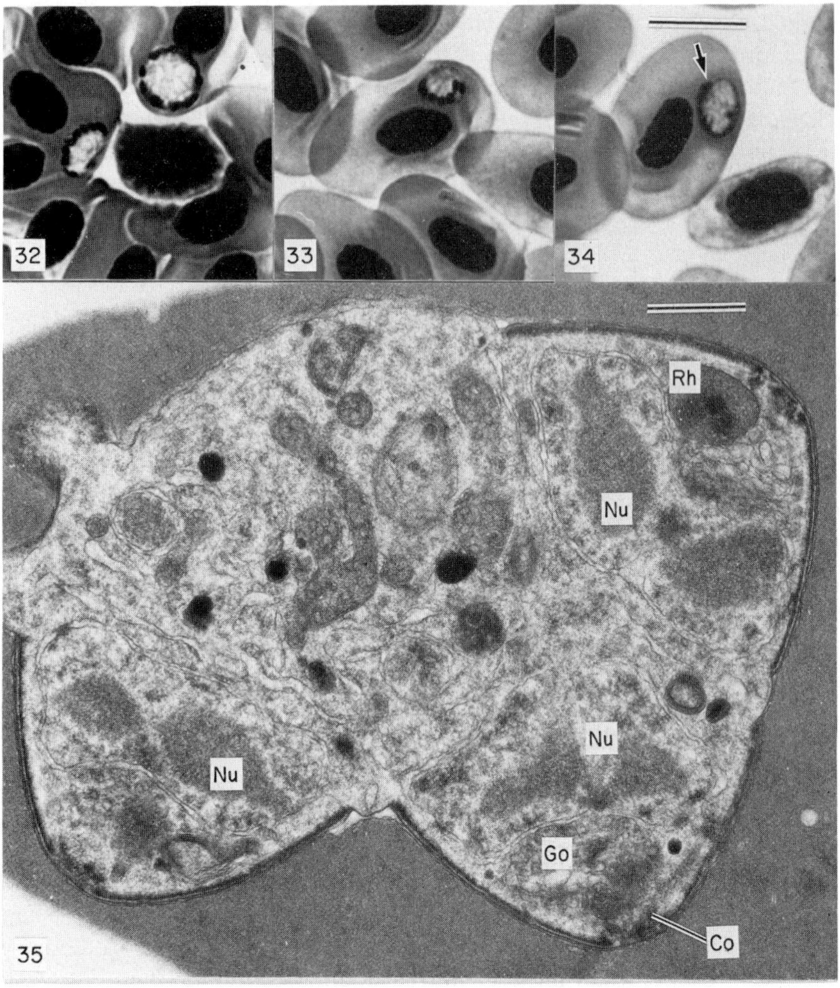

FIGS. 32–34. Light micrographs of intra-erythrocytic development of *Dactylosoma ranarum*. Bar = 10 µm. Fig. 32. Large, palely stained primary meronts that produce up to 16 merozoites. Fig. 33. Smaller, more densely stained secondary meronts that produce only six merozoites. Fig. 34. Typically recurved intra-erythrocytic gamont with the narrower posterior end indicated by an arrow. FIG. 35. Early merozoite development of *Dactylosoma ranarum* in a primary meront. Note the absence of amylopectin inclusions and the large, immature rhoptries (Rh). Co, conoid; Go, Golgi body; Nu, nucleus. Bar = 0.5 µm.

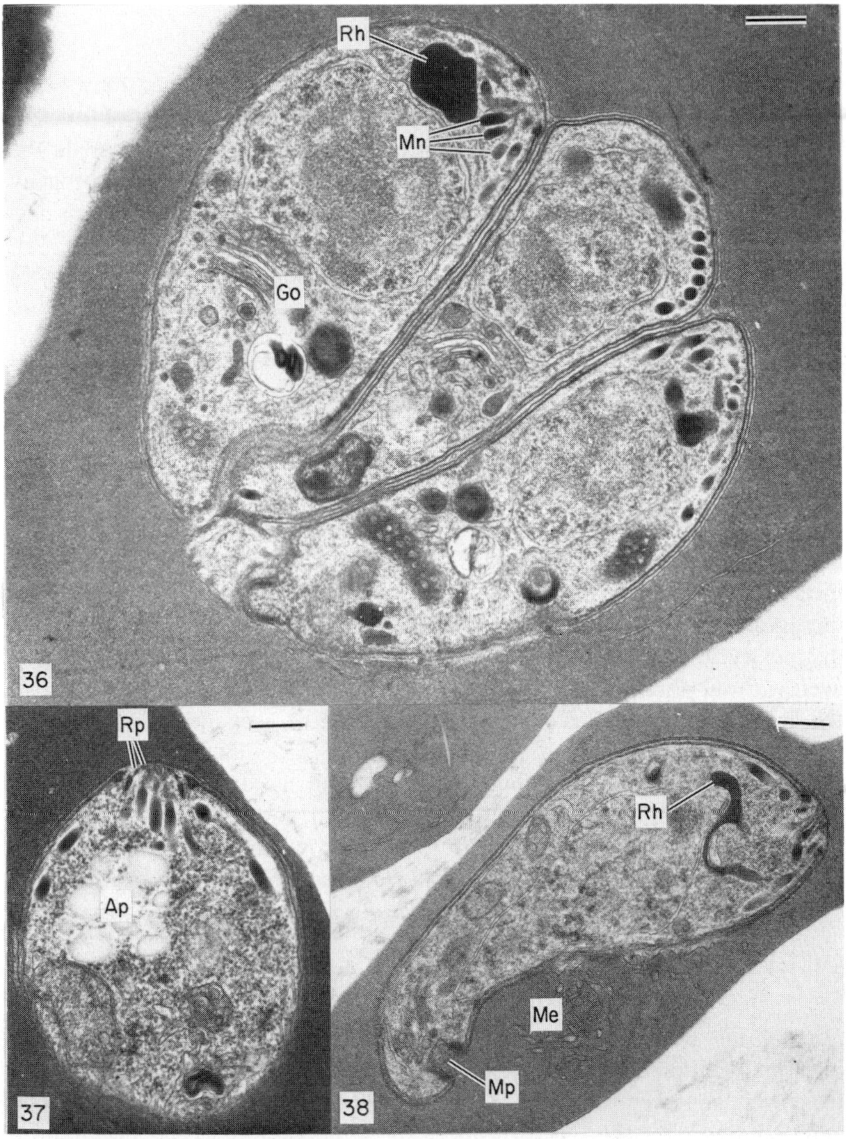

FIG. 36. Advanced secondary meront of *Dactylosoma ranarum*. A prominent rhoptry (Rh) and clustered micronemes (Mn) are found at the apex of a developing merozoite. Go, Golgi. Bar = 0.25 μm (from Barta *et al.*, 1987. Reproduced with the permission of the American Society of Parasitologists.) FIG. 37. Anterior region of a secondary merozoite of *Dactylosoma ranarum* demonstrating an unusual concentration of amylopectin (Ap). The apical complex is complete. Rp, preconoidal rings. Bar = 0.25 μm. FIG. 38. Young gamont of *Dactylosoma ranarum* demonstrating the typical narrow posterior region, which in this specimen has a prominent micropore (Mp). The organism develops in a parasitophorous vacuole that is characterized by numerous membraneous extensions (Me) into the cytoplasm of infected cells. Rh, rhoptry. Bar = 0.25 μm.

Merozoites initiate further cycles of primary merogony or may begin secondary merogonic replication (Figs 33, 36). The process of nuclear division and merozoite development is identical to that which occurs in the primary cycles of merogony except for the smaller number of progeny formed. Merozoites produced during secondary merogony, which are destined to become gamonts, typically have a narrow posterior region (Fig. 38).

Mature gamonts (Figs 39–42) of *D. ranarum* are found in erythrocytes after the appearance of the smaller secondary meronts. Sexual stages are typically recurved within erythrocytes (Fig. 39); the narrow posterior portion of the gamont extends back approximately 80% of the length of the anterior portion of the cell (Fig. 41). Gamonts of *D. ranarum* are much smaller than those of *Babesiosoma* species (see Becker and Katz, 1965; Baker, 1960; Hoare, 1930; Nigrelli, 1930). *Dactylosoma* species, like *Babesiosoma* species, are isogamous; some reports (Hoare, 1930; Nigrelli, 1930; Tanabe, 1931; Baker, 1960; Manwell, 1964) claimed to be able to distinguish micro- and macrogamonts. Indeed, Tanabe (1931) proposed three types of intra-erythrocytic merogony for *D. ranarum*; in addition to the primary merogonic development described above, he suggested that two additional, distinct cycles gave rise to recognizable micro- and macrogamonts. However, without observing syzygy and syngamy in the appropriate annelid host, the question of whether or not *D. ranarum* is isogamous will remain unanswered.

The anterior end of gamonts of *D. ranarum* has typical apicomplexan features (Fig. 42), including ca. 50 subpellicular microtubules; none of the subpellicular microtubules extends into the narrow recurved tail. As with the merogonic stages, few amylopectin inclusions have been observed within the cytoplasm of sexual stages of *D. ranarum*.

2. *Development in an experimental leech host*

D. ranarum was found to undergo sporogonic development within the intestinal epithelium of an experimental leech host, *Dess. picta*. Infected frogs were captured on the island of Corsica, France (see Barta *et al.*, 1989) and fed upon by this North American glossiphoniid leech. The natural definitive host and vector for *D. ranarum* in Corsica is probably the local glossiphoniid leech, *Batracobdella algira*, which is known to feed upon frogs on this island (Sawyer, 1986).

No observation has been made on the development of gametes or zygote formation. At 14 days after feeding, oocysts of *D. ranarum* are found within swollen epithelial cells of the leech intestine. The oocysts appear polysporoblastic and produce 30 or more sporozoites by a process of exogenous budding directly into the cytoplasm of the epithelial cell (Figs 43, 44). Unlike

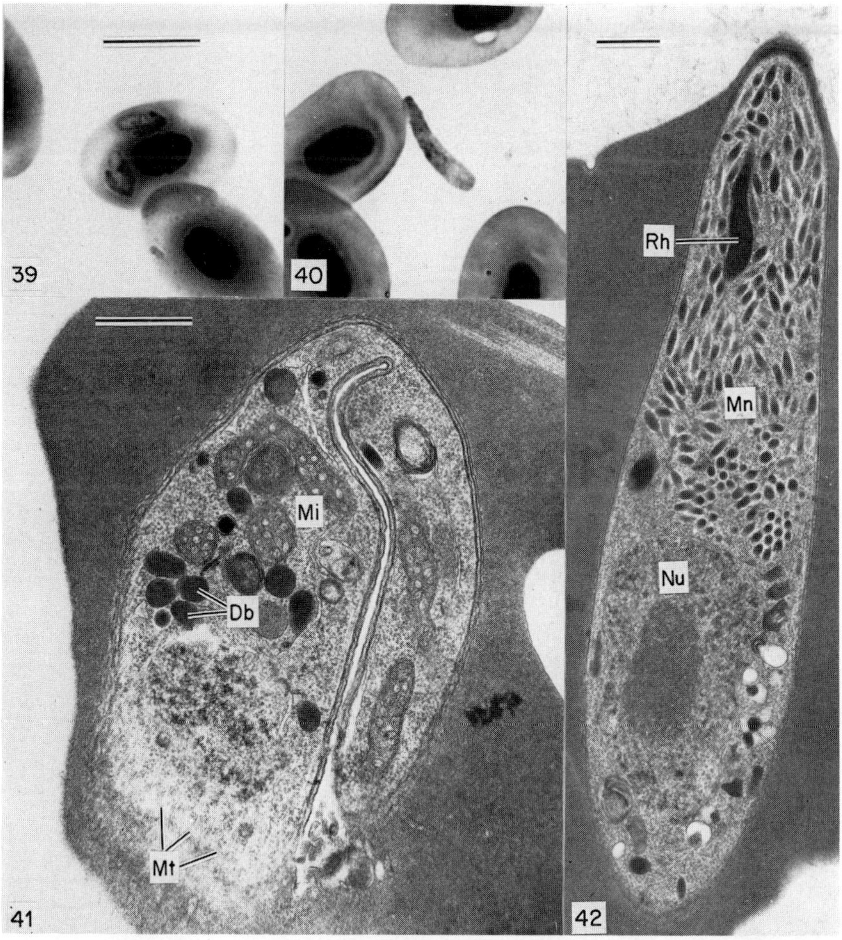

FIGS 39–40. Light micrographs of gamonts of *Dactylosoma ranarum*. Bar = 10 µm. Fig. 39. Typically recurved, mature gamonts of *D. ranarum*. Fig. 40. When outside the host erythrocytes, gamonts extend fully and are highly motile. FIG. 41. Perinuclear region of a mature gamont of *Dactylosoma ranarum*. Note the distinct bend in the organism posterior to the nucleus and the striking absence of amylopectin inclusions which were characteristic of gamonts of *Babesiosoma stableri*. About 50 subpellicular microtubules (Mt) surround the gamonts of *D. ranarum*; these microtubules are not found in the posterior, folded region of the gamonts. Db, dense bodies; Mi, mitochondria. Bar = 0.5 µm (from Barta *et al*., 1987. Reproduced with the permission of the American Society of Parasitologists). FIG. 42. The region anterior to the nucleus of a mature gamont of *Dactylosoma ranarum* is filled with numerous micronemes (Mn) and a few, large rhoptries (Rh). Bar = 0.5 µm.

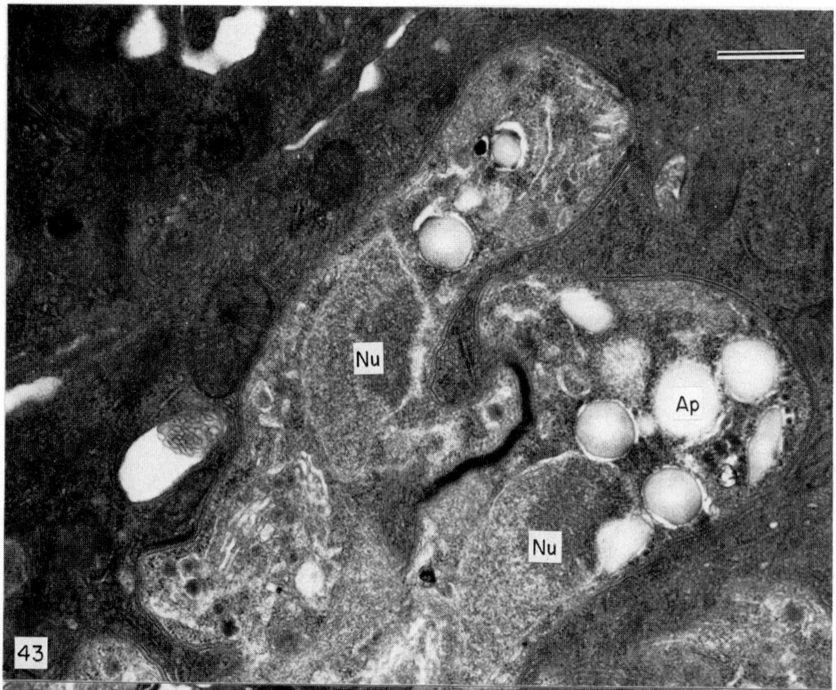
Nu
Ap
Nu
43

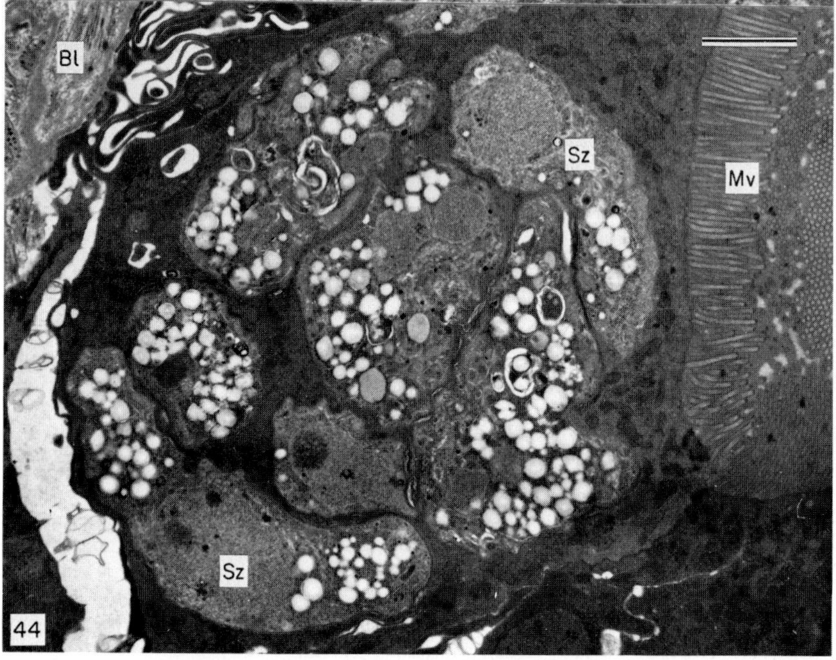
Bl
Sz
Mv
Sz
44

B. stableri (see Barta and Desser, 1989), the production of sporozoites is not synchronized within an oocyst; infected cells demonstrated fully formed sporozoites and multinucleate sporonts at the same time (Fig. 45). Like piroplasms and other adeleid sporozoa, the sporozoites of *D. ranarum* and *B. stableri* possess an electron-lucent polar ring structure (the polar ring complex *sensu* Paterson *et al.*, 1988). The fate of sporozoites within the tissues of this experimental leech host has not been determined and experimental transmission could not be attempted.

Experimental transmission of *D. ranarum* using glossiphoniid leeches was attempted by Nöller (1913), using the glossiphoniid leech *Hemiclepsis marginata*; however, he was unable to demonstrate transmission of the parasite despite repeated attempts. The failure to detect transmission of *D. ranarum* may have been related to the protracted period of development in the annelid definitive host and the lengthy pre-patent period observed for the only dactylosomatid parasite that has been transmitted experimentally (Barta and Desser, 1989).

3. *Epizootiology*

Dactylosoma species have generally been found at low prevalences within their vertebrate hosts. The highest recorded prevalence is 63.6%, observed for *D. ranarum* in *Rana esculenta* collected in the Fium 'Orbo river on the island of Corsica (Barta *et al.*, 1989). Other *Dactylosoma* species have generally been observed at lower prevalences (see, e.g. Nöller, 1913; Tanabe, 1931; Manwell, 1964). The dactylosomatids may not require a high prevalence in the vertebrate hosts to survive. Instead, the potential for extended merogonic replication demonstrated by *B. stableri* in its leech host suggests that the leech host(s) for *Dactylosoma* species may play an important role in the life-cycle of these parasites as a persistent source of infection for their poikilothermic vertebrate hosts.

FIG. 43. Sporogonic development of *Dactylosoma ranarum* within the intestinal epithelium of an experimentally infected annelid host, *Desserobdella picta*, 14 days after the leech had fed on an infected *Rana esculenta*. Sporozoites are formed by peripheral budding like those of *Babesiosoma stableri*. However, more numerous (ca. 30) sporozoites are produced by oocysts of *D. ranarum*. Ap, amylopectin; Nu, nucleus. Bar = 0.25 μm. FIG. 44. Large oocysts of *Dactylosoma ranarum* developing directly within the cytoplasm of epithelial cells between the microvilli (Mv) and basal lamina (Bl). Sz, immature sporozoites. Bar = 2 μm.

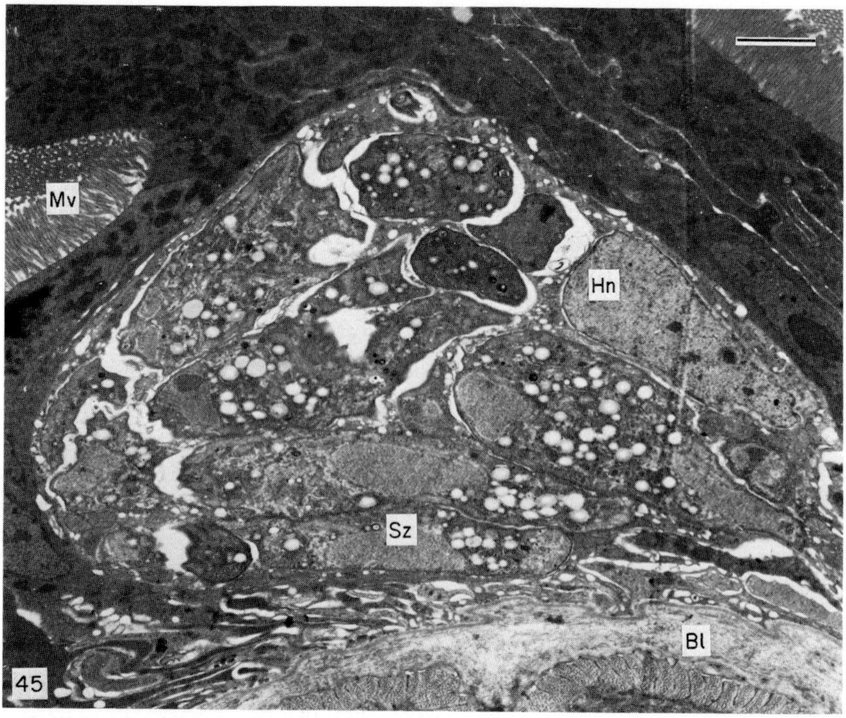

FIG. 45. A maturing oocyst of *Dactylosoma ranarum* contains numerous sporozoites (Sz) which displace the host cell nucleus (Hn) and enlarge the leech epithelial cell significantly. Note that several multinucleate sporonts remain in the host cell. Bl, basal lamina; Mv, microvilli. Bar = 2 μm.

IV. CONCLUSIONS

1. Members of the genera *Babesiosoma* and *Dactylosoma* (family Dactylosomatidae Jakowska and Nigrelli, 1955) are parasites of cosmopolitan distribution which infect the erythrocytes of a wide range of aquatic poikilothermic vertebrates.
2. The genus *Babesiosoma* Jakowska and Nigrelli, 1956 is not synonymous with the genus *Haemohormidium* Henry, 1910.
3. Continued separation of the genera *Babesiosoma* and *Dactylosoma* must be based on the higher number of progeny produced during replicative cycles by members of the latter genus. *Dactylosoma* species produce 6–16 merozoites during intra-erythrocytic replication by simultaneous peripheral budding, whereas *Babesiosoma* species produce exactly 4 merozoites in the same manner.

4. The definitive hosts and vectors of all dactylosomatid parasites recognized in the present chapter are probably rhynchobdellid leeches. The two dactylosomatid parasites for which observations exist undergo sporogonic development directly within the cytoplasm of intestinal epithelial cells of a glossiphoniid leech. Marine species may develop within piscicolid leeches.
5. *Babesiosoma* and *Dactylosoma* species are intriguing parasites that are biologically intermediate between the more commonly encountered adeleid parasites belonging to the genera *Haemogregarina*, *Karyolysus* or *Cyrilia* and the piroplasms of the genera *Babesia* or *Theileria*.

Acknowledgement

This research was supported by the Natural Sciences and Engineering Research Council of Canada (NSERCC).

References

Aikawa, M., Miller, L. H., Johnson, J. and Rabbege, J. R. (1978). Erythrocyte entry by malarial parasites. *Journal of Cell Biology* **77,** 72–82.

Awerinzew, S. V. (1914). Contributions to the morphology and development of parasites from German East Africa [in Russian]. *Zhurnal Microbiologia* **1,** 1–10.

Baker, J. R. (1960). Trypanosomes and dactylosomes from the blood of fresh-water fish in East Africa. *Parasitology* **50,** 515–525.

Barta, J. R. (1989). Phylogenetic analysis of apicomplexans of the class Sporozoea (Phylum Apicomplexa Levine, 1970): evidence for the independent evolution of heteroxenous life cycles. *Journal of Parasitology* **75,** 195–206.

Barta, J. R. and Desser, S. S. (1984). Blood parasites of amphibians from Algonquin Park, Ontario. *Journal of Wildlife Diseases* **20,** 180–189.

Barta, J. R. and Desser, S. S. (1986). Light and electron microscopic observations on the intraerythrocytic development of *Babesiosoma stableri* (Apicomplexa, Dactylosomatidae) in frogs from Algonquin Park, Ontario. *Journal of Protozoology* **36,** 359–368.

Barta, J. R. and Desser, S. S. (1989). Development of *Babesiosoma stableri* (Dactylosomatidae; Adeleida; Apicomplexa) in its leech vector (*Batracobdella picta*) and the relationship of the dactylosomatids to the piroplasms of higher vertebrates. *Journal of Protozoology* **36,** 241–253.

Barta, J. R. and Sawyer, R. T. (1990). Definition of a new genus of glossiphoniid leech and a redescription of the type species, *Clepsine picta* Verrill, 1872. *Canadian Journal of Zoology* **68,** 1942–1950.

Barta, J. R., Boulard, Y. and Desser, S. S. (1987). Ultrastructural observations on secondary merogony and gametogony of *Dactylosoma ranarum* Labbé, 1894. *Journal of Parasitology* **73,** 1019–1029.

Barta, J. R., Boulard, Y. and Desser, S. S. (1989). Blood parasites of *Rana esculenta* from Corsica: comparison of its parasites with those of eastern North American ranids in the context of host phylogeny. *Transactions of the American Microscopical Society* **108,** 6–20.

Becker, C. D. (1970). Haematozoa of fishes with emphasis on North American records. *In* "A Symposium on Diseases of Fishes and Shellfishes" (S. F. Snieszko, ed.), pp. 82–100. Special publication No. 5 of the American Fisheries Society, Washington D.C.

Becker, C. D. and Katz, M. (1965). *Babesiosoma tetragonis* n. sp. (Sporozoa: Dactylosomatidae) from a California teleost. *Journal of Protozoology* **12,** 189–193.

Billet, A. (1904). Sur le *Trypanosoma inopinatum* de la grenouille verte d'Algérie et sa relation possible avec les *Drepanidium. Comptes Rendus des Séances de la Société de Biologie et de ses Filiales* **57,** 161–165.

Boulard, Y., Vivier, E. and Landau, I. (1982). Ultrastructure de *Dactylosoma ranarum* (Kruse, 1890); affinités avec les coccidies; révision du statut taxonomique des dactylasomides. *Protistologica* **18,** 103–121.

Büscher, G., Friedhoff, K. T. and El-Allawy, T. A. A. (1988). Quantitative description of the development of *Babesia ovis* in *Rhipicephalus bursa* (hemolymph, ovary, eggs). *Parasitology Research* **74,** 331–339.

Chaudhuri, S. R. and Choudhury, A. (1983). *Babesiosoma batrachi* n. sp. from *Clarias batrachus* (L.) with a note on the genus *Babesiosoma* (Apicomplexa: Sporozoea). *Archiv für Protistenkunde* **127,** 335–341.

Chobotar, B. and Scholtyseck, E. (1982). Ultrastructure. *In* "The Biology of the Coccidia" (P. L. Long, ed.), pp. 101–165. University Park Press, Baltimore, Maryland.

Davies, A. J. (1980). Some observations on *Haemohormidium cotti* Henry 1910, from the marine fish *Cottus bubalis* Euphrasen. *Zeitschrift für Parasitenkunde* **62,** 31–38.

Desser, S. S. and Weller, I. (1973). Structure, cytochemistry, and locomotion of *Haemogregarina* sp. from *Rana berlandieri. Journal of Protozoology* **20,** 65–73.

Durham, H. E. (1902). Report of the yellow fever expedition to Para of the Liverpool School of Tropical Medicine and Medical Parasitology: *Drepanidium* in the toad. *Liverpool School of Tropical Medicine Memoirs* **7,** 485–563.

Du Toit, P. J. (1938). A new piroplasm (*Sauroplasma thomasi* n. g., n. sp.) of a lizard (*Zonurus giganteus*, Smith). *Onderstepoort Journal of Veterinary Science* **9,** 289–299.

Dutton, J. E., Todd, J. L. and Tobey, E. N. (1907). Concerning certain parasitic Protozoa observed in Africa. Part II. *Annals of Tropical Medicine and Parasitology* **1,** 285–370.

El-Naffer, M. K., Abdel-Rahman, A. M. and Khalifa, R. (1979). A review of the genus *Babesiosoma* Jakowska and Nigrelli, 1956 with a description of the first species from the gecko *Pterodactylus hasselquistii. Journal of the Egyptian Society of Parasitologists* **9,** 305–316.

Fahmy, M. A. M., Arafa, M. S., Mandour, A. M., Khalifa, R. and Abdel-Salem, E. A. (1979). *Babesiosoma gallinarum* n. sp.; a blood protozoan investigated from fowls in Assiut Province. *Journal of the Egyptian Veterinary Medical Association* **39,** 251–263.

Fantham, H. B. (1905). *Lankesterella tritonis* n. sp., a haemogregarine from the blood of the newt, *Triton cristatus (Molge cristata*). *Zoologischer Anzeiger* **29,** 257–263.

Fantham, H. B., Porter, A. and Richardson, L. R. (1942). Some haematozoa observed in vertebrates in eastern Canada. *Parasitology* **34,** 199–226.

Finkelstein, N. J. (1908). Parasites endoglobulaires du sang chez les animaux à sang froid du Caucase. *Archives des Sciences Biologique, St. Pétersburg* **13,** 137–168.

França, C. (1908). Quelques notes sur l'*Haemogregarina splendens* Labbé. *Archives de l'Institut Royale de Bacteriologie (Camara Prestana)* **2,** 123–131.

Haiba, M. H. and El-Shabrawy, M. N. (1967). *Babesiosoma anseris* (n. sp.) investigated in the goose, *Cygnoides cygnoides* in Egypt. *Journal of Veterinary Science of the United Arab Republic* **4,** 189–194.

Haldar, D. P., Misra, K. K. and Chakravarty, M. M. (1971). Observations on *Babesiosoma hareni* n. sp. (Protozoa: Sporozoa) from a freshwater teleost. *Archiv für Protistenkunde* **113,** 1–6.

Henry, H. (1910). On the Haemoprotozoa of British sea fish (a preliminary note). *Journal of Pathology and Bacteriology* **14,** 463–465.

Henry, H. (1913). An intracorpuscular parasite in the blood of *Cottus bubalis* and *Cottus scorpius*. *Journal of Pathology and Bacteriology* **18,** 224–227.

Hintze, R. (1902). Lebensweise und Entwicklung von *Lankesterella minima* (Chaussat). *Zoologische Jahrbücher, Abteilung für Anatomie und Ontogonie der Tiere* **15,** 693–750

Hoare, C. A. (1930). On a new *Dactylosoma* occurring in fish of Victoria Nyanza. *Annals of Tropical Medicine and Parasitology* **24,** 241–248.

Jakowska, S. and Nigrelli, R. F. (1955). A taxonomic re-evaluation of *Dactylosoma* Labbé, 1894, a babesioid of cold blooded vertebrates. *Journal of Protozoology* **2,** supplement, 8.

Jakowska, S. and Nigrelli, R. F. (1956). *Babesiosoma* gen. nov. and other babesioids in erythrocytes of cold-blooded vertebrates. *Annals of the New York Academy of Sciences* **64,** 112–127.

Jones, S. R. M. and Woo P. T. K. (1990). Redescription of the leech, *Desserobdella phalera* (Graf 1899) n. comb. (Rhynchobdellida:Glossiphoniidae), with notes on its biology and occurrence on fishes. *Canadian Journal of Zoology* **68,** 1951–1955.

Khan, R. A. (1980). The leech as a vector of a fish piroplasm *Canadian Journal of Zoology* **58,** 1631–1637.

Kruse, W. (1890). Über blutparasiten. *Virchows Archiv für Pathologische Anatomie und Physiologie und für Klinische Medizin* **120,** 541–560.

Kudo, R. R. (1966). "Protozoology", 5th edn. Charles C. Thomas, Springfield, Illinois.

Kundu, T. K. and Haldar, D. P. (1984). *Dactylosoma notopterae* n. sp. from the freshwater teleost *Notopterus notopterus* (Pallas). *Rivista di Parassitologia* **1,** 391–398.

Labbé, A. (1894). Recherches zoologiques et biologiques sur les parasites endoglobulaires du sang des vertébrés. *Archives de Zoologie Expérimentale et Génerale* **2,** 55–252.

Laird, M. and Bullock, W. L. (1969). Marine fish haematozoa from New Brunswick and New England. *Journal of the Fisheries Research Board of Canada* **26,** 1075–1102.

Lankester, E. R. (1871). On *Undulina*, the type of a new group of Infusoria. *Quarterly Journal of Microscopical Science* **11,** 387–389.

Lankester, E. R. (1882). On *Drepanidium ranarum*, the cell-parasite of the frog's blood and spleen (Gaule's Würmschen). *Quarterly Journal of Microscopical Science* **22,** 56–65.

Laveran, A. (1898). Contribution à l'étude de *Drepanidium ranarum* (Lankester). *Comptes Rendus des Séances de la Société de Biologie et de ses Filiales* **1,** 977–980.
Levine, N. D. (1971). Taxonomy of the piroplasms. *Transactions of the American Microscopical Society* **90,** 2–33.
Levine, N. D. (1984). Nomenclatural corrections and new taxa in the apicomplexan protozoa. *Transactions of the American Microscopical Society* **103,** 195–204.
Levine, N. D. (1985). Phylum II. Apicomplexa Levine, 1970. *In* "An Illustrated Guide to the Protozoa" (J. J. Lee, S. H. Hutner and E. C. Bovee, eds), pp. 322–374. Society of Protozoologists and Allen Press, Lawrence, Kansas.
Levine, N. D. (1988). "The Protozoan Phylum Apicomplexa", 2 vols. CRC Press, Boca Raton, Florida.
Mackerras, I. M. and Mackerras, M. I. (1925). The haematozoa of Australian marine Teleostei. *Proceedings of the Linnean Society of New South Wales* **50,** 359–366.
Mackerras, I. M. and Mackerras, M. I. (1961). The haematozoa of Australian frogs and fish. *Australian Journal of Zoology* **9,** 123–139.
Mandal, A. K. (1979). Studies on the haematozoa of some catfishes belonging to the genus *Mystus* Scopoli from India. *Bulletin of the Zoological Survey of India* **2,** 17–23.
Manwell, R. D. (1964). On the genus *Dactylosoma. Journal of Protozoology* **11,** 526–530.
Martof, B. S., Palmer, W. M., Bailey, J. R., Harrison, J. R., III and Dermid, J. (1980). "Amphibians and Reptiles of the Carolinas and Virginia". University of North Carolina Press, Chapel Hill, North Carolina.
Mathis, C. and Léger, M. (1912). "Recherches de Parasitologie et de Pathologie Humaines et Animales au Tonkin". Masson, Paris.
Mehlhorn, H. and Schein, E. (1984). The piroplasms: life cycle and sexual stages. *Advances in Parasitology* **23,** 37–103.
Mehlhorn, H., Peters, W. and Haberkorn, A. (1980). The formation of kinetes and oocyst in *Plasmodium gallinaceum* (Haemosporidia) and considerations on phylogenetic relationships between Haemosporidia, Piroplasmida and other coccidia. *Protistologica* **16,** 135–154.
Misra, K. K. and Nigrelli, R. F. (1973). On *Babesiosoma*: Haemosporidia of cold-blooded vertebrates. *Acta Protozoologica* **12,** 107–110.
Misra, K. K., Haldar, D. P. and Chakravarty, M. M. (1969). *Babesiosoma ophicephali* n. sp. from the freshwater teleost *Ophicephalus punctatus* Bloch. *Journal of Protozoology* **16,** 446–449.
Nigrelli, R. F. (1929). *Dactylosoma jahni* sp. nov., a sporozoan parasite of the erythrocytes and erythroplastids of the newt (*Triturus viridescens*). *Journal of Parasitology* **16,** 102.
Nigrelli, R. F. (1930). *Dactylosoma jahni* sp. nov., a sporozoan of the erythrocytes and erythroplastids of the newt (*Triturus viridescens*). *Annales de Parasitologie Humaine et Comparée* **3,** 1–7.
Nöller, W. (1913). Die Blutprotozoen des Wasserfrosches und ihre Übertragung. II. Gattung *Dactylosoma* Labbé, 1894. *Archiv für Protistenkunde* **31,** 169–240.
Oosthuizen, J. H. (1989). Redescription of the African fish leech *Batracobdelloides tricarinata* (Blanchard, 1897). *Hydrobiologica* **184,** 153–164.
Paperna, I. (1981). *Dactylosoma hannesi* n. sp. (Dactylosomatidae, Piroplasmia) found in the blood of grey mullets (Mugilidae) from South Africa. *Journal of Protozoology* **28,** 486–491.

Paterson, W. B., Desser, S. S. and Barta, J. R. (1988). Ultrastructural features of the apical complex, pellicle, and membranes investing the gamonts of *Haemogregarina magna* (Apicomplexa: Adeleina). *Journal of Protozoology* **35,** 73–80.

Poisson, R. (1953). Protistes parasites, intra- ou extracellulaires, d'affinités incertaines. *In* "Traité de Zoologie" (P. P. Grassé, ed.), pp. 976–1005. Masson, Paris.

Ruiz, A. (1959). Sobre la presencia de un *Dactylosoma* en *Bufo marinus*. *Revista de Biologia Tropical* **7,** 113–117.

Sarkar, N. K. and Haldar, D. P. (1979). *Dactylosoma striata* n. sp. (Piroplasmida: Dactylosomatidae) from an Indian fish. *Indian Journal of Parasitology* **3,** 99–106.

Saunders, D. S. (1960). A survey of the blood parasites in the fishes of the Red Sea. *Transactions of the American Microscopical Society* **79,** 239–252.

Sawyer, R. T. (1986). "Leech Biology and Behaviour". Oxford University Press, Oxford.

Schmittner, S. M. and McGhee, R. B. (1961). The intra-erythrocytic development of *Babesiosoma stableri* n. sp. in *Rana pipiens pipiens*. *Journal of Protozoology* **8,** 381–386.

Sinden, R. E. (1985). Gametocytogenesis in *Plasmodium* spp. and observations on the meiotic division. *Annales de la Société Belge de Médecine Tropicale* **65,** 21–33.

Tanabe, M. (1931). Studies on blood-inhabiting protozoa of the frog. *Keijo Journal of Medicine* **2,** 53–71.

Vivier, E. (1979). Données nouvelles sur les sporozoaires. Cytologie–cycles–systématique. *Bulletin de la Société Zoologique de France* **104,** 345–381.

Vivier, E. (1982). Réflexions et suggestions à propos de la systématique des sporozoaires: création d'une classe des Hematozoa. *Protistologica* **18,** 449–457.

Walton, A. C. (1946). Protozoan parasites of the Bufonidae (Amphibia). *Transactions of the Illinois State Academy of Science* **39,** 143–147.

Walton, A. C. (1947). Parasites of the Ranidae (Amphibia). *Journal of Parasitology* **33,** supplement, abstract no. 68.

Walton, A. C. (1948). Parasites of the Ranidae (Amphibia). *Journal of Parasitology* **34,** supplement, abstract no. 78.

Walton, A. C. (1949). Parasites of the Ranidae (Amphibia). *Transactions of the American Microscopical Society* **68,** 49–54.

Walton, A. C. (1950). Parasites of the Ranidae (Amphibia). *Anatomical Record* **108,** abstract no. 238.

Weynon, C. M. (1926). "Protozoology. A Manual for Medical Men, Veterinarians and Zoologists". Baillière, Tindall and Cox, London.

Amino Acid Metabolism in Helminths

J. BARRETT

Department of Biological Sciences, University College of Wales, Aberystwyth, UK

I. Introduction

Amino acids are the building blocks of proteins and the amino acid sequence determines biological activity. There are 20 common or standard amino acids (alanine, arginine, asparagine, aspartic acid, cysteine, glutamine, glutamic acid, glycine, histidine, isoleucine, leucine, lysine, methionine, phenylalanine, proline, serine, threonine, tryptophan, tyrosine and valine). All the amino acids (except proline, which is an imino acid) have as a common denominator a free unsubstituted amino group and a free carboxyl group. They differ from one another, however, in the distinctive nature of their side chains.

ADVANCES IN PARASITOLOGY VOL. 30
ISBN 0-12-031730-3

TABLE 1 *Total amino acid analyses of some helminths*

Species	*Reference(s)*
Nematodes	
Anisakis sp.	Freeman *et al.* (1963), Okuno (1968, 1969)
Ascaridia dissimilis	Ossikovski and Khlebarov (1983)
Ascaridia galli	Monteoliva Hernandez (1962), Monteoliva (1963), Dubinsky and Rybos (1978a,b)
Ascaris lumbricoides	Okuno (1968, 1969), Krvavica *et al.* (1964a,b)
Cooperia sp.	Herlich (1966)
Heterakis kotwardensis	Malhotra and Rautela (1986)
Neoascaris vitulorum	Krvavica *et al.* (1964a)
Nippostrongylus brasiliensis	Friedman and Kagan (1958)
Oesophagostomum radiatum	Herlich (1966)
Ostertagia ostertagia	Herlich (1966)
Parascaris equorum	Krvavica *et al.* (1964a)
Trichostrongylus sp.	Herlich (1966)
Digeneans	
Echinoparyphium sp.	Cheng (1963)
Fasciola hepatica	Bankov *et al.* (1978), Bankov and Khlebarov (1987)
Gastrothylax crumenifer	Gupta and Agrawal (1977)
Glypthelmins quieta,	Cheng (1963)
Glypthelmins amplicava	Cheng (1963)
Posthodiplostomum minimum	Lynch and Bogitsh (1962)
Schistosoma mansoni	Robinson (1961), Senft *et al.* (1972), Chappell and Walker (1982)
Monogeneans	
Capsola laevis	Ramalingam (1973a)
Diclodophora merlangi	Arme and Whyte (1975)
Cestodes	
Amoebotaenia cuneata	Bhalya *et al.* (1983c)
Cittotaenia perplexa	Campbell (1960b)
Cotugnia columbae	Bhalya *et al.* (1986)
Davinia hewetensis	Bhalya *et al.* (1984)
Dipylidium caninum	Goodchild and Dennis (1966)
Echinococcus granulosus	Agosin and Repetto (1963)
Gangesia sp.	Nanda *et al.* (1987)
Hymenolepis diminuta	Kent (1957), Goodchild and Wells (1957), Foster and Daugherty (1959), Campbell (1963b), Goodchild and Dennis (1965, 1966)
Hymenolepis microstoma	Litchford (1970)
Hymenolepis palmarum	Bhalya *et al.* (1985)
Introvertus raipurensis	Niyogi and Agarwal (1983)
Ligula intestinalis plerocercoid	Dabrowski (1980), Soutter *et al.* (1980)
Lucknowia indica,	Niyogi and Agarwal (1983)
Lytocestus indicus	Niyogi and Agarwal (1983)
Moniezia expansa	Campbell (1960b), Lopez Gorge and Monteoliva Hernandez (1964), Goodchild and Dennis (1966)
Raillietina cesticillus	Foster and Daugherty (1959), Goodchild and Dennis (1966), Bhalya *et al.* (1983a, 1984)
Raillietina echinobothrida	Il'yasov (1978)
Raillietina penetrans	Bhalya *et al.* (1983a)

TABLE 1 (*continued*)

Species	*Reference(s)*
Raillietina saharanpurensis	Malhotra (1981), Bhalya *et al.* (1983a, 1984)
Raillietina simmonsi,	Bhalya *et al.* (1983a, 1984)
Raillietina tetragona	Bhalya *et al.* (1983a, 1984)
Staphylepis rustica	Bhalya *et al.* (1983b)
Taenia crassiceps larvae	Taylor and Haynes (1966)
Taenia pisiformis, Taenia taeniaeformis, Taeniarhynchus saginatum (= *Taenia saginata*)	Goodchild and Dennis (1966)
Thysanosoma actinioides	Campbell (1960b)

In addition to the 20 common amino acids, there are some 150 known non-protein amino acids, compounds such as citrulline, ornithine, taurine and 4-aminobutyrate. These amino acids do not occur naturally in proteins and are often metabolic intermediates or, in some cases, neurotransmitters. Many of these non-protein amino acids are particularly abundant in plants but they occasionally occur in physiologically active peptides. There are also a number of rare amino acids such as 4-hydroxyproline, 5-hydroxylysine and 6-*N*-methyl-lysine, which are only found in specialized proteins.

Organisms vary in their ability to synthesize amino acids and in the forms of nitrogen used for this purpose. On this basis, amino acids can be classified into two types, the essential and the non-essential. Man and the albino rat can only synthesize 10 of the 20 common amino acids (alanine, asparagine, aspartate, cysteine, glutamine, glutamate, glycine, proline, serine and tyrosine). The remainder of nutritionally indispensable amino acids cannot be synthesized or are synthesized at such a low rate (e.g. arginine and histidine) that they cannot supply nutritional requirements and have to be obtained from the diet.

The common amino acids can also be classified either on the nature of their side chains (neutral, basic and acidic; or hydrophilic and hydrophobic) or on the basis of the end-products of their catabolism (glucogenic, ketogenic or both). The naturally occurring amino acids in eukaryotes are all L-amino acids. However, in invertebrates, and especially marine invertebrates, there may be significant amounts of D-amino acids, in particular D-alanine (Preston, 1987). The physiological role (if any) of D-amino acids in invertebrates is unknown. The tegumental transport mechanisms of cestodes have a similar affinity for D- and L-amino acids, but amino acid racemases have not been found in either cestodes or nematodes.

II. Amino Acid Composition

Amino acids occur in tissues either as free amino acids or as components of proteins and peptides. *In vivo* proteins are constantly turning over, with half-lives varying from minutes to several months. In general, there are usually hardly any qualitative differences between the total amino acid composition of different species, and total amino acid composition has not proved particularly useful as an aid to taxonomy. However, the discovery of unusual amino acids during the analysis of tissue extracts can indicate the presence of novel enzymes or pathways.

A. Total Amino Acids

There have been numerous studies on the total amino acid composition of helminths, particularly cestodes (Table 1). In practically all cases, the majority of the standard amino acids have been found and the apparent absence of a particular amino acid in one or two instances may be due to differences in preparation or analytical technique. Amino acid analyses have also be carried out on semi-purified preparations, including nematode cuticles (Bird, 1954, 1956, 1957; Bird and Rogers, 1956; Simmonds, 1958; Watson and Silvester, 1959; Jaskowski and Ozuk, 1977; Leushner *et al.*, 1979; Cox *et al.*, 1981a,b), nematode egg shells (Kreuzer, 1953; Jaskowski, 1962; Clarke *et al.*, 1967), nematode egg and sperm inclusions (Ebel and Colas, 1954; Klass and Hirsh, 1981; Abbas and Cain, 1984), cestode hooks (Gallagher, 1964; Dvorak, 1969; Pearson *et al.*, 1985), cestode egg membranes (Morseth, 1966; Lethbridge, 1971), monogenean sclerites (Lyons, 1966), monogenean egg shells (Ramalingam, 1973c), digenean egg shells and cysts (Toro-Goyco and del Valle, 1970; Rainsford, 1972; Ramalingam, 1973b; Byram and Senft, 1979; Robbins *et al.*, 1979; Waite and Rice-Ficht, 1987), digenean spines (Pearson *et al.*, 1985) and semi-characterized glycoprotein fractions (Machnicka-Roguska, 1965; Fujimoto and Iizuka, 1972; Peczon *et al.*, 1977; Cossey *et al.*, 1979; Caulfield *et al.*, 1987).

There have been a number of studies on the amino acid composition of purified proteins from helminths, particularly collagens (Table 2). Again, the amino acid composition of isolated parasite proteins is usually very similar to their mammalian counterparts. There are, however, some interesting differences such as the high cysteine content of nematode cuticle collagens and the occurrence of di- and trityrosine in the structural proteins of nematodes, cestodes and digeneans (Ramalingam, 1973b; Muthukrishnan, 1975; Nellaiappan and Ramalingam, 1980; Fujimoto *et al.*, 1981). A further interesting development in amino acid sequencing of parasite proteins is the prediction of amino acid sequence from the DNA sequence (Files *et al.*, 1983; McLachlan and Karn, 1983; Lanar *et al.*, 1986; Bobeck *et al.*, 1986, 1988; Zurita *et al.*, 1987).

TABLE 2 *Amino acid analyses of some isolated helminth proteins*

Species	*Reference(s)*
Collagen	
Ascaris lumbricoides	Josse and Harrington (1964), Fujimoto and Adams (1964), McBride and Harrington (1967a,b), Fujimoto (1968, 1975a) Peczon *et al.* (1975), Evans *et al.* (1976), Hung *et al.* (1977, 1980)
Caenorhabditis elegans	Ouazana and Gilbert (1979), Ouazana (1981), Ouazana and Herbage (1981), Ouazana *et al.* (1984)
Calicophoron erschowi, *Eurytrema pancreaticum*, *Bothriocephalus scorpii*, *Nybelinia* sp. larvae	Yarygina *et al.* (1982)
Fasciola hepatica	Nordwig and Hayduk (1969)
Macracanthorhynchus hirudinaceous	Cain (1970)
Onchocerca volvulus	Titanji *et al.* (1988)
Taenia solium	Torre-Blanco and Toledo (1981)
Cuticulin	
Ascaris lumbricoides	Fujimoto and Kanaya (1973), Fujimoto (1975b)
Actin	
Ascaris lumbricoides	Dedman and Harris (1975), Nakamura *et al.* (1979)
Paramyosin	
Ascaris lumbricoides, *Hymenolepis diminuta*	Winkelman (1976)
Fasciola hepatica	Ishii and Sano (1980)
Schistosoma mansoni	Cohen *et al.* (1987)
Troponin	
Ascaris lumbricoides	Kimura *et al.* (1987)
Calmodulin	
Ascaris lumbricoides	Masaracchia *et al.* (1986)
Haemoglobin	
Ascaris lumbricoides	Wittenberg *et al.* (1965), Okazaki *et al.* (1967), Darawshe *et al.* (1987)
Fasciolopsis buski	Cain (1969)
Cytochrome *c*	
Ascaris lumbricoides	Hill *et al.* (1971), Dedman and Harris (1975), Nakamura *et al.* (1979)
Histones	
Caenorhabditis elegans	Vanfleteren *et al.* (1986, 1987a,b, 1988)
Proteases	
Ancylostoma caninum	Hotez *et al.* (1985)
Schistosoma mansoni	Landsperger *et al.* (1982)
Protease inhibitors	
Ascaris lumbricoides	Fraefel and Acher (1968), Pudles *et al.* (1967) Peanasky *et al.* (1984), Babin *et al.* (1984)
Glutathione *S*-transferase	
Schistosoma mansoni	Taylor *et al.* (1988)

B. FREE AMINO ACIDS

Compared with the protein amino acids, the free amino acid fraction in tissues is small and probably represents several independent pools. In vertebrates, free amino acid levels range from 10 to 50 mg per 100 g fresh weight; in invertebrates, the figures are usually much higher (300–3000 mg per 100 g fresh weight), with the highest values being found in marine invertebrates (Awapara, 1962). Parasitic helminths show a similar wide range of free amino acid levels, with the highest levels being found in marine monogeneans (Arme, 1977). In vertebrates, the free and protein amino acid compositions are usually fairly similar. The free amino acid pools of invertebrates, however, are often dominated by one or two principal amino acids. In free-living invertebrates, amino acids are important in the regulation of intracellular osmotic pressure and this usually involves the synthesis and degradation of a non-essential amino acid such as alanine or proline. Free amino acids may also be an important energy source in free-living invertebrates, e.g. proline in a number of insects and glutamate and aspartate during the onset of anaerobiosis in molluscs (Hochachka and Somero, 1984). In addition, individual amino acids may have specific functions as neurotransmitters (glutamate, 4-aminobutyrate), phosphagens (arginine) or excretory products (see Section III).

In cestodes, the size of the free amino acid pool ranges from 100 to 400 mg per 100 g fresh weight, the highest values being found in the tapeworms of sharks (Simmons, 1969). In *Hymenolepis diminuta*, the major free amino acid is alanine (Daugherty, 1952b; Foster and Daugherty, 1959; Campbell, 1963b; Chappell and Read, 1973; Wack *et al.*, 1983; Webb, 1986), although the free amino acid pool of *H. diminuta* seems to be, at least to a certain extent, in equilibrium with the external amino acid pool (Arme and Read, 1969; Hopkins and Callow, 1965; Hopkins, 1969). The free amino acid pool of *H. diminuta* may undergo diurnal variations (Page *et al.*, 1978) and there is some evidence that free amino acids may be involved in osmotic regulation (Lussier *et al.*, 1978; Wack *et al.*, 1983). In *H. diminuta*, glutamate may also be an excitatory neurotransmitter (Webb, 1986).

Data have been published on the free amino acid pools of a range of cestodes, including *Davinea hewetensis* (Bhalya *et al.*, 1984), *Echinococcus* cysts (Krvacica *et al.*, 1959b), *Hydatigera taeniaeformis* larvae (Gaur and Agarwal, 1981), *Hymenolepis microstoma* (Litchford, 1970), *H. palmarum* (Bhalya *et al.*, 1985), *Ligula intestinalis* plerocercoids (Soutter *et al.*, 1980), *Moniezia expansa* (Campbell, 1960b; Lopez Gorge and Monteoliva Hernandez, 1964), *Raillietina cesticillus* (Foster and Daugherty, 1959; Bhalya *et al.*, 1984). *R. saharanpurensis* and *R. tetragona* (Bhalya *et al.*, 1984), *Taenia hydatigena* (Pathak *et al.*, 1980), *Staphylepis rustica* (Bhalya *et al.*, 1983b), *Triaenophorus nodulosus* procercoids (Guttowa, 1968), *Cittotaenia perplexa* and *Thysanosoma actinioides* (Campbell, 1960b), *Lytocestus indicus, Introvertus raipurensis* and *Lucknowia indica* (Niyogi and Agarwal, 1983) and *Lacistorhynchus tenuis, Calliobothrium verticillatum, Phyllobothrium foliatum, Rhodobothrium pulvinatum* and *Acanthobothrium* sp. (Simmons, 1969). In all these cases, the free amino acid pools are dominated by one or two non-essential amino acids, usually alanine, glycine or proline. A variety of non-protein amino acids has been found in the free amino acid pools of cestodes, including 3-alanine, citrulline, ornithine, 3-aminobutyrate, 3-aminoisobutyrate, 4-aminobutyrate, 4-aminoisobutyrate and taurine. Taurine often occurs in high levels in the tissues of marine animals and high levels are found in some of the shark tapeworms. It is possible that in those shark tapeworms that are impermeable to urea, taurine could have an important osmotic function. 3-Alanine and 3-aminoisobutyrate have been reported from the tissues of free-living as well as parasitic platyhelminths, and are probably derived from the breakdown of pyrimidines (Campbell, 1960a).

In digeneans, like cestodes, alanine, glycine or proline are usually the major free amino acids and the free amino acid pools range in size from 100 to 500 mg per 100 g fresh weight (Barrett, 1981). Proline and alanine are also the principal free amino acids in the parasitic turbellarian *Syndesmis franciscana* (Mettrick and Boddington, 1972a) and alanine and glutamate are the major free amino acids in a number of free-living turbellarians (Mettrick and Boddington, 1972b). Detailed analyses of free amino acid pools are available for the adult and larval stages of several species of digenean, including *Cercaria doricha* sporocysts (Negus, 1968), *Cercaria emasculans* sporocysts (Watts, 1970a), *Cotylophoron orientale* (Gupta and Bahadur, 1985), *Cryptocotyle lingua* rediae (Watts, 1970a), *Echinoparyphium* sp. rediae and cercariae (Cheng, 1963), *Echinostoma revolutum* (Bailey and Fried, 1977), *Fasciola gigantica* (Balogun and Braide, 1972), *Fasciola hepatica* (Daugherty, 1952b; Kurelec and Ehrlich, 1963; Kurelec and Rijavec, 1966; Gundlach *et al.*, 1971; Bankov and Ossikovski, 1976), *Fasciola indica* (Tandon, 1968), *Gastrothylax crumenifer* (Gupta and Agarwal, 1979; Gupta and Bahadur, 1985), *Glypthelmins quieta* and *G. amplicava* sporocysts and cercariae (Cheng, 1963), *Himasthla leptosoma* rediae (Watts, 1970a), *Microphallus pygmaeus* and *M. similis* sporocysts (Richards, 1969, 1970b), *Nanophyetus salmincola* rediae (Porter and Gamble, 1971), *Paramphistomum cervi* (Gaur and Agarwal, 1980; Varma and Sharma, 1984). *Paragonimus westermani* adults, rediae and cercariae (Hamajima, 1966), *Schistosoma mansoni* (Robinson, 1961; Senft *et al.*, 1972; Chappell, 1974; Asch, 1976; Seed *et al.*, 1980) and *Tremiorchis ranarum* (Karyakarte and Baheti, 1980). A number of non-protein amino acids occur in the free amino acid pools of digeneans, including 2-aminobutyrate, 4-aminobutyrate, creatinine, citrulline, homoserine, hydroxytryptophan, hydroxykyneurine, norleucine, norvaline, ornithine, sarcosine and taurine. Cain (1969) found significant amounts of taurine associated with the haemoglobin of *Fasciolopsis buski*. The reason for this is unknown, but it could be acting as an allosteric modifier.

The free amino acid pools of monogeneans again contain high levels of proline. Arme and Whyte (1975) and Arme (1977) examined eight species of monogeneans, and found that the free amino acid pools ranged from 400 mg per 100 g fresh weight for freshwater forms (*Diplozoon paradoxum*, *Discocotyle sagittata* and *Eupolystoma* sp.) up to 2600 mg per 100 g fresh weight for marine species (*Dictyocotyle coeliaca, Diclidophora denticulata, D. merlangi, Entobdella hippoglossi* and *E. soleae*). In the latter, proline may constitute up to 70% of the total free amino acids. There is no evidence that proline is particularly abundant in the diet of these monogeneans, nor could Arme find any evidence that proline was functioning either as an osmotic agent or as an energy source in these parasites. In *Fasciola hepatica*, it seems likely that proline is an end-product of arginine catabolism (see Section V.B.2).

Compared with platyhelminths, there is relatively little information on the free amino acid pools of parasitic nematodes. The overall free amino acid content of *Ascaris lumbricoides* is relatively small for an invertebrate (ca. 20 mg per 100 g fresh weight). In the perienteric fluid, the main amino acids are serine, alanine, glycine, proline and glutamate (Kajihara and Hashimoto, 1952; Cavier and Savel, 1954b,d; Ueno, 1960; Salmenkova, 1962; Jasksowski, 1963; Okuno, 1969; Abbas and Foor, 1978; Cuperlovic *et al.*, 1986); alanine, glutamate and serine are the major amino acids in the ovaries (Pollack and Fairbairn, 1955a; Dubinsky and Rybos, 1979), while lysine predominates in seminal fluid (Abbas and Foor, 1978). High free amino acid levels are also characteristic of the seminal fluids of insects, birds and mammals, but the exact function and source of seminal amino acids is not known.

Additional information on the free amino acid pools of nematodes is available for *Anisakis physeteris* (Viglierchio and Görtz, 1972), *Ascaridia galli* (Monteoliva Hernandez, 1962; Monteoliva Hernandez *et al.*, 1962; Nigam, 1978; Ossikovski, 1983), *Gnathostoma spinigerum* (Ando, 1957), *Haemonchus contortus* (Nigam, 1979; Kapur and Sood, 1984a), *Necator americanus* larvae (Desowitz, 1962), *Oesophagostomum columbianum* (Nigam, 1979; Kaur *et al.*, 1984), *Paranisakis* sp. (Gupta and Garg, 1977), *Setaria cervi* (Gupta and Kalia, 1977), *Toxocara canis* (Learmonth *et al.*, 1987) and *Trichuris ovis* (Nigam, 1979). In *Ascardia galli*, there is evidence that free amino acids may have a role in osmoregulation (Ossikovski, 1984). Among the non-protein amino acids found in nematodes are 3-alanine, 2-aminoisobutyrate, 3-aminoisobutyrate, allo-lysine, allo-leucine, citrulline, ornithine and taurine. 4-Aminobutyrate has also been identified as an inhibitory neurotransmitter in nematodes.

The major free amino acid in the pseudocoelomic fluid of the acanthocephalan *Moniliformis moniliformis* is glutamate, with glycine, alanine and proline being the next most important (Tanaka and MacInnis, 1980; Crompton and Ward, 1984).

In general, the free amino acid pools of parasitic helminths are similar to those of other invertebrates being dominated by a small number of non-essential amino acids. Unfortunately, with the exception of *Ascaris* and *Ascaridia*, there are no data on the amino acid pools of individual tissues in parasitic helminths, nor on how individual amino acid pools vary under different physiological conditions. Similarly, nothing is known about the rate of turnover of individual free amino acids. There is some evidence in parasitic helminths that, as in free-living invertebrates, free amino acids may have an osmotic role. There is always a danger that high levels of a single osmolyte will disrupt protein structure. However, glycine, proline, alanine, taurine and 3-alanine have been found to be compatible with protein function over a wide range of concentrations (Yancey *et al.*, 1986).

III. Excretion of Amino Acids

The excretion of amino nitrogen by invertebrates is widespread, usually comprising about 10% of the total excretory nitrogen, but reaching as much as 30% in echinoderms and some crustaceans. Of the helminths that have so far been studied, most have been found to excrete significant amounts of nitrogen in the form of amino acids, peptides or proteins. For example, amino nitrogen constitutes 35% of the nitrogenous end-products of *Nematodirus* sp. (Rogers, 1952), 49% of *Trichinella spiralis* larvae (Haskins and Weinstein, 1957a) and 28% of *Hymenolepis diminuta* (Fairbairn *et al.*, 1961); in *Nematodirus* sp., most of the amino nitrogen is in the form of peptides, but in *Trichinella* larvae free amino acids constitute 57% of the excreted amino nitrogen. The amino acids excreted by some helminths are summarized in Table 3.

The origin of the excretory amino acids in helminths (and invertebrates generally) is not clear, and they may well come from several different sources. The excretion of amino acids can provide a means of detoxifying ammonia, particularly in organisms that cannot synthesize urea or uric acid, and compounds such as proline, which are neutral in solution, can readily cross membranes. The synthesis of end-products such as urea and uric acid also requires metabolic energy. Some amino acids, in particular alanine and proline, may be true metabolic end-products. Alanine, formed by the transamination of pyruvate, is a major anaerobic end-product of carbohydrate catabolism in free-living invertebrates (Hochachka and Somero,

TABLE 3 *Amino acids excreted by some helminths*

Species	*Amino acid*																							*Reference(s)*
	Alanine	*Arginine*	*Asparagine*	*Aspartate*	*Cysteine*	*Glutamate*	*Glutamine*	*Glycine*	*Histidine*	*Isoleucine*	*Leucine*	*Lysine*	*Methionine*	*Ornithine*	*Phenylalanine*	*Proline*	*Serine*	*Threonine*	*Tryptophan*	*Tyrosine*	*Valine*	*Citrulline*	*4-Aminobutyrate*	
Nematodes																								
Ascaridia galli	+			+		+					+				+	+				+	+			Rogers (1955)
Ascaris lumbricoides	+			+		+					+				+	+					+		+	Flury (1912), Savell (1955), Rogers (1955), Rasero *et al.* (1968)
Caenorhabditis briggsae	+	+		+		+	+	+	+	+	+						+			+	+			Rothstein and Tomlinson (1962), Rothstein and Mayo (1964a,b), Rothstein (1963, 1965)
Ditylenchus dipsaci[a]	+	+	+	+	+	+	+	+		+	+	+	+	+	+		+	+		+	+	+	+	Myers and Krusberg (1965)
Ditylenchus myceliophagus[a]	+	+	+	+		+		+		+	+	+	+	+			+	+		+	+	+		Myers and Krusberg (1965)
Ditylenchus triformis[b]	+	+	+	+	+	+	+	+	+	+	+	+		+	+	+	+	+		+	+	+	+	Myers and Krusberg (1965)
Meloidogyne incognita[a]	+	+	+	+		+	+	+	+	+	+	+		+	+	+	+	+		+	+	+	+	Myers and Krusberg (1965)
Neoaplectana glaseri	+		+													+								Kleshchinova (1980a,b, 1983a,b)
Nematodirus fillicolis, *Nematodirus spathinger*	+			+		+					+				+	+					+			Rogers (1955)
Pratylenchus penetrans[a]	+			+		+		+						+			+	+						Myers and Krusberg (1965)
Trichinella spiralis larvae	+					+		+			+			+	+	+	+			+	+			Haskins and Weinstein (1957a,b)

Digeneans																							
Echinostoma revolutum[c]	+	+				+				+	+		+			+	+			+	+		Bailey and Fried (1977)
Fasciola gigantica	+								+			+			+	+	+						Lutz and Siddiqi (1971)
Fasciola hepatica[c]	+	+	+	+		+	+	+	+	+	+	+	+	+	+	+	+	+	+	+	+		van Grembergen and Pennoit de Cooman (1944), Locatelli and Camerini (1969), Moss (1970)
Microphallus pigmaeus[d] sporocysts	+			+		+		+		+	+	+				+					+		Richards (1970a)
Schistosoma mansoni[a] adults	+									+				+		+							Senft (1963, 1965, 1966), Roth and Hare (1966), Isseroff *et al.* (1983)
cercariae		+																					Granzer and Haas (1986)
Cestodes																							
Anoplocephala magna[e]	+							+				+					+			+	+		Krvavica *et al.* (1959a)
Gyrocotyle fimbriata	+							+															Bishop (1975)
Hymenolepis diminuta[f]	+		+	+		+		+	+	+	+	+	+	+	+	+	+	+		+		+	Fairbairn *et al.* (1961), Campbell (1963b), Webster and Wilson (1970), Zavras and Roberts (1984)
Taenia taeniaeformis larvae	+							+			+		+			+				+	+		Haskins and Olivier (1958)
Acanthocephalans																							
Moniliformis moniliformis	+							+															Crompton and Ward (1984)

[a] Also methionine sulphoxide.
[b] Also cysteic acid, hydroxyproline and methionine sulphoxide.
[c] Also 2-aminobutyrate.
[d] Also contains data on *Himasthla leptosoma* rediae, *Parvatrema homeotecnum* germinal sacs, *Cercaria lebouri* rediae and *Cercaria littorinae* sporocysts.
[e] Also 3-alanine.
[f] Also 3-amino alanine, 3-amino isobutyrate, methylhistidine, taurine, phosphoserine and glucosaminic acid.

1984). In *Hymenolepis diminuta*, alanine constitutes 38% of the total amino acids in protonephridial fluid (Webster and Wilson, 1970), alanine and proline are the major amino acids in the protonephridial fluid of *Fasciola gigantica* (Lutz and Siddiqi, 1971) and all of the helminths that have been studied excrete some alanine. In the digenean *Hirundinella ventricosa*, the major amino acids in the protonephridial fluid were leucine, lysine and valine (Lutz *et al.*, 1981).

Proline is formed from the breakdown of arginine (see Section V.B.2). Again, almost all of the helminths studied excrete proline, but digeneans in particular seem to excrete relatively large amounts. In hosts infected with *Fasciola hepatica*, the levels of biliary proline increase 10 000-fold (Isseroff *et al.*, 1972; Sheers *et al.*, 1980; Campbell *et al.*, 1981; Hudson and Thomas, 1981). This increase may not solely be due to excreted proline from the parasite and there are indications that proline catabolism in the host is impaired during fascioliasis (Chi and Isseroff, 1979). There is considerable evidence that proline is responsible for the bile duct hyperplasia found in fascioliasis and may also induce anaemia (Isseroff *et al.*, 1977, 1979; Sawma *et al.*, 1978; Chi and Isseroff, 1979; Isseroff, 1980; Girotra and Isseroff, 1980; Isseroff and Chi, 1981; Wolf-Spengler and Isseroff, 1983; Modavi and Isseroff, 1984). The excretion of proline by the adults and eggs of schistosomes has also been implicated in causing bile duct hyperplasia and fibrosis around egg granulomata (Bedi and Isseroff, 1979; Isseroff *et al.*, 1983; Kawanaka *et al.*, 1986).

Another possibility is that helminths lack selective transport mechanisms and absorb all available amino acids and then excrete the excess. Excretion of excess dietary amino acids has been well established for insects. The amino nitrogen excreted by helminths could also arise from the partial digestion of proteins (via faeces or vomit) or from the activity of extracorporeal enzymes. In this context, it is perhaps significant that peptide excretion occurs extensively in nematodes and digeneans, both of which possess an intestine, but not in cestodes.

Alternatively, amino nitrogen could be leaking out of moribund helminths and not be a normal excretory product. In many invertebrates and possibly in helminths as well, amino acids are involved in the regulation of intracellular osmotic pressure (particularly proline). Disruption of the osmotic balance when helminths are removed from their hosts and incubated *in vitro* could well lead to amino acid leakage.

The significance of amino acid excretion by helminths is uncertain. Much of the amino nitrogen excreted by helminths probably comes from the partial products of digestion. However, in digeneans and cestodes, at least some of the amino acids are excreted via the protonephridial system and may represent true excretory products. Finally, excreted amino acids could have a signalling function. Arginine released from the post-acetabular glands of *Schistosoma mansoni* cercariae stimulate the attachment response (Granzer and Hass, 1986). Peptide transfer occurs between male and female *S. mansoni* (Popiel and Basch, 1984), but whether amino acid exchange also takes place does not seem to have been investigated. The excretion of a large number of amino acids and amides by plant parasitic nematodes may have something to do with circumventing the plant's defence responses. However, excreted amino acids do not seem to be involved in the crowding effect in *H. diminuta* (Zavras and Roberts, 1984).

In addition to amino acids, some helminths also excrete amines. Adult and juvenile *T. spiralis* excrete a similar range of amines, including methylamine, ethylamine, *n*-propylamine, *n*-butylamine, *n*-pentylamine, *n*-hexylamine, *n*-heptylamine, ethylene diamine, 1,5-pentanediamine (cadaverine), *iso*-propylamine, *sec*-butylamine, *iso*-butylamine, 1-amino, 2-propanol and ethanolamine (Haskins and Weinstein, 1957a,c; Castro *et al.*, 1973). In larval *T. spiralis*, amines may constitute as much as 7% of the total nitrogen excreted. The amines excreted by the infective larvae of *Nippostrongylus brasiliensis* were methylamine, *n*-propylamine, *n*-butylamine, *iso*-butylamine, ethylene diamine, 1,5-pentanediamine and ethanolamine together with unidentified aliphatic monoamines (Weinstein and Haskins, 1955). The infective larvae of *Ascaris lumbricoides* (which are enclosed in a cleidoic egg) produced a similar range of amines to *N. brasiliensis*, with the exception of ethylenediamine which was absent from *A. lumbricoides* and ethylamine and 1-amino,2-propanol which were present (Haskins and Weinstein, 1957c). The plant parasitic nematode *Ditylenchus triformis* produces a number of secondary aliphatic amines (Myers and Krusberg, 1965), but amine excretion could not be detected in the free-living nematode *Caenorhabditis briggsae* (Rothstein, 1963). Amine excretion has only been reported in one non-nematode, the larva of *Taenia taeniaeformis*, which excretes methylamine, ethylamine, butylamine, ethylene diamine, 1,5-pentanediamine and 1-amino,2-propanol (Haskins and Olivier, 1958).

The origin of these amines is obscure, some of them can be formed by the decarboxylation of the appropriate amino acids, glycine, alanine, lysine and threonine, giving rise to methylamine, ethylamine, 1,5-pentanediamine and 1-amino,2-propanol, respectively. The direct decarboxylation of serine to ethanolamine has not been reported in animal tissues. So the synthesis of ethanolamine and of *n*-, *sec*- and *iso*-butylamine, pentylamine, hexylamine, heptylamine, ethylene diamine and isopropylamine must involve specialized and, as yet, unknown pathways. Possible sources of amines are from the transamination of aldehydes, the reduction of nitro- and azo-compounds or the catabolism of amino sugars. Aliphatic amines are often found in the flowering parts of plants, but again the biosynthetic pathways are unknown.

Amines are, of course, strongly alkaline as well as being toxic. Trimethylamine occurs in the tissues of marine organisms, but amine excretion appears to be confined to parasitic helminths. A possible function of amine production in helminths may be to neutralize the acidic end-products from carbohydrate catabolism. Amines also disrupt the keratin layer of the epidermis and so may aid the penetration of infective larvae.

Two compounds related to amines, betaine and creatinine, have been found in *Echinococcus* cysts (Frayha and Haddad, 1980).

IV. Amino Acid Synthesis

There are three basic methods available to study amino acid synthesis: nutrition experiments, labelling experiments and enzyme assays. Each of these methodologies has its attendant pitfalls, and an added problem in all studies on synthetic pathways is the possibility of a contribution from endosymbionts.

A. Dietary Requirements and Utilization

Nutritional studies on parasites *in vivo*, where the hosts are fed different diets, are almost impossible to interpret. Nutrients absent from the host's diet may be available to the parasite via the exocrino-enteric circulation or from the digestion of host tissues. Any effect on a parasite of host dietary restriction may be direct or the indirect result of a change in host metabolism. In theory, if a parasite can be cultured in a defined medium, then classical deletion or supplementation experiments should give unambiguous data on amino acid requirements. However, in practice, the results are rarely clear cut. The *in vitro* culture of parasitic helminths is not easy and cultivation media nearly always contain a number of undefined components such as liver extract, yeast extract or foetal serum. The results of deletion experiments can also be misleading. A non-essential amino acid may be designated essential if it is contaminated with a growth factor not otherwise supplied in the basal diet. An amino acid may also be mistakenly designated essential, when it is in fact only one of several possible alternative metabolites. Growth of a parasite in culture, particularly nematodes, may be restricted by underfeeding if a particular amino acid is required as a phagostimulant, even though it may not be nutritionally required. So again, a phagostimulatory amino acid could be wrongly designated as essential.

Classical deletion and supplementation studies have been carried out on a number of species of nematodes (*Aphelenchoides rutgersi*: Balasubramanian and Myers, 1971; Myers and Balasubramanian, 1973; *Caenorhabditis briggsae*: Vanfleteren, 1973; *Chiloplacus lentus*: Roy, 1975; *Neoaplectana glaseri*: Jackson, 1973; Kleshchinova, 1983b; *Rhabditis maupasi*: Brockelman and Jackson, 1978). The results indicate that nematodes require all 10 essential amino acids and in some cases possibly tyrosine as well. In *A. rutgersi*, tyrosine spares phenylalanine and cysteic acid can replace cysteine, suggesting that phenylalanine can be converted to tyrosine and cysteic acid to cysteine. Similarly, in *C. briggsae* methionine can be replaced by homocysteine.

Schistosoma eggs require exogenous amino acids for development. Deletion experiments with *S. japonicum* eggs have shown that, in addition to the 10 essential amino acids (arginine, histidine, isoleucine, leucine, lysine, methionine, phenylalanine, threonine, tryptophan and valine), cysteine, glycine, serine and tyrosine are also required (Kawanaka *et al.*, 1983). Senft (1963) demonstrated the utilization of arginine, histidine and tryptophan by adult *S. mansoni* during long-term maintenance. There is considerable information available on the uptake of amino acids by parasitic helminths, particularly cestodes and digeneans (for a review see Barrett, 1981), but this gives little indication of synthetic abilities. If a particular amino acid is essential, it will be taken up and utilized. However, non-essential amino acids will also be utilized if available in the diet. Also, during the life-cycle of a parasite, amino acid requirements and hence utilization may vary considerably, and therefore results will depend on the age or developmental stage investigated.

B. ISOTOPE STUDIES

Labelled substrates provide a useful approach for determining amino acid requirements, but again there are limitations to the method. Although such experiments may show the capacity for the synthesis of a particular amino acid, they do not show whether the rate is rapid enough to meet all the requirements of the organism. Contamination with microorganisms can again present a problem, particularly if long incubation times are used. Ideally, specifically labelled precursors should be used and the products should be shown to be labelled in the correct position. Failure to adequately purify either substrates or products can give rise to erroneous results.

The incorporation of label from a variety of substrates into amino acids by helminths is summarized in Table 4. Alanine, aspartate and glutamate are readily synthesized in most cases, as are glycine and serine. The apparent ability of a number of nematodes and of *S. mansoni* and *Echinococcus granulosus* protoscoleces to synthesize essential amino acids is particularly interesting. In the case of *C. briggsae*, Rothstein and Tomlinson (1962) ruled out the involvement of microorganisms in the synthesis of essential amino

TABLE 4 *Biosynthesis of amino acids in helminths*

Species	*Amino acids synthesized*		*Reference(s)*
	Non-essential[a]	*Essential*	
Nematodes			
Aphelenchoides sp.	ala, asp, cys, glu, gly, pro, ser, tyr	arg, ile, leu, lys, phe, thr, val	Balasubramanian and Myers (1971)
Caenorhabditis briggsae	ala, asp, cys, glu, gln, gly, pro, ser, tyr	arg, his, ile, leu, lys, thr, val	Nicholas *et al.* (1960), Rothstein (1963, 1965), Rothstein and Mayo (1964a,b) Rothstein and Tomlinson (1961, 1962)
Cooperia punctata	ala, asp, cys, glu, gly, pro, ser, tyr	arg, his, ile, leu, lys, met, phe, thr, val	Slonka *et al.* (1973)
Ditylenchus triformis	ala, asp, asn, cys, glu, gly, ser		Myers and Krusberg (1965)
Haemonchus contortus	ala, asp, cys, glu, gly, pro, ser, tyr	arg, his, ile, leu, lys, met, val	Kapur and Sood (1984b)
Heligmosomoides polygyrus	ala, asp, glu, gly, ser, pro, tyr	arg, ile, leu, lys, phe, thr, val	Grantham (1986)
Meloidogyne sp.	ala, glu	trp	Myers and Krusberg (1965)
Neoaplectana glaseri	ala, glu, pro, tyr	arg, ile, leu, lys, his, phe	Kleshinova (1983a)

Digeneans			
Fasciola hepatica	ala, asp, glu		Bryant and Smith (1963)
Schistosoma mansoni	ala, asp, glu	arg, leu, met, val	Senft (1963) Chappell and Walker (1982), Rahman *et al.* (1985)
Cestodes			
Echinococcus granulosus	asp, glu	arg, thr	Agosin and Repetto (1963)
Hymenolepis diminuta	ala, asp, gln		Prescott and Campbell (1965), Wack *et al.* (1983), Webb (1986)
Turbellarians			
Bipalium kewense	ala, asp, gln, glu, gly, ser	arg	Campbell (1965)
Acanthocephalans			
Moniliformis moniliformis	ala, asp, asn, glu, ser		Graff, (1964, 1965), Bryant and Nicholas (1965)

[a] In mammals.

TABLE 5 *2-Oxoglutarate/glutamate-linked transaminases in helminths*[a]

Species	*Donor amino acids*	*Reference(s)*
Nematodes		
Ancylostoma caninum	ala	Perez-Gimenez *et al.* (1967)
Ancylostoma ceylanicum	ala, arg, asp, cys, gly, leu, lys, met, orn, phe, ser, thr, trp, tyr, val, 2-aminobutyrate, DOPA	Singh, *et al.* (1987)
Ascaridia galli	ala, arg, asp, ile, leu, met, phe, tyr, val	Govorova (1965, 1968) Roy and Srivastava (1981), Singh *et al.* (1983a,b), Singh and Srivastava (1983)
Ascaris lumbricoides	ala, arg, asp, gly, met, phe, ser, 4-aminobutyrate	Cavier and Savel (1954c), Pollack and Fairbairn (1955b), Rasero *et al.* (1968) Zenka and Prokopic (1983), Dubinsky *et al.* (1984)
Bunostomum trigonocephalum	ala, asp	Trivedi and Gupta (1987)
Caenorhabditis elegans	4-aminobutyrate	Schaeffer and Bergstrom (1988)
Dictyocaulus filaria	ala, arg, asp, leu, lys, met, orn, phe, trp, tyr, val, norleucine, norvaline	Polyakova (1962)
Dipetalonema viteae (now *Acanthocheilonema viteae*)	ala, arg, asp, cys, gly, his, ile, leu, lys, met, orn, phe, ser, thr, val	Singh and Srivastava (1984), Srivastava *et al.* (1987)
Heligmosomoides polygyrus	ala, asn, asp, cys, leu, lys, 4-aminobutyrate, citrulline	Grantham and Barrett (1986a)
Litomosoides carinii	ala, asp	Singh and Srivastava (1984)
Neoaplectana glaseri	ala, arg, asp, gly, ile, leu, lys, pro, val	Kleschinova (1980a)
Nippostrongylus brasiliensis	ala, arg, cys, gly, his, leu, lys, met, orn, phe, ser, thr, val, 2-aminobutyrate, 4-aminobutyrate, DOPA	Watts and Atkins (1983, 1984, 1986), Singh *et al.* (1987)

Panagrellus redivivus	ala, arg, asn, asp, glu, ile, leu, met, orn, phe, pro, thr, tyr, val, 4-aminobutyrate citrulline	Grantham and Barrett (1986a)
Pelodera strongyloides	ala, asp	Scott and Whittaker (1970)
Setaria cervi	ala, asp	Singh and Srivastava (1984)
Stephanurus dentatus	ala, asp	Sharma and Singh (1977)
Trichuris ovis	ala, asp	Trivedi and Gupta (1987)
Digeneans		
Cercaria emasculans sporocyst	ala, asp	Watts (1970b)
Clinostomum complanatum	ala, asp	Sahu *et al.* (1987)
Cryptocotyle lingua rediae	ala, asp	Watts *et al.* (1970b)
Fasciola hepatica	ala, arg, asp, ile, leu, met, phe, orn, pro, tyr, val	Daugherty (1952a), Huang *et al.* (1962), Kurelec and Ehrlich (1963), Connolly and Downey (1968), Ertel and Isseroff (1974, 1976), Kurelec (1975), Sheers *et al.* (1980), Goldberg *et al.* (1980), Campbell *et al.* (1981), Han *et al.* (1983), Park *et al.* (1983), Lee *et al.* (1983)
Isoparorchis hypselobagri	ala,	Gupta and Agarwal (1986)
Microphallus pygmaeus sporocysts	ala, arg, asp, gly, ile, lys	Richards (1970a)
Paragonimus uterobilateralis	ala, asp	Zillman *et al.* (1987)
Schistosoma japonicum	ala, arg, asp	Huang *et al.* (1962), Hsiao and Hsu (1965)
Schistosoma mansoni	ala, arg, asp, gly, orn	Garson and Williams (1957), Coles (1973), Goldberg *et al.* (1979, 1980)
cercariae	ala, asp	Coles (1973)
Cestodes		
Gyrocotyle fimbriata	ala, asp	Bishop (1975)
Hymenolepis citelli	ala, asn, asp	Wertheim *et al.* (1960)

TABLE 5 (*continued*)

Species	*Donor amino acids*	*Reference(s)*
Hymenolepis diminuta	ala, asn, asp, cys, cysteine sulphinate	Aldrich *et al.* (1954), Foster and Daugherty (1959), Wertheim *et al.* (1960), Prescott and Campbell (1965), Gomez-Bautista and Barrett (1988)
Hymenolepis nana	asn, asp	Wertheim *et al.* (1960)
Lytocestus indicus	ala, asp	Rasheed (1981)
Moniezia expansa	ala, asp, 4-aminobutyrate	Rasero *et al.* (1968), Cornish and Bryant (1975)
Raillietina cesticillus	ala, asp	Foster and Daugherty (1959)
Taenia taeniaeformis larvae and adults	asp	Waitz (1963)
Acanthocephalans		
Macracanthorhynchus hirudinaceus	4-aminobutyrate	Rasero *et al.* (1968)

[a] Additional data in Tandon and Misra (1984).

acids and suggested that the inhibition of syntheses caused by the addition of antibiotics was due to a direct effect on the nematode enzymes. However, in none of the cases where essential amino acids were apparently being synthesized has the position of the label been determined and so contamination cannot be ruled out. Grantham (1986; Ph.D. thesis, University of Wales) noted that the apparent incorporation of label from ^{14}C-glucose into leucine, isoleucine and phenylalanine by *Heligmosomoides polygyrus* and *Panagrellus redivivus* dropped by 46% after rechromatography of the amino acids on ion exchange Sephadex, suggesting significant non-amino contamination.

C. SYNTHETIC ENZYMES

The synthetic pathways for the non-essential amino acids are usually relatively short (1–3 steps), whereas the essential amino acids have much longer pathways (5–10 steps). Amino acids can be conveniently divided into six groups, reflecting their synthesis from key intermediates in carbohydrate metabolism.

1. *Group I*

Glutamate, glutamine, proline, hydroxyproline and arginine are all ultimately derived from 2-oxoglutarate. Glutamate is formed by transamination of 2-oxoglutarate and 2-oxoglutarate-linked amino-transferases are probably universally distributed in helminths (Table 5). Glutamine is an important donor of amino groups in a variety of synthetic reactions and is formed from glutamate via the enzyme glutamine synthetase:

$$\text{glutamate} + NH_3 + \text{ATP} \longrightarrow \text{glutamine} + H_2O + \text{ADP} + \text{Pi}$$

Glutamine synthetase has been demonstrated in *H. polygyrus* and *P. redivivus* (Grantham and Barrett, 1988). The incorporation of label into glutamate has been widely reported in helminths and incorporation into glutamine has been shown in *Caenorhabditis elegans, Hymenolepis diminuta* and *Bipalium kewense* (Table 4).

Proline is synthesized from glutamate by a four-step pathway:

$$\text{glutamate} \xrightarrow{(1)} \text{5-glutamylphosphate} \xrightarrow{(2)} \text{glutamic acid semi-aldehyde} \xrightarrow{(3)}$$

$$\Delta\text{-pyrroline-5-carboxylic acid} \xrightarrow{(4)} \text{proline}$$

where (1) is glutamate kinase, (2) glutamylphosphate dehydrogenase, (3) a spontaneous reaction and (4) Δ-pyrroline-5-carboxylate reductase (proline-5-oxidoreductase).

The digenean *Fasciola hepatica*, which excretes large amounts of proline, has high levels of Δ-pyrroline-5-carboxylate reductase (Rijavec, 1974; Kurelec, 1975; Isseroff and Ertel, 1976; Isseroff, 1980). However, in *F. hepatica*, glutamate is not the principal source of glutamic acid semi-aldehyde; instead, the latter is formed from ornithine by transamination and high levels of ornithine 5-transaminase have been found in this parasite (Rijavec, 1974; Kurelec, 1975; Isseroff, 1980; Ertel and Isseroff, 1974, 1976; Goldberg *et al.*, 1980; Campbell *et al.*, 1981). The source of ornithine in *F. hepatica* is probably the cleavage of arginine by arginase (see Section V.A.3). There appears to be no Δ-pyrroline-5-carboxylate dehydrogenase (Δ-pyrroline-5-carboxylate oxidoreductase) in *F. hepatica* (Isseroff and Ertel, 1976); this is the enzyme which reduces Δ-pyrroline-5-carboxylate to glutamate, providing a route for the conversion of arginine or proline to glutamate. Proline excretion also occurs in *Schistosoma mansoni*, and again high activities of ornithine transaminase and Δ-pyrroline-5-carboxylate reductase have been found in adults and ova (Goldberg *et al.*, 1979, 1980; Isseroff, 1980; Isseroff *et al.*, 1983). Because the conversion of ornithine to proline results in the oxidation of NADH, proline production is redox linked and could act as an anaerobic electron sink (Kurelec, 1975).

Proline is converted to hydroxyproline only after it has first been incorporated into protocollagen. The proline residue is hydroxylated by proline-4-monoxygenase, a complex enzyme that requires as cofactors ascorbate, 2-oxoglutarate, CoA, ferrous ions and molecular oxygen:

$$\text{proline residue} + O_2 + \text{2-oxoglutarate} + \text{CoA} \longrightarrow \text{4-hydroxyproline residue} + \text{succinylCoA} + CO_2 + H_2O$$

Proline 4-monoxygenase has been found in the cuticle, muscles and developing eggs of *Ascaris lumbricoides* (Fujimoto and Prockop, 1969; Chvapil and Ehrlich, 1970; Chvapil *et al.*, 1970; Cain and Fairbairn, 1971). The enzyme from the muscles of *A. lumbricoides* is inhibited by oxygen above 5 vol%, the cuticle and egg enzymes, on the other hand, are stimulated by oxygen. Proline 4-monoxygenase has also been demonstrated in *Panagrellus silusiae* (Leushner and Pasternack, 1978), but could not be detected in the cysticerci of *Taenia solium* (Torre-Blanco and Alvizouri, 1987). The latter parasite appears to be unique in having a collagen that contains no hydroxyproline (Torre-Blanco and Toledo, 1981).

Glutamate is also the starting point for arginine synthesis. However, the details of the pathway differ in different organisms. In microorganisms, glutamate is first acetylated and the resulting *N*-acetylglutamate is converted by reduction and transamination to *N*-acetylornithine. After hydrolysis, ornithine is converted to arginine via a series of reactions that are common with the urea cycle (carbamoylphosphate synthetase, ornithine transcarbamoylase, arginosuccinate synthetase, arginosuccinate lyase). Mammals cannot carry out the conversion of glutamate to ornithine via acetylglutamate. However, they are able to synthesize small amounts of ornithine by transamination from glutamic acid semi-aldehyde; hence in mammals arginine is "semi-essential". In helminths, there is no evidence for the conversion of glutamate to ornithine via *N*-acetylglutamate. In addition, the apparent absence, or very low activities of, the urea cycle enzymes (see Section V.C) makes it unlikely that parasitic helminths can synthesize arginine from ornithine.

2. *Group II*

Aspartate, asparagine, lysine, methionine and threonine form the aspartate group of amino acids. Aspartate is synthesized by the transamination of oxaloacetate. Again, oxaloacetate-linked transaminases are probably universally distributed in helminths and incorporation of label into aspartate has been widely reported (Tables 4, 5 and 6). The biosynthesis of asparagine is catalysed by asparagine synthetase, in a reaction analogous to that catalysed by glutamine synthetase. A major difference, however, is that while ATP is converted to ADP and Pi in the glutamine synthetase reaction, AMP and PPi are formed in the reaction catalysed by asparagine synthetase:

$$\text{aspartate} + NH_3 + \text{ATP} \longrightarrow \text{asparagine} + H_2O + \text{AMP} + \text{PPi}$$

In most animals, glutamine is the nitrogen donor, rather than ammonia. Grantham and Barrett (1988) reported low levels of asparagine synthetase in *Panagrellus redivivus*, but could not detect it in *Heligmosomoides polygyrus*.

In plants and bacteria, the starting point for lysine, methionine and threonine synthesis is aspartate semi-aldehyde. This compound is not synthesized by most higher organisms, and hence all three amino acids are normally essential.

Methionine synthesis involves a series of intermediates, some of which are common to cysteine metabolism. The sequence is:

$$\text{aspartate semi-aldehyde} \longrightarrow \text{homoserine} \longrightarrow \text{cystathionine} \longrightarrow \text{homocysteine} \longrightarrow \text{methionine}$$

TABLE 6 *Pyruvate/alanine-linked transaminases in helminths*

Species	*Donor amino acids*	*Reference(s)*
Nematodes		
Aphelenchoides ritzemabosi	asp, glu	Miller and Roberts (1964)
Ascaridia galli	asp, glu	Singh and Srivastava (1983)
Ascaris lumbricoides	asp, glu, gly, ser	Pollack and Fairbairn (1955b)
Dictyocaulus filaria	asp, glu	Polyakova (1962)
Dipetalonema viteae (now *Acanthocheilonema viteae*)	arg, asp, cys, gly, glu, his, ile, leu, lys, met, orn, phe, ser, thr, val	Srivastava *et al.* (1987)
Heligmosomoides polygyrus	asp, gln, glu, his, try, citrulline	Grantham and Barrett (1986a)
Neoaplectana glaseri	asp, glu	Kleschinova (1980a)
Panagrellus redivivus	arg, asp, gln, glu, lys, met, phe, try	Grantham and Barrett (1986a)
Digeneans		
Cercaria emasculans sporocysts	glu, gly	Watts (1970b)
Cryptocotyle lingua rediae	glu	Watts (1970b)
Fasciola hepatica	glu, orn	Bryant and Smith (1963), Kurelec and Erhlich (1963), Kurelec (1975), Huang *et al.* (1962)
Schistosoma japonicum	glu	Huang *et al.* (1962)
Schistosoma mansoni	glu	Garson and Williams (1957)
Microphallus pygmaeus sporocysts	glu, gly	Richards (1970a)
Cestodes		
Hymenolepis citelli	glu	Wertheim *et al.* (1960)
Hymenolepis diminuta	asn, asp, glu	Foster and Daugherty (1959), Aldrich *et al.* (1954), Wertheim *et al.* (1960)
Hymenolepis nana	glu	Wertheim *et al.* (1960)
Raillietina cesticillus	glu	Foster and Daugherty (1959)

Despite some evidence from dietary experiments that *Caenorhabditis briggsae* could convert homocysteine to methionine (Vanfleteren, 1973), neither methionine synthase nor betaine : homocysteine transmethylase could be demonstrated in *Brugia pahangi* and both *B. pahangi* and *Dirofilaria immitis* appear to be unable to synthesize methionine (Jaffe, 1980; Jaffe and Chrin, 1979, 1981).

Cell-free extracts of *A. lumbricoides* can convert 2-hydroxy-4-methylthiobutyrate and 2-oxo-4-methylthiobutyrate to methionine (Langer *et al.*, 1971). The probable reactions are:

$$\text{2-hydroxy-4-methylthiobutyrate} + NAD^+ \longrightarrow \text{2-oxo-4-methylthiobutyrate} + NADH + H^+ \quad (\textit{2-hydroxyacid dehydrogenase})$$

$$\text{2-oxo-4-methylthiobutyrate} + \text{glutamate} \longrightarrow \text{methionine} + \text{2-oxoglutarate} \quad (\textit{transaminase})$$

These two steps, however, are not part of the normal pathway either for methionine synthesis or methionine catabolism (Livesey, 1984).

3. *Group III*

These are the amino acids formed from pyruvate, namely alanine, leucine, valine and isoleucine. Alanine is formed by the direct transamination of pyruvate and these transaminases have been found in all the parasites that have been studied (Table 5).

Part of the carbon skeleton of isoleucine comes from threonine, the remainder from pyruvate. Pyruvate also supplies the carbon atoms of valine and leucine. These pathways do not occur in animals and despite isotope studies suggesting that valine, leucine and isoleucine synthesis might occur in some parasites (see Section IV.B), none of the enzymes involved has ever been investigated. Mammals and helminths do contain transaminases capable of interconverting all three of the branched chain amino acids with their corresponding 2-oxoacids (see Section V.B.3). This explains the ability of oxo acids to replace their corresponding amino acids in dietary experiments.

4. *Group IV*

The three aromatic amino acids phenylalanine, tyrosine and tryptophan are all formed from phosphoenolpyruvate. The three have a common pathway via shikimic acid to chorismic acid. In phenylalanine and tyrosine synthesis, chorismate is rearranged to give prephenic acid which is then decarboxylated and reduced. The resulting phenylpyruvate is transaminated to yield phenylalanine, which can be hydroxylated to tyrosine. The synthesis of tryptophan is more complicated and proceeds via anthranilic acid and indole-glycerophosphate. Neither of these pathways operates in mammals; the last enzyme in the tryptophan pathway, tryptophan synthetase, could not be detected in *Nippostrongylus brasiliensis* (Walker and Barrett, unpublished), but none of the other enzymes involved has been studied in helminths. Mammals can readily convert phenylalanine to tyrosine via a biopterin-dependent phenylalanine hydroxylase and there is some evidence from nutritional studies that *Aphelenchoides rutgersi* can do the same (see Section IV.A).

5. *Group V*

The carbon chain of serine, glycine and cysteine are all derived from the glycolytic intermediate, 3-phosphoglyceric acid. Oxidation and transamination of 3-phosphoglycerate yields phosphoserine which is then cleaved by a phosphatase:

$$\text{3-phosphoglycerate} \xrightarrow{(1)} \text{3-hydroxyphosphopyruvate} \xrightarrow{(2)}$$
$$\text{3-phosphoserine} \xrightarrow{(3)} \text{serine}$$

where (1) is phosphoglycerate dehydrogenase, (2) amino transferase and (3) phosphoserine phosphatase. There is also a minor, non-phosphorylating pathway for serine synthesis in animals:

$$\text{3-phosphoglycerate} \xrightarrow{(1)} \text{2-phosphoglycerate} \xrightarrow{(2)} \text{glycerate} \xrightarrow{(3)}$$
$$\text{3-hydroxypyruvate} \xrightarrow{(4)} \text{serine}$$

where (1) is phosphoglyceromutase, (2) glycerate-2-phosphohydrolase, (3) glycerate dehydrogenase and (4) amino transferase.

Although serine is readily synthesized by helminths from labelled precursors (Table 4), the enzymes involved have not been investigated. Glycine and serine can be interconverted via serine hydroxymethyl transferase:

$$\text{serine} + \text{tetrahydrofolate} \rightleftharpoons \text{glycine} + \text{methylenetetrahydrofolate} + H_2O$$

Glycine is again readily synthesized from labelled precursors in helminths and serine hydroxymethyltransferase has been demonstrated in *N. brasiliensis*

(Walker and Barrett, unpublished), *Brugia pahangi* and *Dirofilaria immitis* (Jaffe and Chrin, 1980). Glycine and serine are readily interconverted in most tissues and are important contributors to the 1 carbon pool.

In mammals, cysteine is synthesized from serine and homocysteine (the latter being an intermediate in methionine metabolism):

$$\text{serine} + \text{homocysteine} \longrightarrow \text{cystathionine} \qquad (\textit{cystathionine-}\beta\textit{-synthase})$$

$$\text{cystathionine} + H_2O \longrightarrow \text{cysteine} + NH_3 + \text{2-oxobutyrate} \; (\gamma\textit{-cystathionase})$$

So, in mammals, the sulphur atom of cysteine comes from methionine, which is one of the essential amino acids. In some tissues, cystathionine-β-synthase can also catalyse the direct displacement of the hydroxyl group of serine by inorganic sulphide (Thong and Coombs, 1985):

$$\text{serine} + H_2S \rightleftarrows \text{cysteine} + H_2O \qquad (\text{“}\textit{serine sulphydrase}\text{”})$$

In plants and microorganisms, the incorporation of inorganic sulphide into serine involves two steps:

$$\text{serine} + \text{acetylCoA} \longrightarrow O\text{-acetylserine} + \text{CoA} \qquad (\textit{acetyl transferase})$$

$$O\text{-acetylserine} + H_2S \longrightarrow \text{cysteine} + \text{acetate} \qquad (\text{O-}\textit{acetylserine sulphydrase})$$

Both cystathionine-β-synthase and γ-cystathionase have been detected in *Hymenolepis diminuta* (Gomez-Bautista and Barrett, 1988) and in a range of nematodes (*Cooperia oncophora, Panagrellus redivivus, Ostertagia circumcincta, Trichostrongylus colubriformis*: Walker and Barrett, unpublished). Crompton and Ward (1984) reported low rates of conversion of labelled serine to cystathionine in *Moniliformis moniliformis*. Cystathionine-β-synthase is a multifunctional enzyme catalysing the synthesis of cystathionine from homocysteine and serine or cysteine, the synthesis of thio-ethers from cysteine and thiol compounds and of cysteine from serine and inorganic sulphide. The enzymes from different sources differ in their substrate specificities and in contrast to the enzyme from mammals the enzymes so far studied from helminths show considerable "serine sulphydrase" activity. γ-Cystathionase is again a multifunctional enzyme catalysing a variety of 2- and 3-elimination reactions, the cleavage of cystathionine to cysteine and 2-oxobutyrate, the cleavage of cystine to thiocysteine, ammonia and pyruvate and of homoserine to 2-oxobutyrate and ammonia. This enzyme may also have threonine dehydratase activity. Again, the relative specificity of the enzyme for different substrates may differ in different organisms.

Jaffe (1980) has reported that *B. pahangi* and *D. immitis* possess the pathways to convert methionine via *S*-adenosylmethionine and *S*-adenosylhomocysteine to homocysteine and hence via cystathionine to cysteine. 5-Adenosylmethionine decarboxylase has been characterized from *Ascaris suum* and *Onchocerca volvulus* (Rathaur *et al.*, 1988); this enzyme is, however, primarily involved in polyamine synthesis.

6. *Group VI*

Histidine arises from phosphoribosylpyrophosphate and part of the purine ring of ATP. The pathway of histidine synthesis is long and complicated and histidine is an essential amino acid in animals.

Studies on synthetic enzymes, like dietary and isotope studies, need to be interpreted with caution. The presence of an enzyme does not necessarily mean that a particular pathway is active and care must be taken not to "construct" pathways from unrelated enzymes. Synthetic pathways are nearly always under tight feedback control and many synthetic enzymes may be inducible.

V. Amino Acid Catabolism

Amino acids are not an important energy source in parasitic helminths, only in digenean rediae and sporocysts and in some plant parasitic nematodes is there any evidence for appreciable amino acid catabolism (Van Gundy *et al.*, 1967; Friedl, 1961; Vernberg and Hunter, 1963; Richards, 1970b; Pascoe, 1970; Pascoe and Richards, 1970). In organisms that are heavily committed to synthesis, such as dividing bacteria and growing plants, amino acid degradation is not a significant process and the same may apply to parasites.

Amino acid catabolism can be considered in two phases, the initial removal of the amino group, followed by catabolism of the carbon skeleton. The pathways of amino acid catabolism are not the reverse of the synthetic pathways, although common steps may occur. Several of the intermediates from amino acid catabolism are also important precursors for other synthetic pathways.

A. Removal of the Amino Group

The first step in amino acid catabolism is the removal of the 2-amino group. There are two main routes for this—transamination and oxidative deamination. In addition, there are also a number of non-oxidative deaminases for specific amino acids.

1. *Transamination*

The transaminases are a series of pyridoxal phosphate-dependent enzymes that catalyse the transfer of the 2-amino group of an amino acid to the 2 carbon of a 2-oxo acid. The reaction is freely reversible, with an equilibrium constant of approximately 1. Of the 20 common protein amino acids, the catabolism of at least 12 of them (alanine, arginine, aspartate, asparagine, cysteine, isoleucine, leucine, lysine, phenylalanine, tryptophan, tyrosine and valine) starts with transamination.

A large number of different aminotransferases have been described (at least 40). Most of them are specific for 2-oxoglutarate as the oxo-acid, but show a rather wider specificity for the amino donor. A small group of aminotransferases appear to be specific for pyruvate as the oxo-acid. However, there is almost certainly not a separate class of transferases specific for oxaloacetate, and the reactions so described probably represent the reverse reaction of aspartate : oxoglutarate and aspartate : pyruvate transaminases.

2-Oxoglutarate-linked transaminases are widely distributed in parasitic helminths (Table 5) and, as in vertebrates, aspartate : oxoglutarate and alanine : oxoglutarate transaminases are usually the most active. Pyruvate-linked transaminases have also been demonstrated in helminths (Table 6), but the number of amino acids that can act as donors in this system is very limited. In general, compared with mammals, relatively few amino acids seem to be able to act as co-substrates in helminth transaminase reactions (Grantham and Barrett, 1986a).

The synthesis of amino sugars also involves aminotransferases, and a glutamine : fructose-6-phosphate aminotransferase has been demonstrated in *Ascaris lumbricoides* (Dubinsky *et al.*, 1985).

2. *Oxidative deamination*

The amino groups from the different amino acids are collected via transaminase reactions into glutamate, which then acts as the amino donor for the formation of the different nitrogenous end-products. The enzyme glutamate dehydrogenase catalyses the oxidative deamination of glutamate to release ammonia:

$$\text{L-glutamate} + H_2O + NAD(P)^+ \rightleftharpoons \text{2-oxoglutarate} + NH_4^+ + NAD(P)H$$

Glutamate dehydrogenase from animal sources can use either $NADP^+$ or NAD^+ as the co-factor. In mammals, the enzyme is primarily mitochondrial, with separate isoenzymes existing in the cytoplasm and mitochondrion. Mammalian glutamate dehydrogenase is an allosteric enzyme, it is activated by ADP, GDP and certain amino acids and inhibited by ATP, GTP and NADH. The enzyme is also affected by hormones. Glutamate dehydrogenase has been demonstrated in a wide range of helminths, including *Hymenolepis diminuta* (Daugherty, 1955; Read, 1953; Mustafa *et al.*, 1978), *Moniezia benedeni* (Van Grembergen, 1944), *M. expansa* (Barrett, unpublished), *Microphallus pygmaeus* (Pascoe, 1970), *M. similis* (McManus and James, 1975), *Ancylostoma caninum* (Perez-Gimenez *et al.*, 1967), *Ascaridia galli* (Singh *et al.*, 1983a,b), *Ascaris lumbricoides* (Pollack and Fairbairn, 1955b; Pollack 1957a,b), *Dirofilaria immitis* (McNeill and Hutchison, 1971; Turner *et al.*, 1986), *Haemonchus contortus* (Rhodes and Ferguson, 1973), *Heligmosomoides polygyrus* (Grantham and Barrett, 1986a) and *Panagrellus redivivus* (Wright, 1975; Grantham and Barrett, 1986a). The regulatory properties of glutamate dehydrogenase have been studied in detail in four of the species—*H. contortus*, *H. polygyrus*, *P. redivivus* and *H. diminuta*. The enzyme from these helminths shows both similarities to and differences from their mammalian counterpart. In *H. diminuta*, the enzyme was primarily cytosolic and the nucleotides AMP, ADP, ATP, GDP and GTP have little effect on activity (Mustafa *et al.*, 1978). The mitochondrial glutamate dehydrogenase from *H. contortus* was specific for NAD and was inhibited by AMP, ADP, ATP, aspartate and thyroxine (Rhodes and Ferguson, 1973). In *H. polygyrus* and *P. redivivus*, glutamate dehydrogenase activity was found in both the cytoplasm and the mitochondria. When NAD(H) was the co-factor, the mitochondrial enzymes showed similar properties to mammalian glutamate dehydrogenase, being activated by AMP and ADP, and inhibited by ATP, GTP and GDP. If NADP(H) was the co-factor, the mitochondrial glutamate dehydrogenases from these two nematodes showed anomalous properties (Grantham and Barrett, 1986a). Turner *et al.* (1986) have looked at the properties of the mitochondrial and cytoplasmic glutamate dehydrogenases of *D. immitis*; both enzymes were inhibited by ATP, but differed almost totally from each other in their responses to other nucleotides.

Studies on the regulatory properties of glutamate dehydrogenase are difficult to interpret, because the response of the enzyme depends very much on the conditions (pH, metal ions) and on the other co-factors present ($NAD^+/NADP^+$). Glutamate dehydrogenase has a central role linking amino acid and carbohydrate metabolism. Because it is a reversible enzyme, glutamate dehydrogenase can be involved both in amino acid synthesis and amino acid catabolism, depending on the intracellular conditions. The primary metabolic role of glutamate dehydrogenase (catabolic or synthetic) has not yet been established and may differ with the different isoenzymes. For example, those isoenzymes that show maximum activity with NAD^+

may function primarily in catabolism, whereas those that show maximum activity with $NADP^+$ may be mostly synthetic.

Alanine could have an important role in invertebrates as a nitrogen carrier. There have been a number of claims that there is a specific alanine dehydrogenase in invertebrates which catalyses the oxidative deamination of alanine in a reaction analogous to glutamate dehydrogenase (most glutamate dehydrogenases also show some activity with alanine as substrate). Such an enzyme might be expected to have a key role in the anaerobic production of alanine (Hochachka and Somero, 1984). However, although an alanine dehydrogenase has been isolated from microorganisms, there is no evidence that such an enzyme occurs in eukaryotes. The apparent reductive amination of pyruvate reported in *H. diminuta* (Daugherty, 1954), *A. lumbricoides* (Pollack and Fairbairn, 1955b; Pollack, 1957a,b), *A. galli* (Singh *et al.*, 1983b) and *D. immitis* (Polyakova, 1962) is probably the result of the sequential activities of glutamate dehydrogenase and pyruvate: glutamate transaminase. A similar mechanism would also account for the reductive amination of oxaloacetate reported in *H. diminuta* (Daugherty, 1954).

A minor pathway for the oxidative deamination of amino acids is via the flavin-linked L-amino acid oxidase:

$$\text{L-amino acid} + H_2O + FMN \longrightarrow \text{2-oxoacid} + NH_3 + FMNH_2$$
$$FMNH_2 + O_2 \longrightarrow FMN + H_2O_2$$

L-Amino acid oxidases have been found in *H. diminuta* (Daugherty, 1955), *A. galli* (Rogers, 1952), the muscles and intestine of *A. lumbricoides* (Cavier and Savel, 1954c; Pollack and Fairbairn, 1955b; Lopez Gorge, 1969), *H. polygyrus* (Grantham and Barrett, 1986a), *Nematodirus* sp. (Rogers, 1952) and *P. redivivus* (Grantham and Barrett, 1986a). Low activities of the corresponding D-amino acid oxidase have also been demonstrated in *H. polygyrus* and in the free-living nematodes *P. redivivus* and *T. aceti* (Aueron and Rothstein, 1974; Grantham and Barrett, 1986a). In addition to the general D- and L-amino acid oxidases, mammals have a separate flavoprotein glycine oxidase. The latter may be an important alternative to glutamate dehydrogenase as a route for oxidative deamination.

3. *Specific deaminases*

A number of amino acids—arginine, asparagine, histidine, serine, threonine and glutamine—can be non-oxidatively deaminated by specific deaminases. All of these enzymes have now been demonstrated in helminths.

Arginase cleaves arginine into ornithine and urea:

$$\text{arginine} + H_2O \longrightarrow \text{ornithine} + \text{urea}$$

Arginase has been widely reported from parasitic helminths, including nematodes (Rogers, 1952; Cavier and Savel, 1954a; Lopez-Gorge and Monteoliva, 1969; Paltridge and Janssens, 1971; Wright, 1975; Sokhina, 1976; Shishkova-Kasatochkina *et al.*, 1976, 1986; Sokhina and Koloskova, 1978; Sokhina and Shishkova-Kasatochkina, 1979; Roy and Srivastava, 1981; Grantham and Barrett, 1986a), digeneans (Van Grembergen and Pennoit de Cooman, 1944; Campbell and Lee, 1963; Tao and Huang, 1965; Rijavec, 1965, 1974; Rijavec and Kurelec, 1966; Senft, 1966; Read, 1968; Janssens and Bryant, 1969; Kurelec, 1964a, 1974a, 1975; Sokhina, 1975) and cestodes (Van Grembergen and Pennoit de Cooman, 1944; Campbell, 1963a; Campbell and Lee, 1963; Rijavec, 1965; Read, 1968; Janssens and Bryant, 1969; Dubovskaya, 1979, 1982a,b, 1984; Fukase *et al.*, 1984; Celik, 1986a). Arginase may be involved in at least two different pathways—the conversion of arginine to proline (see Section III.C) and in the urea cycle (see Section V.C). In addition, in free-living organisms, arginase may play a role in maintaining homeostasis during long-term anaerobiosis (Wieser and Platzer, 1983).

L-Serine dehydratase and L-threonine dehydratase catalyse the deamination of serine and threonine, respectively:

$$\text{serine} + H_2O \longrightarrow \text{pyruvate} + NH_3 + H_2O$$

$$\text{threonine} + H_2O \longrightarrow \text{2-oxobutyrate} + NH_3 + H_2O$$

Both of these enzymes have been detected in *Heligmosomoides polygyrus* and *Panagrellus redivivus* (Grantham and Barrett, 1986a) and in *Fasciola indica* (Tandon and Misra, 1980). Serine dehydratase has also been demonstrated in *Gyrocotyle fimbriata* (Bishop, 1975) and threonine dehydratase in *Ascaridia galli* (Roy and Srivastava, 1981; Singh *et al.*, 1983b) and *Hymenolepis diminuta* (J. Barrett, unpublished). There is some evidence that threonine dehydratase may be identical with γ-cystathionase, an important enzyme of cysteine metabolism. It has also been suggested that in some cases serine and threonine dehydratase are the same enzyme (Keleti *et al.*, 1987). Histidase, the first enzyme in the catabolism of histidine, has been found in *A. galli* (Berdyeva and Dryuchenko, 1975; Dryuchenko, 1979; Roy and Srivastava, 1981; Singh *et al.*, 1983b), *Ascaris lumbricoides* (Berdyeva and Dryuchenko, 1975; Dryuchenko and Berdyeva, 1975, 1976), *H. polygyrus* (Grantham and Barrett, 1986a), *P. redivivus* (Grantham and Barrett, 1986a), *G. fimbriata*

(Bishop, 1975), *Fasciola hepatica* (Berdyeva and Dryuchenko, 1975) and *Schistosoma mansoni* (Saber and Wu, 1985):

$$\text{histidine} \longrightarrow \text{urocanic acid} + NH_3$$

Finally, glutaminase and asparaginase have been shown in *H. polygyrus* and *P. redivivus* (Wright, 1975; Grantham and Barrett, 1986a):

$$\text{glutamine} + H_2O \longrightarrow \text{glutamic acid} + NH_3$$
$$\text{asparagine} + H_2O \longrightarrow \text{aspartic acid} + NH_3$$

Surprisingly high levels of glutamine (0.5 mM) occur in mammalian tissues and could be catabolized by tissue parasites. Aspartase, the enzyme which cleaves aspartate to fumarate and ammonia, has not been found in animal tissues.

B. METABOLISM OF THE CARBON SKELETON

Following removal of the amino group, the carbon skeletons of amino acids are converted to glycolytic or tricarboxylic acid cycle intermediates and ultimately to carbon dioxide. Amino acids can be grouped according to their catabolic end-products: some give rise to oxaloacctate or pyruvate, others to 2-oxoglutarate, a third class gives rise to succinlyCoA and a fourth to acetylCoA or acetoacetylCoA. In vertebrates, there are some 20 multi-enzyme sequences for the degradation of the different amino acids.

1. *Conversion to oxaloacetate and pyruvate*

Seven amino acids—aspartate, asparagine, alanine, serine, glycine, cystine and cysteine—are eventually broken down to oxaloacetate or pyruvate. The removal of the amino groups from asparagine and aspartate gives oxalo-acetate, whereas the deamination of alanine and serine yields pyruvate (see Section V.A.1). The catabolism of labelled aspartate, serine and alanine to primarily ethanol and carbon dioxide has been demonstrated in *Moniliformis moniliformis* (Crompton and Ward, 1984; Ward and Crompton, 1986, 1987). Aspartate, serine and alanine have also been shown to be catabolized in *Ancylostoma ceylanicum*, *Ascaridia galli*, *Dipetalonema viteae* and *Nippostrongylus brasiliensis* (Singh *et al.*, 1983a,b, 1985, 1987; Srivastava *et al.*, 1987) and alanine and aspartate in *Schistosoma mansoni*, *S. japonicum*, *Hymenolepis diminuta* and *Setaria cervi* (Bruce *et al.*, 1972; Wack *et al.*, 1983; Singh *et al.*, 1984).

There are several possible pathways for the breakdown of glycine. In mammals, the major route is direct enzymatic cleavage to carbon dioxide and the methylene carbon of $N^{5,10}$ methylene tetrahydrofolate. Other routes are conversion to serine, deamination to glyoxylate or condensation with acetylCoA to give aminoacetone. The pathway of glycine catabolism in helminths has not been investigated, but interconversion with serine has been demonstrated (see Section IV.C). The production of carbon dioxide from labelled glycine has been shown in *A. galli*, *Ancylostoma ceylanicum*, *D. viteae* and *N. brasiliensis* (Singh *et al.*, 1983a, 1987; Srivastava *et al.*, 1987). In *M. moniliformis*, although glycine is taken up, it does not appear to be catabolized (Ward and Crompton, 1987). Similarly, no significant catabolism of glycine was found in *Brugia pahangi*, although alanine was readily converted to carbon dioxide (Srivastava *et al.*, 1988).

The catabolism of cystine first involves cleavage to cysteine. Helminths appear to be unable to reduce cystine directly to cysteine (Munir and Barrett, 1985), but a glutathione-cystine transhydrogenase system has been demonstrated in *H. diminuta* (Gomez-Bautista and Barrett, 1988):

$$\text{cystine} + 2\text{GSH} \longrightarrow 2\text{ cysteine} + \text{GSSG}$$
$$(\textit{glutathione cysteine transhydrogenase})$$
$$\text{GSSG} + \text{NADPH} + \text{H}^+ \longrightarrow 2\text{GSH} + \text{NADP}^+ \ (\textit{glutathione reductase})$$

The glutathione transhydrogenase system is probably responsible for the reduction of a wide variety of sulphydryl links in tissues. Cystine can also be cleaved by γ-cystathionase to pyruvate, thiocysteine and ammonia (see Section IV.C).

There are at least three different pathways leading to pyruvate from the breakdown of cysteine:

$$\text{cysteine} \xrightarrow{(1)} \text{cysteine sulphinic acid} \xrightarrow{(2)} \text{3-sulphinylpyruvate} \xrightarrow{(3)} \text{pyruvate} + SO_2$$

where (1) is cysteine dioxygenase, (2) cysteine sulphinate transaminase and (3) is non-enzymic,

$$\text{cysteine} \xrightarrow{(4)} \text{3-mercaptopyruvate} \xrightarrow{(5)} \text{pyruvate} + H_2S$$

where (4) is cysteine transaminase and (5) 3-mercaptopyruvate sulphotransferase, and

$$\text{cysteine} \xrightarrow{(6)} \text{pyruvate} + H_2S + NH_3$$

where (6) is γ-cystathionase.

Cysteine transaminase has been reported in a number of helminths (Table 5), cysteine sulphinate transaminase has been demonstrated in *N. brasiliensis* and *Fasciola hepatica* (Walker and Barrett, unpublished) and cysteine transaminase, cysteine sulphinate transaminase and cysteine dioxygenase have all been reported from *H. diminuta* (Gomez-Bautista and Barrett, 1988). However, no 3-mercaptopyruvate sulphotransferase activity could be detected in *H. diminuta* and the γ-cystathionase from both *H. diminuta* and *N. brasiliensis* does not cleave cysteine. In mammals, the dioxygenase pathway is probably the principal route for cysteine catabolism.

2. *Conversion to 2-oxoglutarate*

This group is composed of five amino acids—glutamate, glutamine, proline, arginine and histidine. Deamination of glutamine and glutamate yields 2-oxoglutarate and the enzymes responsible—glutaminase, 2-oxoglutarate transaminase and glutamate dehydrogenase—are widely distributed in helminths (see Section V.A). The catabolism of labelled glutamate has been demonstrated in a number of helminths (Singh *et al.*, 1983a,b, 1984, 1987; Srivastava *et al.*, 1987, 1988; Bruce *et al.*, 1969, 1972; Stjernholm and Warren, 1974). However, Ward and Crompton (1987) reported that labelled glutamate, like glycine, was readily taken up by *Moniliformis moniliformis*, but not significantly metabolized.

An alternative pathway for the catabolism of glutamate is the 4-aminobutyrate bypass. This pathway, which occurs in the vertebrate brain and in some plants and microorganisms, provides a route for the conversion of glutamate to succinate bypassing 2-oxoglutarate decarboxylase:

$$\text{Glutamate} \xrightarrow{(1)} \text{4-aminobutyrate} \xrightarrow{(2)} \text{succinic semi-aldehyde} \xrightarrow{(3)} \text{succinate}$$

where (1) is glutamate decarboxylase, (2) 4-aminobutyrate transaminase and (3) succinate semi-aldehyde dehydrogenase.

Glutamate decarboxylase has been reported from *Ascaridia galli*, *Ascaris lumbricoides*, *Macracanthorhynchus hirudinaceus*, *Moniezia expansa* and *Taenia solium*, while 4-aminobutyrate transaminase has been found in *A. lumbricoides*, *Heligmosomoides polygyrus*, *Nippostrongylus brasiliensis*, *Panagrellus redivivus*, *M. expansa* and *M. hirudinaceus* (Monteoliva *et al.*, 1965; Rasero *et al.*, 1968; Watts and Atkins, 1983, 1984, 1986; Grantham and Barrett, 1986a). However, Cornish and Bryant (1975) were unable to detect glutamate decarboxylase, 4-aminobutyrate transaminase or succinic semi-aldehyde dehydrogenase in *M. expansa* and isotope studies provided no evidence for the presence of a functional 4-aminobutyrate bypass in this parasite.

The major pathway for the catabolism of arginine is hydrolysis to ornithine (via arginase; see Section V.A.3) followed by transfer of the 5-amino group to a suitable acceptor, usually 2-oxoglutarate, to give glutamic-4-semi-aldehyde, which spontaneously cyclizes to Δ-pyrroline-5-carboxylic acid. The latter can be oxidized to glutamate via Δ-pyrroline-5-carboxylate dehydrogenase. In *Fasciola hepatica*, the dehydrogenase is absent and Δ-pyrroline-5-carboxylate is instead reduced to proline (see Section IV.C). In *F. hepatica* and *Schistosoma mansoni*, proline appears to be the major end-product of arginine catabolism. The situation in other helminths has not been investigated, but the production of labelled carbon dioxide from labelled arginine has been described in *A. galli*, *Ancylostoma ceylanicum*, *D. viteae*, *N. brasiliensis*, *S. mansoni* and *S. japonicum* (Singh *et al.*, 1983a, 1987; Srivastava *et al.*, 1987; Bruce *et al.*, 1972; Stjernholm and Warren, 1974). An alternative route for arginine catabolism in animals which lack arginase is oxidative decarboxylation to 4-guanidobutyramide.

The catabolism of proline in mammals involves initial oxidation by proline oxidase to Δ-pyrroline-5-carboxylate, followed by reduction to glutamate as described above. Proline oxidase, however, could not be demonstrated in *F. hepatica*, *S. mansoni* or *A. lumbricoides* (Ertel and Isseroff, 1974; Isseroff, 1980; Barrett, unpublished). However, low activities of proline oxidase have been found in *B. pahangi*, *P. redivivus* and *H. diminuta* (Srivastava *et al.*, 1988; Butterworth, unpublished). The situation in other helminths has again not been investigated, but the production of carbon dioxide from labelled proline has been noted in *A. ceylanicum*, *B. pahangi*, *D. viteae*, *N. brasiliensis*, *S. mansoni* and *S. japonicum* (Srivastava *et al.*, 1987, 1988; Singh *et al.*, 1987; Bruce *et al.*, 1972).

Histidine catabolism is fairly straightforward, involving deamination to urocanic acid followed by hydration to give formaminoglutamate. The formamino group is then removed by reaction with tetrahydrofolate to yield glutamate. The initial deamination step in the pathway has been demonstrated in several helminths (see Section V.3), but none of the other enzymes has been investigated.

3. *Conversion to succinylCoA*

Methionine, threonine, isoleucine and valine are all ultimately degraded to succinylCoA. The initial steps of methionine catabolism are identical with the pathway for the conversion of methionine to cysteine discussed in Section IV.C, namely conversion to cystathionine then cleavage by γ-cystathionase to cysteine, ammonia and 2-oxobutyrate. 2-Oxobutyrate is oxidatively decarboxylated to give propionylCoA, which can then be metabolized via methylmalonylCoA to succinylCoA. The catabolism of 2-oxobutyrate has not been investigated in helminths, but the interconversion of propionyl- and succinylCoA is widespread in parasites as part of carbohydrate catabolism (Barrett, 1981). In mammals, a major route for methionine catabolism is amino transfer to yield 2-oxo-4-methylthiobutyrate. Methionine-linked aminotransferases have been described in several nematodes (Table 5), but little is known about the further metabolism of 2-oxo-4-methylthiobutyrate in nematodes other than reduction to the corresponding hydroxy compound (Langer *et al.*, 1971).

Deamination of threonine by threonine dehydratase (see Section V.A.3) yields 2-oxobutyrate, which is metabolized as above. In mammals, however, the major route for threonine catabolism is conversion to amino acetone by an NAD-linked threonine dehydrogenase. Threonine dehydrogenase could not, however, be detected in *Panagrellus redivivus* (Barrett, unpublished).

The breakdown of the branched chain amino acids leucine, isoleucine and valine seems to be identical in all organisms which have been so far studied. Valine is ultimately catabolized to succinylCoA and isoleucine to succinyl- and acetylCoA. Leucine, which is catabolized to acetylCoA and acetoacetate, will be considered in detail in the next section. The first three steps in catabolism are common to all three branched chain amino acids, namely transamination to the corresponding 2-oxo-acid, oxidative decarboxylation to the acylCoA derivative followed by dehydrogenation to the 2,3 unsaturated acylCoA derivative:

Leucine	Isoleucine	Valine
↓ (1)	↓ (1)	↓ (1)
2-oxoisocaproate	2-oxo-3-methylvalerate	2-oxoisovalerate
↓ (2)	↓ (2)	↓ (2)
isovalerylCoA	2-methylbutyrylCoA	isobutyrylCoA
↓ (3)	↓ (3)	↓ (3)
2-methylcrotonylCoA	tiglylCoA	methacrylCoA

where (1) is aminotransferase, (2) a branched chain oxoacid dehydrogenase and (3) acylCoA dehydrogenase.

The further metabolism of isoleucine involves the hydration of tiglylCoA by crotonase to give 2-methyl-3-hydroxybutyrylCoA. This is then oxidized by an NAD-dependent dehydrogenase to 2-methylacetoacylCoA, which is then cleaved by a CoA lyase to give acetylCoA and propionylCoA.

The catabolism of methacrylCoA from the breakdown of valine is slightly more complex. Hydration of methacrylCoA by crotonase yields 3-hydroxyisobutrylCoA, which is then hydrolysed to the free acid. Free 3-hydroxy-isobutyrate is then oxidized to methylmalonic semi-aldehyde. In bacteria, methylmalonic semi-aldehyde can be oxidatively decarboxylated to propionylCoA, but in eukaryotes it is usually dehydrogenated and condensed with CoA to form methylmalonylCoA. MethylmalonylCoA mutase then catalyses the conversion of methylmalonylCoA to succinylCoA.

Valine is converted to 2-oxoisovalerate by the tapeworm *Calliobothrium verticillatum* (Fisher and Starling, 1970) and to isobutyrate in *Ancylostoma caninum* and *Fasciola hepatica* (Warren and Poole, 1970; Lahoud *et al.*, 1971). The latter species also converts isoleucine to 2-methylbutyrate. These acids are all intermediates in valine or isoleucine catabolism. The breakdown of valine to carbon dioxide has been described in *Ancylostoma ceylanicum*, *Ascaridia galli*, *Dipetalonema viteae*, *Nippostrongylus brasiliensis* and *Setaria cervi* (Singh *et al.*, 1983a, 1984, 1987; Srivastava *et al.*, 1987) and the complete catabolism of both valine and isoleucine has been demonstrated in *Heligmosomoides polygyrus* and *Panagrellus redivivus* (Grantham and Barrett, 1986b). The complete sequence of catabolic enzymes for valine and isoleucine has also been demonstrated in the last two species.

4. *Conversion to acetylCoA and acetoacylCoA*

There are five amino acids in this class—leucine, lysine, tryptophan, phenylalanine and tyrosine. 2-MethylcrotonylCoA, the unsaturated acylCoA derivative from leucine (see above), is carboxylated in an ATP-utilizing biotin-dependent reaction to give 2-methylglutaconylCoA, which is then hydrated by crotonase to 2-hydroxy-2-methylglutaconylCoA. The latter can be used in the synthesis of mevalonic acid and hence as a precursor for isoprenoids, but its major metabolic fate is hydrolysis to acetylCoA and acetoacetate. Carbon dioxide production from labelled leucine has again been reported in a number of nematodes (Singh *et al.*, 1983a, 1987; Srivastava *et al.*, 1987; Grantham and Barrett, 1986b) and the complete sequence of catabolic enzymes has been demonstrated in *H. polygyrus* and *P. redivivus* (Grantham and Barrett, 1986b).

Lysine is an exception to the general rule that the first step in amino acid catabolism is the removal of the amino group; in lysine, neither the 2 nor the 6 amino group undergoes transamination. The catabolism of lysine is complicated with a number of possible pathways. In mammalian liver, lysine is degraded via saccharopine to 2-aminoadipate; the further metabolism of 2-aminoadipate involves transamination to 2-oxoadipate followed by oxidative decarboxylation to glutarylCoA. The subsequent catabolism of glutarylCoA in mammalian tissues is not known with any certainty. An alternative mammalian pathway for lysine breakdown leads to pipecolic acid (a cyclic imino acid resembling proline). A number of other pathways of lysine catabolism occur in microorganisms. Nothing is known about lysine catabolism in helminths, although carbon dioxide production from lysine has been noted in *D. viteae* (Srivastava *et al.*, 1987).

The catabolism of phenylalanine involves conversion to tyrosine via a biopterin-dependent hydroxylase. Phenylalanine can also undergo transamination to phenylpyruvate (Table 5), but the Km of the transferase is relatively high and so it is probably only a minor route. A five-enzyme sequence is involved in the catabolism of tyrosine. Transamination yields *p*-hydroxyphenylpyruvate, which is then metabolized via homogentisic acid and maleylacetoacetic acid to fumarylacetoacetic acid. The latter is then hydrolysed to acetoacetic and fumaric acids. The enzymes involved in phenylalanine and tyrosine catabolism have not been studied in helminths, although carbon dioxide production from phenylalanine has again been reported in several nematodes (Singh *et al.*, 1983a, 1987; Srivastava *et al.*, 1987).

The final amino acid is tryptophan, which is notable for its variety of important metabolic reactions and intermediates. The first step in tryptophan catabolism is the oxidative opening of the heterocyclic ring catalysed by tryptophan oxygenase to yield *N*-formyl kynurenine. This is then metabolized via kynurenine and 3-hydroxykynurenine to 3-hydroxyanthranilic acid. 3-Hydroxyanthranilic acid has a number of metabolic fates: it can be decarboxylated to *o*-aminophenol or completely catabolized to acetylCoA; it is also the starting point for NAD synthesis. Kynurenine and 3-hydroxykynurenine are also precursors for other synthetic pathways. A number of intermediates in the tryptophan pathway—kynurenine, 3-hydroxykynurenine, anthranilic acid and *o*-aminophenol—have been detected in *Caenorhabditis elegans*, as have two of the enzymes—kynurenine hydroxylase and kynurenine hydrolase (Babu, 1974; Siddiqui and Babu, 1980; Siddiqui and von Ehrenstein, 1980). However, tryptophan oxygenase could not be demonstrated in *Schistosoma mansoni* (Brown and Smith, 1973).

C. UREA CYCLE AND RELATED ENZYMES

In vertebrates, the 2-amino groups removed from amino acids during their oxidative degradation are ultimately excreted as ammonia, uric acid or urea. In parasitic helminths, some 2–10% of the total nitrogenous end-products is urea. However, the mechanism of urea formation in helminths is unclear and it is doubtful if a classical urea cycle is present in any of the parasitic forms.

Limited evidence for a functional urea cycle has been presented for *Fasciola hepatica* (Ehrlich *et al.*, 1963, 1968; Kurelec, 1964b; Rijavec and Kurelec, 1965), *Hymenolepis diminuta* (Campbell, 1963a), *Ascaridia galli* (Rogers, 1952) and *Ascaris lumbricoides* (Rogers, 1952; Cavier and Savel,

1954e; Savel, 1955). However, the work of Janssens and Bryant (1969), Paltridge and Janssens (1971), Kurelec (1972, 1973, 1974b), Dragojevic *et al.* (1974) and Celik (1986b) have failed to substantiate these earlier claims. There is, in contrast, better evidence for a urea cycle in the free-living planarian *Bipalium kewense* (Campbell, 1965) and in the free-living nematode *Panagrellus redivivus* (Wright, 1975; Grantham and Barrett, 1986a).

Two of the enzymes of the urea cycle are widely distributed in parasites: arginase (see Section V.A.3) and ornithine carbamoyltransferase (Campbell, 1963a; Campbell and Lee, 1963; Rijavec and Kurelec, 1966; Lopez-Gorge and Monteoliva, 1969; Grantham and Barrett, 1986a). The other enzymes of the cycle, carbamoylphosphate synthetase, arginosuccinate synthetase and arginosuccinate lyase, have been reported only sporadically (Kurelec, 1964a; Rijavec and Kurelec, 1965; Janssens and Bryant, 1969; Grantham and Barrett, 1986a). These enzymes are, however, rather labile in tissue extracts. Three different classes of carbamoylphosphate synthetases are known—an ammonia-dependent enzyme that requires *N*-acetylglutamate as an activator, a glutamine-dependent enzyme and a glutamine plus *N*-acetylglutamine-dependent enzyme. The ammonia-dependent enzyme is mitochondrial and functions in the urea cycle; the other variants are cytoplasmic and are involved in pyrimidine and arginine synthesis. In contrast to the ammonia-dependent enzyme, low levels of glutamine-dependent carbamoylphosphate synthetase have been detected in a wide range of helminths (Aoki *et al.*, 1975, 1980; Kobayashi *et al.*, 1978, 1984; Aoki and Oya, 1979; Hill *et al.*, 1981).

Urease, the enzyme which cleaves urea into ammonia and carbon dioxide, occurs sporadically in invertebrates but not in vertebrates. Low activities of urease have been found in nematodes (Rogers, 1952; Sokhina, 1976; Shishkova-Kasatochkina *et al.*, 1976, 1986; Sokhina and Koloskova, 1978), but not in digeneans (van Grembergen and Pennoit de Cooman, 1944; Tao and Huang, 1965). Urease was not detected in cyclophyllidean or tetraphyllidean cestodes, but was found in some, but not all, trypanorhynchid cestodes (van Grembergen and Pennoit de Cooman, 1944; Simmons, 1961). The presence of a urease in two trypanorhynchid cestodes is intriguing, as these worms live in the intestines of sharks and are exposed to high levels of urea (0.2–0.5 M). The metabolic significance of urease in these cestodes is obscure (Pappas, 1978). Possibly, the carbon dioxide released by the urease reaction is utilized in acid/base regulation.

VI. Amino Acid Derivatives

Many of the intermediates in amino acid synthesis or catabolism are precursors for other pathways. In addition, there are two other important reactions which amino acids can undergo—decarboxylation and hydroxylation.

A. AMINO ACID DECARBOXYLATION

A wide range of amino acid decarboxylases have been described in helminths. Arginine, glutamate, histidine, lysine, ornithine, tryptophan, 5-hydroxytryptophan, phenylalanine and tyrosine decarboxylation have been described in adult *Ascaris lumbricoides* (Cavier and Savel, 1954c; Savel, 1955; Monteoliva *et al.*, 1965; Rasero *et al.*, 1968; Lopez-Gorge and Monteoliva, 1969; Lopez-Gorge *et al.*, 1968, 1969; Chaudhuri *et al.*, 1988). The decarboxylation of lysine, glutamate and 5-hydroxytryptophan has been described in *Ascaridia galli* (Monteoliva *et al.*, 1965; Dryuchenko and Kulikovo, 1984; Smart, 1987, and *Moniezia expansa* (Monteoliva *et al.*, 1965; Rasero *et al.*, 1968; Lopez-Gorge *et al.*, 1968, 1969) and glutamate decarboxylation in *Taenia solium* (Monteoliva *et al.*, 1965). Histidine decarboxylase has been identified in *Mesocoelium monodi* (Mettrick and Telford, 1963, 1965), 5-hydroxytryptophan decarboxylase in *M. corti* (Hariri, 1975) and, finally, histidine and 5-hydroxytryptophan decarboxylase in *Fasciola hepatica* (Mansour *et al.*, 1957; Kurelec *et al.*, 1969; Mansour and Stone, 1970) and *Schistosoma mansoni* (Hillman and Senft, 1973; Bennett and Beuding, 1973; Catto, 1981; Saber and Wu, 1985). Aromatic L-amino acid decarboxylase and glutamate decarboxylase have also been shown in the free-living *Caenorhabditis elegans* (Sulston *et al.*, 1975; Shaeffer and Bergstrom, 1988).

However, other studies have failed to detect such a wide range of decarboxylase activities. Mettrick and Telford (1965) were unable to detect histidine decarboxylase in a range of species (*F. hepatica*, *Macracanthorhynchus hirudinaceus*, *Oochoristica ameivae*, *Stephanurus dentatus*, *Syndesmis franciscana*) and Catto (1981) could not detect histidine decarboxylase activity in *S. mansoni*. No ornithine, lysine, arginine or histidine decarboxylase could be detected in *Nippostrongylus brasiliensis* (Walker and Barrett, unpublished). Cornish and Bryant (1975) were unable to confirm the presence of glutamate decarboxylase activity in *Moniezia expansa*. Wittich *et al.*, (1987) and Walter (1988) reported no ornithine or arginine decarboxylation by *Brugia patei*, *Dirofilaria immitis*, *L. carinii* or *Onchocerca volvulus*, and were also unable to detect ornithine decarboxylase activity in *Ascaris lumbricoides* or *Paragonimus uterobilateralis*. Finally, Nations *et al.* (1973) have shown that the apparent decarboxylation of aspartate by *Hymenolepis diminuta* was in fact due to the decarboxylation of oxaloacetate formed by transamination.

The number of amino acid decarboxylases present in helminths may eventually turn out to be relatively limited and, so far, the only one which has been consistently identified is the aromatic L-amino acid decarboxylase that is responsible for the decarboxylation of 5-hydroxytryptophan. In mammals, amino acid decarboxylation is quantitatively a very minor pathway, but it is often a major route in bacteria.

B. AMINO ACID HYDROXYLATION

The hydroxylation of proline and phenylalanine has been discussed in connection with amino acid synthesis. Two other important hydroxylation reactions in amino acid metabolism are tryptophan and tyrosine hydroxylase, both important steps in the formation of biogenic amines. Despite indications that tryptophan hydroxylase might be absent from helminths (Mansour and Stone, 1970; Hillman and Senft, 1973; Bennett and Beuding, 1973; Hariri, 1975; Mansour, 1979), the enzyme has now been characterized in *Ascaris lumbricoides* (Chaudhuri *et al.*, 1988). Labelling experiments have indicated the presence of tyrosine hydroxylase in *Hymenolepis diminuta* (Ribeiro and Webb, 1983) and *Caenorhabditis briggsae* (Kisiel *et al.*, 1976), while low activities of this enzyme have been demonstrated in *C. elegans* (Sulston *et al.*, 1975) and *Ascaridia galli* (Smart, 1987, 1988). Tyrosine hydroxylase requires a reduced pteridine co-factor, and Smart (1987) has also isolated a dihydropteridine reductase system from *A. galli*.

VII. Summary and Conclusions

Amino acids are major constituents of biological material. Chemically they are extremely stable and combine a relatively simple molecular structure with a wide range of properties and functions. In general, amino acid metabolism in helminths has been relatively neglected and the information available is often uneven and of uncertain quality. However, the search for new target sites for anthelmintic development has led to a renewed interest in this area.

The amino acid composition of helminths is similar to that of other invertebrates and no unique amino acids have been reported. With the possible addition of tyrosine, helminths seem to require the same 10 essential amino acids as mammals and, where studied in detail, the pathways of amino acid synthesis in helminths are similar to those of mammals. Although amino acids are not a significant energy source in parasites, helminths are able to catabolize amino acids by pathways which, again, appear identical to those found in mammals. Helminths have also been shown to carry out a number of oxidative reactions associated with amino acid metabolism, including cysteine dioxygenase, proline hydroxylase and tryptophan hydroxylase.

There are, however, differences in detail between the pathways of amino acid metabolism in helminths and mammals, particularly in the metabolism of the sulphur amino acids and arginine and proline. These differences may be exploitable in anthelmintic design and proline analogues and proline biosynthesis inhibitors show some potential as fasciolicides (Sheers *et al.*, 1982). Differences in metabolism between parasites and their hosts may be the result of parasitic adaptation or they may merely reflect general features of the invertebrate phyla as a whole. Thus a comparison of amino acid metabolism in parasitic helminths with that of their free-living relatives may give some insight into the biochemical basis of parasitism.

References

Abbas, M. K. and Cain, G. D. (1984). Amino acid and lipid composition of refringent granules from the ameboid sperm of *Ascaris suum* (Nematoda). *Histochemistry* **81,** 59–65.

Abbas, M. and Foor, W. E. (1978). *Ascaris suum*: free amino acids and proteins in the pseudocoelom, seminal vesicle and glandular vas deferens. *Experimental Parasitology* **45,** 263–273.

Agosin, M. and Repetto, Y. (1963). Studies on the metabolism of *Echinococcus granulosus—VII.* Reactions of the tricarboxylic acid cycle in *E. granulosus* scolices. *Comparative Biochemistry and Physiology* **8,** 245–261.

Aldrich, D. V., Chandler, A. C. and Daugherty, J. W. (1954). Intermediary protein metabolism in helminths. II. Effect of host castration on amino acid metabolism in *Hymenolepis diminuta. Experimental Parasitology* **3,** 173–184.

Ando, T. (1957). A study of *Gnathostoma spinigerum. Acta Medica (Fukuoka)* **27,** 2342–2359.

Aoki, T. and Oya, H. (1979). Glutamine dependent carbamoyl-phosphate synthetase and control of pyrimidine biosynthesis in the parasitic helminth *Schistosoma mansoni. Comparative Biochemistry and Physiology* **63B,** 511–515.

Aoki, T., Oya, H., Mori, M. and Tatibana, M. (1975). Glutamine-dependent carbamyl phosphate synthetase in *Ascaris* ovary and its regulatory properties. *Proceedings of the Japan Academy* **51,** 733–736.

Aoki, T., Oya, H., Mori, M. and Tatibana, M. (1980). Control of pyrimidine biosynthesis in the *Ascaris* ovary: Regulatory properties of glutamine dependent carbamoyl-phosphate synthetase and copurification of the enzyme with aspartate carbamoyltransferase and dihydroorotase. *Molecular and Biochemical Parasitology* **1,** 55–68.

Arme, C. (1977). Amino acids in eight species of Monogenea. *Zeitschrift für Parasitenkunde* **51,** 261–263.

Arme, C. and Read, C. P. (1969). Fluxes of amino acids between the rat and a cestode symbiote. *Comparative Biochemistry and Physiology* **29,** 1135–1147.

Arme, C. and Whyte, A. (1975). Amino acids of *Diclidophora merlangi* (Monogenea). *Parasitology* **70,** 39–46.

Asch, H. L. (1976). Amino acid pools of *Schistosoma mansoni* and mouse hepatic portal serum. *Rice University Studies* **62,** 35–42.

Aueron, F. and Rothstein, M. (1974). Nematode biochemistry—XIII. Peroxisomes in the free-living nematode, *Turbatrix aceti*. *Comparative Biochemistry and Physiology* **49B,** 261–271.

Awapara, J. (1962). Free amino acids in invertebrates: A comparative study of their distribution and metabolism. *In* "Amino Acid Pools" (J. T. Holden, ed.), pp. 158–175. Elsevier, Amsterdam.

Babin, D. R., Peanasky, R. J. and Goos, S. M. (1984). The isoinhibitors of chymotrypsin/elastase from *Ascaris lumbricoides*: The primary structure. *Archives of Biochemistry and Biophysics* **232,** 143–161.

Babu, P. (1974). Biochemical genetics of *Caenorhabditis elegans*. *Molecular and General Genetics* **135,** 39–44.

Bailey, R. S. and Fried, B. (1977). Thin layer chromatographic analyses of amino acids in *Echinostoma revolutum* (Trematoda) adults. *International Journal of Parasitology* **7,** 497–499.

Balasubramanian, M. and Myers, R. F. (1971). Nutrient media for plant-parasitic nematodes. II. Amino acid requirements of *Aphelenchoides* sp. *Experimental Parasitology* **29,** 330–336.

Balogun, R. A. and Braide, E. I. (1972). Studies on the amino acid pool of the liver fluke *Fasciola gigantica* (Cobbold). *Nigerian Journal of Science* **5,** 191–195.

Bankov, I. and Khlebarov, Z. (1987). Biochemical studies of *Fasciola hepatica* L. and of its pathogenic effect. XII. Amino acid composition of fractions from the water soluble proteins obtained after gelfiltration on Sephadex G-100. *Bulgarian Academy of Sciences* **23,** 24–29.

Bankov, I. and Ossikovski, E. (1976). Biochemical studies on *Fasciola hepatica* L. and its pathogenic effect. II. Studies in the free amino acids and other ninhydrin-positive compounds in *Fasciola hepatica* L. *Bulgarian Academy of Sciences: Helminthology* **2,** 13–19.

Bankov, I., Ossikovski, E. and Khlebarov, Z. (1978). Biochemical studies of *Fasciola hepatica* L. and its pathogenic action. V. Amino acid composition of the water-soluble proteins from *Fasciola hepatica* L. *Bulgarian Academy of Sciences: Helminthology* **5,** 3–9.

Barrett, J. (1981). "Biochemistry of Parasitic Helminths". Macmillan, London.

Bedi, A. J. K. and Isseroff, H. (1979). Bile duct hyperplasia in mice infected with *Schistosoma mansoni*. *International Journal for Parasitology* **9,** 401–404.

Bennett, J. L. and Bueding, E. (1973). Uptake of 5-hydroxytryptamine by *Schistosoma mansoni*. *Molecular Pharmacology* **9,** 311–319.

Berdyeva, G. T. and Dryuchenko, E. A. (1975). Histidine metabolism in *Ascaris suum, Ascaridia galli* and *Fasciola hepatica*. *Parazitologiya* **9,** 28–30.

Bhalya, A., Seth, A., Malhotra, S. K. and Capoor, V. N. (1983a). Chemotaxonomic differentiation of 3 subgenera *Paroniella, Raillietina* and *Skrjabinia* of cestode genus *Raillietina*. *Advances in Biosciences* **2,** 93–96.

Bhalya, A., Seth, A., Malhotra, S. K. and Capoor, V. N. (1983b). Chemotaxonomic studies of hymenolepidid tapeworms *Staphylepis rustica* (Meggitt, 1926). *Allahabad University Studies 15 (N.S.)* **1,** 73–77.

Bhalya, A., Seth, A., Malhotra, S. K. and Capoor, V. N. (1983c). Amino-acids of *Amoebotaenia cuneata* (Cestoda : Dilepidoidea Wardle McLeod & Radinovsky, 1974). *Journal of Helminthology* **57,** 9–10.

Bhalya, A., Seth, A., Capoor, V. N. and Malhotra, S. K. (1984). Chemotaxonomic studies on devaineid tapeworms based on amino acid analysis. *Journal of Helminthology* **58,** 325–326.

Bhalya, A., Seth, A., Malhotra, S. K. and Capoor, V. N. (1985). Amino acids of *Hymenolepis palmarum* (Johri, 1956) and chemotaxonomic studies on hymenolepidid cestodes. *Journal of Helminthology* **59,** 39–42.

Bhalya, A., Seth, A., Malhotra, S. K. and Capoor, V. N. (1985). Chemotaxonomic patterns of amino acids in *Cotugnia columbae* Malviya and Dutta (1969). *In* "Proceedings of the National Symposium on New Dimensions in Parasitology" (V. N. Capoor, S. C. Misra and S. K. Malhotra, eds), pp. 89–93. Allahabad University Press, Allahabad.

Bird, A. F. (1954). The cuticle of nematode larvae. *Nature (London)* **174,** 362.

Bird, A. F. (1956). Chemical composition of the nematode cuticle: Observations on the whole cuticle. *Experimental Parasitology* **5,** 320–358.

Bird, A. F. (1957). Chemical composition of the nematode cuticle: Observations on individual layers and extracts from these layers in *Ascaris lumbricoides* cuticle. *Experimental Parasitology* **6,** 383–403.

Bird, A. F. and Rogers, W. P. (1956). Chemical composition of the cuticle of third stage nematode larvae. *Experimental Parasitology* **5,** 449–457.

Bishop, S. H. (1975). Ammonia formation and amino acid excretion by *Gyrocotyle fimbriata* (Cestoidea). *Journal of Parasitology* **61,** 79–88.

Bobek, L., Rekosh, D. M., van Keulen, H. and Lo Verde, P. T. (1986). Characterisation of a female-specific cDNA derived from a developmentally regulated mRNA in the human blood fluke *Schistosoma mansoni. Proceedings of the National Academy of Sciences (USA)* **83,** 5544–5548.

Bobek, L. A., Rekosh, D. M. and Lo Verde, P. T. (1988). Small gene family encoding an eggshell (chorion) protein of the human parasite *Schistosoma mansoni. Molecular and Cellular Biology* **8,** 3008–3016.

Brockelman, C. R. and Jackson, G. J. (1978). Amino acid, heme and sterol requirements of the nematode, *Rhabditis maupasi. Journal of Parasitology* **64,** 803–809.

Brown, J. N. and Smith, T. M. (1973). Effects of *Schistosoma mansoni* infection on induction of tryptophan oxygenase in mouse livers. *Comparative Biochemistry and Physiology* **45B,** 487–489.

Bruce, J. I., Weiss, E., Stirewalt, M. A. and Lincicome, D. R. (1969). *Schistosoma mansoni*: Glycogen content and utilization of glucose, pyruvate, glutamate and citric acid cycle intermediates by cercariae and schistosomules. *Experimental Parasitology* **26,** 29–40.

Bruce, J. I., Ruff, M. D., Belusko, R. J. and Werner, J. K. (1972). *Schistosoma mansoni* and *Schistosoma japonicum*: Utilization of amino acids. *International Journal for Parasitology* **2,** 425–430.

Bryant, C. and Nicholas, W. L. (1965). Intermediary metabolism in *Moniliformis dubius* (Acanthocephala). *Comparative Biochemistry and Physiology* **15,** 103–112.

Bryant, C. and Smith, M. J. H. (1963). Some aspects of intermediary metabolism in *Fasciola hepatica* and *Polycelis nigra. Comparative Biochemistry and Physiology* **9,** 189–194.

Byram, J. E. and Senft, A. W. (1979). Structure of the schistosome eggshell: Amino acid analysis and incorporation of labelled amino acids. *American Journal of Tropical Medicine and Hygiene* **28,** 539–547.

Cain, G. D. (1969). Purification and properties of hemoglobin from *Fasciolopsis buski. Journal of Parasitology* **55,** 311–320.

Cain, G. D. (1970). Collagen from the giant acanthocephalan *Macracanthorhynchus hirudinaceous. Archives of Biochemistry and Biophysics* **141,** 264–270.

Cain, G. D. and Fairbairn, D. (1971). Protocollagen proline hydroxylase and collagen synthesis in developing eggs of *Ascaris lumbricoides*. *Comparative Biochemistry and Physiology* **40B,** 165–179.

Campbell, J. W. (1960a). The occurrence of β-alanine and β-aminoisobutyric acid in flatworms. *Biological Bulletin* **119,** 75–79.

Campbell, J. W. (1960b). Nitrogen and amino acid composition of three species of anoplocephalid cestodes: *Moniezia expansa, Thysanosoma actinioides* and *Cittotaenia perplexa*. *Experimental Parasitology* **9,** 1–8.

Campbell, J. W. (1963a). Urea formation and urea cycle enzymes in the cestode, *Hymenolepis diminuta*. *Comparative Biochemistry and Physiology* **8,** 13–27.

Campbell, J. W. (1963b). Amino acids and nucleotides of the cestode *Hymenolepis diminuta*. *Comparative Biochemistry and Physiology* **8,** 181–185.

Campbell, J. W. (1965). Arginine and urea biosynthesis in the land planarian: Its significance in biochemical evolution. *Nature (London)* **208,** 1299–1301.

Campbell, J. W. and Lee, T. W. (1963). Ornithine transcarbamylase and arginase activity in flatworms. *Comparative Biochemistry and Physiology* **8,** 29–38.

Campbell, A. J., Sheers, M., Moore, R. J., Edwards, S. R. and Montague, P. E. (1981). Proline biosynthesis by *Fasciola hepatica* at different developmental stages *in vivo* and *in vitro*. *Molecular and Biochemical Parasitology* **3,** 91–101.

Castro, G. A., Ferguson, J. D. and Gorden, C. W. (1973). Amine excretion in excysted larvae and adults of *Trichinella spiralis*. *Comparative Biochemistry and Physiology* **45A,** 819–828.

Catto, B. A. (1981). *Schistosoma mansoni*: Decarboxylation of 5-hydroxytryptophan, L-dopa and L-histidine in adult and larval schistosomes. *Experimental Parasitology* **51,** 152–157.

Caulfield, J. P., Cianci, C. M. L., McDiarmid, S. S., Suyemitsu, T. and Schmidt, K. (1987). Ultrastructure, carbohydrate and amino acid analysis of two preparations of the cercarial glycocalyx of *Schistosoma mansoni*. *Journal of Parasitology* **73,** 514–522.

Cavier, R. and Savel, J. (1954a). L'uréogénèse chez l'ascaris du porc (*Ascaris lumbricoides* Linné 1758). *Bulletin de la Societé de Chimie Biologique* **36,** 1425–1431.

Cavier, R. and Savel, J. (1954b). Contribution à l'étude de l'azote aminé chez l'ascaris du porc (*Ascaris lumbricoides* Linné, 1758). *Bulletin de la Société de Chimie Biologique* **36,** 1433–1438.

Cavier, R. and Savel, J. (1954c). Etude de quelques aspects du métabolisme intermédiaire des acides aminés chez l'ascaris du porc (*Ascaris lumbricoides* Linné, 1758). *Bulletin de la Société de Chimie Biologique* **36,** 1631–1639.

Cavier, R. and Savel, J. (1954d). Les effets du jeûne sur les constituants azotés du liquide périviscéral d'*Ascaris lumbricoides* (Linné 1758). *Comptes Rendus des Séances de l'Académie des Sciences* **238,** 2035–2037.

Cavier, R. and Savel, J. (1954e). Le métabolisme protéique de l'ascaris du porc, *Ascaris lumbricoides* Linné, 1758, est-il ammoniotélique ou uréotélique? *Comptes Rendus des Séances de l'Académie des Sciences* **238,** 2448–2450.

Celik, C. (1986a). Investigation of the kinetic characteristics of arginase activity in the hydatid cyst layers (*Stratum germinativum, Stratum cuticularis*) of cattle and sheep. *Veteriner Fakultesi Dergisi Ankara Universitesi* **33,** 63–75.

Celik, C. (1986b). The urea biosynthesis in hydatid cysts of *Echinococcus granulosus DOGA, Turk Veterinerlik ve Hayvancilik* **10,** 130–136.

Chappell, L. H. (1974). Methionine uptake by larval and adult *Schistosoma mansoni. International Journal for Parasitology* **4,** 361–369.

Chappell, L. H. and Read, C. P. (1973). Studies on the free pool of amino acids of the cestode *Hymenolepis diminuta. Parasitology* **67,** 289–305.

Chappell, L. H. and Walker, E. (1982). *Schistosoma mansoni:* Incorporation and metabolism of protein amino acids *in vitro. Comparative Biochemistry and Physiology* **73B,** 701–707.

Chaudhuri, J., Martin, R. E. and Donahue, M. J. (1988). Tryptophan hydroxylase and aromatic L-amino acid decarboxylase activities in the tissues of adult *Ascaris suum. International Journal for Parasitology* **18,** 341–346.

Cheng, T. C. (1963). Biochemical requirements of larval trematodes. *Annals of the New York Academy of Sciences* **113,** 289–321.

Chi, C.-W. and Isseroff, H. (1979). Fascioliasis: Nitrogen balance studies and the disposition of excessive proline in rats. *Journal of Nutrition* **109,** 1299–1306.

Chvapil, M. and Ehrlich, E. (1970). Effect of increased oxygen on hydroxyproline synthesis in the cuticle and body wall of *Ascaris lumbricoides. Biochimica et Biophysica Acta* **208,** 467–474.

Chvapil, M., Boucek, M. and Ehrlich, E. (1970). Differences in the protocollagen hydroxylase activities from *Ascaris* muscle and hypodermis. *Archives of Biochemistry and Biophysics* **140,** 11–18.

Clarke, A. J., Cox, P. M. and Shepherd, A. M. (1967). The chemical composition of the egg shells of the potato cyst-nematode, *Heterodera rostochiensis* Woll. *Biochemical Journal* **104,** 1056–1060.

Cohen, C., Lanar, D. E. and Parry, D. A. D. (1987). Amino acid sequence and structural repeats in schistosome paramyosin match those of myosin. *Bioscience Reports* **7,** 11–16.

Coles, G. C. (1973). Enzyme levels in cercariae and adult *Schistosoma mansoni. International Journal for Parasitology* **3,** 505–510.

Connolly, J. F. and Downey, N. E. (1968). Glutamate transaminase activities of the liver fluke, *Fasciola hepatica* (Linnaeus, 1758). *Research in Veterinary Science* **9,** 248–250.

Cornish, R. A. and Bryant, C. (1975). Studies of regulatory metabolism in *Moniezia expansa*: Glutamate and the absence of the γ-aminobutyrate pathway. *International Journal for Parasitology* **5,** 355–362.

Cossey, A., Dimichiel, A. and Dunstone, J. (1979). The use of equilibrium density-gradient ultracentrifugation in the isolation and characterisation of glycoproteins with blood group P_1 activity from sheep hydatid-cyst fluid. *European Journal of Biochemistry* **98,** 53–60.

Cox, G. N., Kusch, M. and Edgar, R. S. (1981a). Cuticle of *Caenorhabditis elegans*: Its isolation and partial characterization. *Journal of Cell Biology* **90,** 7–17.

Cox, G. N., Staprans, S. and Edgar, R. S. (1981b). The cuticle of *Caenorhabditis elegans* II. Stage-specific changes in ultrastructure and protein composition during postembryonic development. *Developmental Biology* **86,** 456–470.

Crompton, D. W. T. and Ward, P. F. V. (1984). Selective metabolism of L-serine by *Moniliformis* (Acanthocephala) *in vitro. Parasitology* **89,** 133–144.

Cuperlovic, M., Movesijan, M. and Milosevic, Z. (1986). Free amino acids in the tissues of *Ascaris suum* (Nematoda). *Acta Veterinaria Beograd* **36,** 187–196.

Dabrowski, K. R. (1980). Amino-acid composition of *Ligula intestinalis* (L.) (Cestoda) plerocercoids and of the host parasitized by these cestodes. *Acta Parasitologica Polonica* **27,** 45–48.

Darawshe, S., Tsafadyah, Y. and Daniel, E. (1987). Quaternary structure of erythrocruorin from the nematode *Ascaris suum. Biochemical Journal* **242,** 689–694.

Daugherty, J. W. (1952a). Intermediary protein metabolism in helminths. I. Transaminase reactions in *Fasciola hepatica. Experimental Parasitology* **1,** 331–338.

Daugherty, J. W. (1952b). Studies on the protein metabolism of certain helminth parasites. *Journal of Parasitology* **38,** supplement 32.

Daugherty, J. W. (1954). Synthesis of amino nitrogen from ammonia in *Hymenolepis diminuta. Proceedings of the Society for Experimental Biology and Medicine* **85,** 288–291.

Daugherty, J. W. (1955). Intermediary protein metabolism in helminths. III: L-amino acid oxidases in *Hymenolepis diminuta* and some effects of changes in host physiology. *Experimental Parasitology* **4,** 455–463.

Dedman, J. R. and Harris, B. G. (1975). *Ascaris suum* actin: Properties and similarities to rabbit actin. *Biochemical and Biophysical Research Communications* **65,** 170–175.

Desowitz, R. S. (1962). The amino acids of *Necator americanus* filariform larvae. *Transactions of the Royal Society of Tropical Medicine and Hygiene* **56,** 257.

Dragojevic, M., Rezic, I. and Kurelec, B. (1974). Metabolic fate of citrulline in the liver fluke (*Fasciola hepatica* L.). *Acta Parasitologica Jugoslavica* **5,** 25–31.

Dryuchenko, E. A. (1979). Study of the histidine metabolism in *Ascaridia galli* and host tissues. *Trudy Gel'mintologicheskoi Laboratorii (Gel'minty zhivotnykh i rastenii)* **29,** 41–45.

Dryuchenko, E. A. and Berdyeva, G. T. (1975). Some properties of histidine ammonia-lyase in *Ascaris suum* and host liver. *In* "Problemy parazitologii. Materialy VIII nauchnoi konferentsii parazitologov Uk SSR", pp. 160–162. Chast'I, Kiev.

Dryuchenko, E. A. and Berdyeva, G. T. (1976). Comparative study of histidine ammonia-lyase (4.3.1.3) in the intestine of *Ascaris suum* and in the liver of pig. *Izvestiya Akademii Nauk Turkmenskoi SSR Biologicheskie Nauki* **3,** 24–27.

Dryuchenko, E. A. and Kulikovo, M. N. (1984). Cadaverine and its possible role in host–parasite relationships in ascaridiasis. *Parazitologiya* **18,** 291–295.

Dubinsky, P. and Rybos, M. (1978a). Participation of amino acids in the post-invasion developmental stages of *Ascaridia galli. Biologia* (*Bratislava*) **33,** 657–662.

Dubinsky, P. and Rybos, M. (1978b). On the differences in amino acid contents in the reproductive organs of *Ascaris suum* Goeze, 1782 and *Ascaridia galli* Shrank, 1788; Freeborn, 1923. *Zoologischer Anzeiger (Jena)* **201,** 391–396.

Dubinsky, P. and Rybos, M. (1979). Quantitative changes of free amino acids in unfertilized and fertilized eggs of *Ascaris suum. Helminthologia* **16,** 299–305.

Dubinsky, P., Rybos, M. and Turcekova, L. (1984). Subcellular distribution and properties of aspartate aminotransferase and alanine aminotransferase in the reproductive organs and muscles of *Ascaris suum* females. *Biologia (Bratislava)* **39,** 557–565.

Dubinsky, P., Rybos, M. and Turcekova, L. (1985). Enzymes regulating glucosamine 6-phosphate synthesis in the zygote of *Ascaris suum. International Journal for Parasitology* **15,** 415–419.

Dubovskaya, A. Ya. (1979). A study of the arginase activity in cestode plerocercoids and adults. *Trudy Gel'mintologicheskoi Laboratorii (Gel'minty zhivotnykh i rastenii)* **29,** 46–49.

Dubovskaya, A. Ya. (1982a). Temperature dependence of arginase activity in *Triaenophorus nodulosus* and *Diphyllobothrium latum*. *Parasitologiya* **16,** 494–497.

Dubovskaya, A. Ya. (1982b). Study of the properties of arginase in adult and larval cestodes, parasitic in freshwater fish. *In* "Gel'minty v presnovodnykh biotsenozakh", pp. 85–90. Moscow.

Dubovskaya, A., Ya. (1984). Thermostability of arginase from cestodes and the temperature homeostasis of the host. *In* "Gel'minty sel'skokhozyaistvennykh i okhotnich'e-promyslovykh zhivotnykh" (M. D. Sonin, ed.), pp. 5–10. Gelan, Moscow.

Dvorak, J. A. (1969). *Hydatigera taeniaeformis*: Strobilocerci hooks I. Collection and preparation; elemental, amino acid and infra red spectrophotometric analyses. *Experimental Parasitology* **26,** 111–121.

Ebel, J. P. and Colas, J. (1954). Propriétés des réserves protéiques de l'oocyte de *Parascaris equorum*. *Comptes Rendus des Séances de la Societé de Biologie, Paris* **148,** 1580–1583.

Ehrlich, I., Rijavec, M. and Kurelec, B. (1963). Urea synthesis in the liver fluke (*Fasciola hepatica* L.). *Bulletin Scientifique Conseil des Académies de la RFP de Yougoslavie* **8,** 133.

Ehrlich, I., Rijavic, M. and Kurelec, B. (1968). Urea excretion and synthesis in the liver fluke (*Fasciola hepatica* L.). *Veterinarski* **38,** 146–151.

Ertel, J. and Isseroff, H. (1974). Proline in fascioliasis: I. Comparative activities of ornithine-δ-transaminase and proline oxidase in *Fasciola* and in mammalian livers. *Journal of Parasitology* **60,** 574–577.

Ertel, J. C. and Isseroff, H. (1976). Proline in fascioliasis. II. Characteristics of a partially purified ornithine-δ-transaminase from *Fasciola*. *Rice University Studies* **62,** 97–109.

Evans, H. J., Sullivan, C. E. and Piez, K. A. (1976). The resolution of *Ascaris* cuticle collagen into three chain types. *Biochemistry* **15,** 1435–1439.

Fairbairn, D., Wertheim, G., Harpur, R. P. and Schiller, E. L. (1961). Biochemistry of normal and irradiated strains of *Hymenolepis diminuta*. *Experimental Parasitology* **11,** 248–263.

Files, J. G., Carr, S. and Hirsh, D. (1983). Actin gene family of *Caenorhabditis elegans*. *Journal of Molecular Biology* **164,** 355–375.

Fisher, F. M. and Starling, J. A. (1970). The metabolism of L-valine by *Calliobothrium verticillatum* (Cestoda: Tetraphyllidea): Identification of α-ketoisovaleric acid. *Journal of Parasitology* **56,** 103–107.

Flury, F. (1912). Zur Chemie and Toxikologie der Ascariden. *Archiv. für experimentelle Pathologie und Pharmakologie* **67,** 275–392.

Foster, W. B. and Daugherty, J. W. (1959). Establishment and distribution of *Raillietina cesticillus* in the fowl and comparative studies on amino acid metabolism of *R. cesticillus* and *Hymenolepis diminuta*. *Experimental Parasitology* **8,** 413–426.

Fraefel, W. and Archer, R. (1968). The amino acid sequence of a trypsin inhibitor isolated from ascaris (*Ascaris lumbricoides* var. *suum*). *Biochimica et Biophysica Acta* **154,** 615–617.

Frayha, G. J. and Haddad, R. (1980). Comparative chemical composition of protoscolices and hydatid cyst fluid of *Echinococcus granulosus* (Cestoda). *International Journal for Parasitology* **10,** 359–364.

Freeman, H. C., Hoogland, P. L. and Odense, P. H. (1963). The nature of an unusual amino acid–pyrimidine complex from the cod parasite. *Porrocaecum decipiens*. *Journal of the Fisheries Research Board of Canada* **20,** 1–11.

Friedl, F. E. (1961). Studies on larval *Fascioloides magna*. 1. Observations on the survival of rediae *in vitro*. *Journal of Parasitology* **47,** 71–75.
Friedman, F. and Kagan, I. G. (1958). A paper chromatographic analysis of *Nippostrongylus muris* larvae. *Transactions of the American Microscopical Society* **77,** 365–372.
Fujimoto, D. (1968). Isolation of collagens of high hydroxyproline, hydroxylysine and carbohydrate content from muscle layer of *Ascaris lumbricoides* and pig kidney. *Biochimica et Biophysica Acta* **168,** 537–543.
Fujimoto, D. (1975a). Action of bacterial collagenase on *Ascaris* cuticle collagen. *Journal of Biochemistry (Japan)* **78,** 905–909.
Fujimoto, D. (1975b). Occurrence of dityrosine in cuticlin, a structural protein from *Ascaris* cuticle. *Comparative Biochemistry and Physiology* **51B,** 205–207.
Fujimoto, D. and Adams, E. (1964). Intraspecies composition difference in collagen from cuticle and body of *Ascaris* and *Lumbricus*. *Biochemical* and *Biophysical Research Communications* **17,** 437–442.
Fujimoto, D. and Iizuka, K. (1972). Isolation of noncollagen protein associated with collagen from the muscle layer of *Ascaris lumbricoides*. *Journal of Biochemistry (Japan)* **71,** 1089–1091.
Fujimoto, D. and Kanaya, S. (1973). Cuticlin: A noncollagen structural protein from *Ascaris* cuticle. *Archives of Biochemistry and Biophysics* **157,** 1–6.
Fujimoto, D. and Prockop, D. J. (1969). Protocollagen proline hydroxylase from *Ascaris lumbricoides*. *Journal of Biological Chemistry* **244,** 205–210.
Fujimoto, D., Horiuchi, K. and Hirama, M. (1981). Isotrityrosine, a new crosslinking amino acid isolated from *Ascaris* cuticle collagen. *Biochemical and Biophysical Research Communications* **99,** 637–643.
Fukase, T., Matsuda, Y., Akihama, S. and Itagaki, H. (1984). Some hydrolysing enzymes, especially arginine amidase, in plerocercoids of *Spirometra erinacei* (Cestoda : Diphyllobothridae). *Japanese Journal of Parasitology* **33,** 283–290.
Gallagher, I. H. C. (1964). Chemical composition of hooks isolated from hydatid scolices. *Experimental Parasitology* **15,** 110–117.
Garson, S. and Williams, J. S. (1957). Transamination in *Schistosoma mansoni*. *Journal of Parasitology* **43,** supplement, 27–28.
Gaur, A. S. and Agarwal, S. M. (1980). Biochemical studies on *Paramphistomum cervi* (Shrank, 1790) (Trematoda : Paramphistomatidae). *Indian Journal of Experimental Biology* **18,** 1518–1519.
Gaur, A. S. and Agarwal, S. M. (1981). Studies on amino acids and sugars in larval *Hydatigera taeniaeformis* (Batsch, 1758). *Proceedings of the Indian Academy of Parasitology* **2,** 4–6.
Girotra, K. L. and Isseroff, H. (1980). *Fasciola hepatica*: Azetidine inhibition of bile duct hyperplasia in the infected rat. *Experimental Parasitology* **49,** 41–46.
Goldberg, M., Flescher, E. and Lengy, J. (1979). *Schistosoma mansoni*: Partial purification and properties of ornithine-δ-transaminase. *Experimental Parasitology* **47,** 333–341.
Goldberg, M., Flescher, E., Gold, D. and Lengy, J. (1980). Ornithine-δ-transaminase from the liver fluke *Fasciola hepatica* and the blood fluke *Schistosoma mansoni*: A comparative study. *Comparative Biochemistry and Physiology* **65B,** 605–613.
Gomez-Bautista, M. and Barrett, J. (1988). Cysteine metabolism in the cestode *Hymenolepis diminuta*. *Parasitology* **97,** 149–159.
Goodchild, C. G. and Dennis, E. S. (1965). Effects of casein and zein diets on *Hymenolepis diminuta* in rats and quantitation of nitrogen and amino acids in parasite and host. *Journal of Parasitology* **51,** 253–259.

Goodchild, C. G. and Dennis, E. S. (1966). Amino acids in seven species of cestodes. *Journal of Parasitology* **52,** 60–62.
Goodchild, C. G. and Wells, O. C. (1957). Amino acids in larval and adult tapeworms (*Hymenolepis diminuta*) and in the tissue of their rat and beetle hosts. *Experimental Parasitology* **6,** 575–585.
Govorova, S. V. (1965). Transamination reactions in *Ascaridia galli. Materialy Nauchnoi Konferentsii Vsesoyuznogo Obschestva Gel'mintologov*, 60–64.
Govorova, S. V. (1968). Localization of transaminases in *Ascaridia galli* homogenate fractions. *Materialy Nauchnoi Konferentsii Vsesoyuznogo Obschestva Gel'mintologov*, 87–90.
Graff, D. J. (1964). Metabolism of C^{14}-glucose by *Moniliformis dubius* (Acanthocephala). *Journal of Parasitology* **50,** 230–234.
Graff, D. J. (1965). The utilization of $C^{14}O_2$ in the production of acid metabolites by *Moniliformis dubius* (Acanthocephala). *Journal of Parasitology* **51,** 72–75.
Grantham, B. D. (1986). A comparative study of amino acid metabolism in nematodes. Ph.D. thesis, University of Wales, Aberystwyth.
Grantham, B. D. and Barrett, J. (1986a). Amino acid catabolism in the nematodes *Heligmosomoides polygyrus* and *Panagrellus redivivus.* 1. Removal of the amino group. *Parasitology* **93,** 481–493.
Grantham, B. D. and Barrett, J. (1986b). Amino acid catabolism in the nematodes *Heligmosomoides polygyrus* and *Panagrellus redivivus.* 2. Metabolism of the carbon skeleton. *Parasitology* **93,** 495–504.
Grantham, B. and Barrett, J. (1988). Glutamine and asparagine synthesis in the nematodes *Heligmosomoides polygyrus* and *Panagrellus redivivus. Journal of Parasitology* **74,** 1052–1053.
Granzer, M. and Haas, W. (1986). The chemical stimuli of human skin surface for the attachment response of *Schistosoma mansoni* cercariae. *International Journal for Parasitology* **16,** 575–579.
Gundlach, J. L., Glinski, Z. and Rzedzicki, J. (1971). Electrochromatography of amino acids in *Fasciola hepatica* L. *Acta Parasitologica Polonica* **19,** 375–384.
Gupta, A. K. and Agarwal, S. M. (1986). Biochemical investigations on *Isoparorchis hypselobagri* from the swim bladder of *Wallagonia attu* and body cavities of *W. attu* and *Channa punctatus. Indian Journal of Parasitology* **10,** 47–51.
Gupta, N. K. and Garg, V. K. (1977). Free amino acids in *Paranisakis* sp. *Indian Journal of Parasitology* **1,** 103.
Gupta, N. K. and Kalia, D. C. (1977). Free amino acids of *Setaria cervi* (Rud., 1819) Baylis. 1939. *Indian Journal of Parasitology* **1,** 101–102.
Gupta, P. C. and Bahadur, R. (1985). Free amino acid composition of rumen trematodes, *Gastrothylax crumenifer* and *Cotylophoron orientale. Indian Journal of Parasitology* **9,** 155–157.
Gupta, V. and Agrawal, S. K. (1977). Amino acid composition in *Gastrothylax crumenifer* (Trematoda). *Indian Journal of Helminthology* **29(2),** 140–143.
Guttowa, A. (1968). Amino acids in the tissues of procercoids of *Triaenophorus nodulosus* (Cestoda) and in the coelomic fluids of their host, *Eudiaptomus gracilis* (Copepoda) before and after infection. *Acta Parasitologica Polonica* **15,** 313–320.
Hamajima, F. (1966). Studies on metabolism of lung flukes genus *Paragonimus.* I. Paper chromatographic analyses of free amino acids and aminosugar in uterine eggs, larvae and adults. *Japanese Journal of Parasitology* **15,** 124–127.
Han, I. S., Lee, D. W., Kwon, N. S. and Lee, H. S. (1983). Studies on the Krebs cycle and α-glycerophosphate shuttle in *Fasciola hepatica. Chung-Ang Journal of Medicine* **8,** 139–150.

Hariri, M. (1975). Uptake of 5-hydroxytryptamine by *Mesocestoides corti* (Cestoda). *Journal of Parasitology* **61,** 440–448.
Haskins, W. T. and Olivier, L. (1958). Nitrogenous excretory products of *Taenia taeniaeformis* larvae. *Journal of Parasitology* **44,** 569–573.
Haskins, W. T. and Weinstein, P. P. (1957a). Nitrogenous excretory products of *Trichinella spiralis* larvae. *Journal of Parasitology* **43,** 19–24.
Haskins, W. T. and Weinstein, P. P. (1957b). Amino acids excreted by *Trichinella spiralis* larvae. *Journal of Parasitology* **43,** 25–27.
Haskins, W. T. and Weinstein, P. P. (1957c). The amine constituents from the excretory products of *Ascaris lumbricoides* and *Trichinella spiralis* larvae. *Journal of Parasitology* **43,** 28–32.
Herlich, H. (1966). Amino acid composition of some strongyle parasites of cattle. *Proceedings of the Helminthological Society of Washington* **33,** 103–105.
Hill, B., Kilsby, J., Rogerson, G. W., McIntosh, R. T. and Ginger, C. D. (1981). The enzymes of pyrimidine biosynthesis in a range of parasitic protozoa and helminths. *Molecular and Biochemical Parasitology* **2,** 123–134.
Hill, G. C., Perkowski, C. A. and Mathewson, N. W. (1971). Purification and properties of cytochrome C_{550} from *Ascaris lumbricoides* var. *suum. Biochimica et Biophysica Acta* **236,** 242–245.
Hillman, G. R. and Senft, A. W. (1973). Schistosome motility measurements: response to drugs. *Journal of Pharmacology and Experimental Therapeutics* **185,** 177–184.
Hochachka, P. W. and Somero, G. N. (1984). "Biochemical Adaptation". Princeton University Press, Princeton, N.J.
Hopkins, C. A. (1969). The influence of dietary methionine on the amino acid pool of *Hymenolepis diminuta* in the rat's intestine. *Parasitology* **59,** 407–427.
Hopkins, C. A. and Callow, L. L. (1965). Methionine flux between a tapeworm (*Hymenolepis diminuta*) and its environment. *Parasitology* **55,** 653–666.
Hotez, P. J., Le Trang N., McKerrow, J. H. and Cerami, A. (1985). Isolation and characterization of a proteolytic enzyme from the adult hookworm *Ancylostoma caninum. Journal of Biological Chemistry* **260,** 7343–7348.
Hsiao, S. H. and Hsu, Y.-C. (1965). The effect of some schistosomacides on glutamic-pyruvic transaminase and glutamic-oxalacetic transaminase of *Schistosoma japonicum. Chinese Academy of Medical Science (Shanghai)* **12,** 242–248.
Huang, T. Y., T'ao, Y. H. and Chu, C. H. (1962). Studies on transaminases of *Schistosoma japonicum. Chinese Medical Journal (Peking)* **81,** 79–85.
Hudson, R. E. and Thomas, K. W. (1981). Measurement of free serum proline in the diagnosis of bovine Fascioliasis. *Proceedings of the Australian Biochemical Society* **14,** 46.
Hung, C.-H., Ohno, M., Freytag, J. W. and Hudson, B. G. (1977). Intestinal basement membranes of *Ascaris suum*: analysis of polypeptide components. *Journal of Biological Chemistry* **252,** 3995–4001.
Hung, C-H., Butkowski, R. J. and Hudson, B. G. (1980). Intestinal basement membrane of *Ascaris suum*: Properties of the collagenous domain. *Journal of Biological Chemistry* **255,** 4964–4971.
Il'yasov, I. N. (1978). Amino acid composition of *Raillietina tetragona* (Molin, 1858) and *R. echinobothrida* (Megnin, 1881). *Trudy Nauchno-Issledovatel'skogo Veterinarnogo Instituta Tadzhikskoi SSR* **8,** 75–79.
Ishii, A. I. and Sano, M. (1980). Isolation and identification of paramyosin from liver fluke muscle layer. *Comparative Biochemistry and Physiology* **65B,** 537–541.

Isseroff, H. (1980). The enzymes of proline biosynthesis in *Fasciola* and *Schistosoma* and the possible role of proline in fascioliasis and schistosomiasis. *FEBS Trends in Enzymology* **61,** 303–314.

Isseroff, H. and Chi. C.-W. (1981). Fascioliasis: A model for pathology based on nitrogen balance in rats infused intraduodenally with proline. *Comparative Biochemistry and Physiology* **70A,** 547–550.

Isseroff, H. and Ertel, J. C. (1976). Proline in fascioliasis: III. Activities of pyrroline-5-carboxylic acid reductase and pyrroline-5-carboxylic acid dehydrogenase in *Fasciola. International Journal for Parasitology* **6,** 183–188.

Isseroff, H., Tunis, M. and Read, C. P. (1972). Changes in amino acids of bile in *Fasciola hepatica* infections. *Comparative Biochemistry and Physiology* **41B,** 157–163.

Isseroff, H., Sawma, J. T. and Reino, D. (1977). Fascioliasis: Role of proline in bile duct hyperplasia. *Science (New York)* **198,** 1157–1159.

Isseroff, H., Spengler, R. N. and Charnock, D. R. (1979). Fascioliasis: Similarities of the anaemia in rats to that produced by infused proline. *Journal of Parasitology* **65,** 709–714.

Isseroff, H., Bock, K., Owczarek, A. and Smith, K. R. (1983). Schistosomiasis: Proline production and release by ova. *Journal of Parasitology* **69,** 285–289.

Jackson, G. J. (1973). *Neoaplectana glaseri*: Essential amino acids. *Experimental Parasitology* **34,** 111–114.

Jaffe, J. J. (1980). Folate related metabolism. *In* "Report of the Fifth Meeting of the Scientific Working Group on Filariasis", pp. 19–20. WHO, Geneva.

Jaffe, J. J. and Chrin, L. R. (1979). *De novo* synthesis of methionine in normal and *Brugia*-infected *Aedes aegypti. Journal of Parasitology* **65,** 550–554.

Jaffe, J. J. and Chrin, L. R. (1980). Folate metabolism in filariae: Enzymes associated with 5,10-methylenetetrahydrofolate. *Journal of Parasitology* **66,** 53–58.

Jaffe, J. J. and Chrin, L. R. (1981). Involvement of tetrahydrofolate cofactors in *de novo* purine ribonucleotide synthesis by adult *Brugia pahangi* and *Dirofilaria immitis. Molecular and Biochemical Parasitology* **2,** 259–270.

Janssens, P. A. and Bryant, C. (1969). The ornithine–urea cycle in some parasitic helminths. *Comparative Biochemistry and Physiology* **30,** 261–272.

Jaskowski, B. J. (1962). Paper chromatography of some fractions of *Ascaris suum* eggs. *Experimental Parasitology* **12,** 19–24.

Jaskowski, B. J. (1963). Amino acids in coelomic fluid of *Ascaris suum. Journal of Parasitology* **49,** 50 pp.

Jaskowski, B. J. and Ozuk, B. A. (1977). The cuticular amino acids in the dog heartworm (*Dirofilaria immitis*). *Transactions of the Illinois State Academy of Science* **70,** 363–369.

Josse, J. and Harrington, W. F (1964). Role of pyrrolidine residues in the structure and stabilization of collagen. *Journal of Molecular Biology* **9,** 269–287.

Kajihara, S. and Hashimoto, N. (1952). Paper chromatography of the coelomic fluid of *Ascaris. Medicine and Biology* **25,** 108–110.

Kapur, Y. and Sood, M. L. (1984a). Amino acid composition of the adults of *Haemonchus contortus* (Nematoda : Trichostrongylidae). *Helminthologia* **21,** 267–273.

Kapur, J. and Sood, M. L. (1984b). Amino acid biosynthesis in *Haemonchus contortus* from C^{14}-labelled precursors, *in vitro. Veterinary Parasitology* **15,** 293–299.

Karyakarte, P. P. and Baheti, S. P. (1980). Studies on free amino acids in *Tremiorchis ranarum* Mehra and Negi, 1926 (Trematoda : Digenea). *Rivista di Parassitologia* **41,** 81–83.

Kaur, D., Johal, M. and Johri, S. (1984). Quantitative and qualitative free-amino acid (F.A.A.) composition of *Oesophagostomum columbianum*, a nematode parasite of sheep and goat. *Indian Journal of Helminthology* **34,** 118–122.

Kawanaka, M., Hayashi, S. and Ohtomo, H. (1983). Nutritional requirements of *Schistosoma japonicum* eggs. *Journal of Parasitology* **69,** 857–861.

Kawanaka, M., Hayashi, S. and Carter, C. E. (1986). Uptake and excretion of amino acids and utilization of glucose by *Schistosoma japonicum* eggs. *Japanese Journal of Medical Science and Biology* **39,** 199–206.

Keleti, T., Leoncini, R., Paganini, R. and Marinello, E. (1987). A kinetic method for distinguishing whether an enzyme has one or two active sites for two different substrates. Rat liver L-threonine dehydratase has a single active site for threonine and serine. *European Journal of Biochemistry* **170,** 179–183.

Kent, H. N. (1957). Biochemical studies on the proteins of *Hymenolepis diminuta*. *Experimental Parasitology* **6,** 351–357.

Kimura, K., Tanaka, T., Nakae, H. and Obinata, T. (1987). Troponin from nematode: Purification and characterization of troponin from *Ascaris* body wall muscle. *Comparative Biochemistry and Physiology* **88B,** 399–407.

Kisiel, M. J., Deubert, K. H. and Zuckerman, B. M. (1976). Biogenic amines in the free-living nematode *Caenorhabditis briggsae*. *Experimental Ageing Research* **2,** 37–44.

Klass, M. R. and Hirsh, D. (1981). Sperm isolation and biochemical analysis of the major sperm protein from *Caenorhabditis elegans*. *Developmental Biology* **84,** 299–312.

Kleshchinova, E. A. (1980a). The activity of aminotransferases in the insect nematode *Neoaplectana glaseri*. *Byulleten' Vsesoyuznogo Instituta Gel'mintologii im K.I. Skryabina* **25,** 19–23.

Kleshchinova, E. A. (1980b). The utilization of amino acids by the nematode *Neoaplectana glaseri* during its development in a synthetic medium. *Byulleten' Vsesoyuznogo Instituta Gel'mintologii im K.I. Skryabina* **25,** 24–28.

Kleschinova, E. A. (1983a). Biosynthesis of amino acids by the nematode *Neoaplectana glasseri*. *Byulleten' Vsesoyuznogo Instituta Gel'mintologii im K.I. Skryabina* **35,** 21–24.

Kleshchinova, E. A. (1983b). Absorption of amino acids by the insect nematode *Neoaplectana glaseri* during its development in a culture medium. *Parazitologiya* **17,** 459–463.

Kobayashi, M., Yokogawa, M., Mori, M. and Tatibana, M. (1978). Pyrimidine nucleotide biosynthesis in *Clonorchis sinensis* and *Paragonimus ohirai*. *International Journal for Parasitology* **8,** 471–477.

Kobayashi, M., Asai, T. and Yokogawa, M. (1984). Enzymes of *de novo* pyrimidine biosynthesis in *Paragonimus ohirai*. *Japanese Journal of Parasitology* **33,** 203–210.

Kreuzer, L. (1953). Zur Kenntnis des chemischen Aufbaus der Eihulle von *Ascaris lumbricoides*. 1. Mitteilung. *Zeitschrift für vergleichende Physiologie* **35,** 13–26.

Krvavica, S., Martincic, T. and Asaj, R. (1959a). Metabolism of amino acids in some parasites. 1. Absorption and excretion of amino acids in the tapeworm *Anoplocephala magna*. *Veterinarski Arhiv* **29,** 305–313.

Krvavica, S., Martincic, T. and Asaj, R. (1959b). Metabolism of amino acids in some parasites. II. Amino acids in the hydatid fluid and germinal layer of *Echinococcus*. *Veterinarski Arhiv* **29,** 314–321.

Krvavica, S., Maloseja, Z. and Lui A. (1964a). The amino acid composition of the eggs of *Ascaris suum*, *Neoascaris vitulorum* and *Parascaris equorum*. *Veterinarski Arhiv* **34,** 167–169.

Krvavica, S., Maloseja, Z., Wagner, H. and Martincic, T. (1964b). The amino acid composition of the proteins and the distribution of fatty acids in the lipids of eggs of *Fasciola hepatica* and *Ascaris suum*. *Veterinarski Arhiv* **34,** 165–166.

Kurelec, B., (1964a). Urea synthesis in the liver fluke (*Fasciola hepatica* L.). I. Krebs-Henseleit ornithine cycle enzymes. *Veterinarksi Arhiv* **34,** 193–201.

Kurelec, B. (1964b). Urea synthesis in the liver fluke (*Fasciola hepatica* L.). II. Functional link of the urea cycle with the tricarboxylic acid cycle. *Veterinarski Arhiv* **34,** 221–227.

Kurelec, B. (1972). Lack of carbamyl phosphate synthesis in some parasitic platyhelminths. *Comparative Biochemistry and Physiology* **43B,** 769–780.

Kurelec, B. (1973). Initial steps in the *de novo* pyrimidine biosynthesis in *Ascaris suum*. *Journal of Parasitology* **59,** 1006–1011.

Kurelec, B. (1974a). Die physiologische Funktion der Arginase im Leberegel (*Fasciola hepatica* L.). *Acta Parasitologica Iugoslavica* **5,** 33–43.

Kurelec, B. (1974b). Aspartate transcarbamylase in some parasitic helminths. *Comparative Biochemistry and Physiology* **47B,** 33–40.

Kurelec, B. (1975). Catabolic path of arginine and NAD regeneration in the parasite *Fasciola hepatica*. *Comparative Biochemistry and Physiology* **51B,** 151–156.

Kurelec, B. and Ehrlich, I. (1963). Über die Natur der von *Fasciola hepatica* (L.). *In vitro* Ausgeschiedenen Amino- und Ketosauren. *Experimental Parasitology* **13,** 113–117.

Kurelec, B. and Rijavec, M. (1966). Amino acid pool of the liver fluke (*Fasciola hepatica* L.). *Comparative Biochemistry and Physiology* **19,** 525–531.

Kurelec, B., Rijavec, M. and Klepac, R. (1969). Metabolic fate of histidine in the parasitic worm *Fasciola hepatica* L. *Comparative Biochemistry and Physiology* **29,** 885–887.

Lahoud, H., Prichard, R. K., McManus, W. R. and Schofield, P. J. (1971). The dissimilation of leucine, isoleucine and valine to volatile fatty acids by adult *Fasciola hepatica*. *International Journal for Parasitology* **1,** 223–233.

Lanar, D. E., Pearce, E. J., James, S. L. and Sher, A. (1986). Identification of paramyosin as schistosome antigen recognized by intradermally vaccinated mice. *Science* (New York) **234,** 593–596.

Landsperger, W. J., Stirewalt, M. A. and Dresden, M. H. (1982). Purification and properties of a proteolytic enzyme from the cercariae of the human trematode parasite *Schistosoma mansoni*. *Biochemical Journal* **201,** 137–144.

Langer, B. W., Smith, W. J. and Theodorides, V. J. (1971). Conversion of the alpha-hydroxy and alpha-keto analogues of methionine to methionine by cell-free extracts of adult female *Ascaris suum*. *Journal of Parasitology* **57,** 836–839.

Learmonth, M. P., Euerby, M. R., Jacobs, D. E. and Gibbons, W. A. (1987). Metabolite mapping of *Toxocara canis* using one- and two-dimensional proton magnetic resonance spectroscopy. *Molecular and Biochemical Parasitology* **25,** 293–298.

Lee, J. H., Lee, D. W., Lee, H. S. and Song, C. Y. (1983). Purification and properties of branched-chain amino acid aminotransferase from *Fasciola hepatica*. *Korean Journal of Parasitology* **21,** 49–57.

Lethbridge, R. C. (1971). The chemical composition and some properties of the egg layers in *Hymenolepis diminuta* eggs. *Parasitology* **63,** 275–288.

Leushner, J. R. A. and Pasternak, J. (1978). Partial purification and characterization of prolyl hydroxylase from the free-living nematode *Panagrellus silusiae*. *Canadian Journal of Zoology* **56,** 159–165.

Leushner, J. R. A., Semple, N. L. and Pasternack, J. (1979). Isolation and characterization of the cuticle from the free-living nematode *Panagrellus silusiae*. *Biochimica et Biophysica Acta* **580,** 166–174.

Litchford, R. G. (1970). Amino acids of *Hymenolepis microstoma* and host bile. *Comparative Biochemistry and Physiology* **32,** 61–67.

Livesey, G. (1984). Methionine degradation: "Anabolic and catabolic". *Trends in Biochemical Sciences* **9,** 27–29.

Locatelli, A. and Camerini, E. (1969). Chomatographic studies of amino acids released in the incubation media of *Fasciola hepatica*. *Italian Journal of Biochemistry* **18,** 376–381.

Lopez-Gorge, J. (1969). Metabolismo de la arginina en *Ascaris lumbricoides*. Actividad L-aminoacido oxidasica. *Revista Iberíca de Parasitologia* **29,** 371–397.

Lopez-Gorge, J. and Monteoliva Hernandez, M. (1964). Algunos aspectos sobre la composicion quimica de *Moniezia expansa* (Rudolphi, 1805): I. Protidos. *Revista Iberíca de Parasitologia* **24,** 183–207.

Lopez-Gorge, J. and Monteoliva, M. (1969). Metabolismo de la arginina en *Ascaris lumbricoides* actividades L-argininacarboxiliasica y arginasica. *Revista Iberíca de Parasitologia* **29,** 3–34.

Lopez-Gorge, J., Monteoliva, M. and Mayor, F. (1968). L-Lysine carboxylase in *Ascaris lumbricoides* and *Moniezia expansa*. *Experimental Parasitology* **23,** 129–133.

Lopez-Gorge, J., Monteoliva, M. and Mayor, F. (1969). Actividad lisinadecarboxilasa en *Ascaris lumbricoides* y *Moniezia expansa*. *Revista Iberíca de Parasitologica* **29,** 219–227.

Lussier, P. E., Podesta, R. B. and Mettrick, D. F. (1978). *Hymenolepis diminuta*: Amino acid transport and osmoregulation. *Journal of Parasitology* **64,** 1140–1141.

Lutz, P. L. and Siddiqi, A. H. (1971). Nonprotein nitrogenous composition of the protonephridial fluid of the trematode *Fasciola gigantica*. *Comparative Biochemistry and Physiology* **40A,** 453–457.

Lutz, P. L., Iversen, E. S. and Tocci, P. M. (1981). Composition of the protonephridial fluid from the giant trematode *Hirundinella ventricosa*. *Journal of Parasitology* **67,** 280–281.

Lynch, D. L. and Bogitsh, B. J. (1962). The chemical nature of metacercarial cysts. II. Biochemical investigations on the cysts of *Posthodiplostomum minimum*. *Journal of Parasitology* **48,** 241–243.

Lyons, K. M. (1966). The chemical nature and evolutionary significance of monogenean attachment sclerites. *Parasitology* **56,** 63–100.

Machnicka-Roguska, B. (1965). Preparation of *Taenia saginata* antigens and chemical analysis of antigenic fractions. *Acta Parasitologica Polonica* **13,** 337–347.

Malhotra, S. K. (1981). Studies on amino acids in *Raillietina (Raillietina) saharanpurensis* (Malhotra and Capoor, 1981) with a note on biochemical variations in cyclophyllidean cestodes. *Comparative Physiology and Ecology* **7,** 207–210.

Malhotra, S. K. and Rautela, A. A. (1986). Amino acids of a poultry nematode *Heterakis kotwardensis* (Malhotra and Rautela) from *Gallus gallus*. *In* "Proceedings of the National Symposium on New Dimensions in Parasitology" (V. N. Capoor, S. C. Misra and S. K. Malhotra, eds), pp. 102–104. Allahabad University Press, Allahabad.

Mansour, T. E. (1979). Chemotherapy of parasitic worms: New biochemical strategies. *Science (New York)* **205,** 462–469.

Mansour, T. E. and Stone, D. B. (1970). Biochemical effects of lysergic acid diethylamide on the liver fluke *Fasciola hepatica. Biochemical Pharmacology* **19,** 1137–1146.

Mansour, T. E., Lago, A. D. and Hawkins, J. L. (1957). Occurrence and possible role of serotonin in *Fasciola hepatica. Federation Proceedings* **16,** 319 pp.

Masaracchia, R. A., Hassell, T. C. and Donahue, M. J. (1986). Structural analysis of the calcium-binding protein calmodulin from *Ascaris suum* obliquely striated muscle. *Journal of Parasitology* **72,** 299–305.

McBride, O. W. and Harrington, W. F. (1967a). *Ascaris* cuticle collagen: On the disulphide cross-linkages and the molecular properties of the subunits. *Biochemistry* **6,** 1484–1498.

McBride, O. W. and Harrington, W. F. (1967b). Helix-coil transition in collagen. Evidence for a single-stranded triple helix, *Biochemistry* **6,** 1499–1514.

McLachlan, A. D. and Karn, J. (1983). Periodic features in the amino acid sequence of nematode rod myosin. *Journal of Molecular Biology* **164,** 605–626.

McManus, D. P. and James, B. L. (1975). Tricarboxylic acid cycle enzymes in the digestive gland of *Littorina saxatilis rudis* (Maton) and in the daughter sporocysts of *Microphallus similis* (Jag) (Digenea: Microphallidae). *Comparative Biochemistry and Physiology* **50B,** 490–495.

McNeill, K. M. and Hutchison, W. F. (1971). The tricarboxylic acid cycle enzymes in the adult dog heartworm *Dirofilaria immitis. Comparative Biochemistry and Physiology*. **38B,** 493–500.

Mettrick, D. F. and Boddington, M. J. (1972a). Amino acid pools of *Syndesmis franciscana* (Turbellaria: Platyhelminthes) and of host coelomic fluid. *Canadian Journal of Zoology* **50,** 411–413.

Mettrick, D. F. and Boddington, M. J. (1972b). The chemical composition of some marine and fresh-water turbellarians. *Caribbean Journal of Science* **12,** 1–7.

Mettrick, D. F. and Telford, J. M. (1963). Histamine in the phylum Platyhelminthes. *Journal of Parasitology* **49,** 653–656.

Mettrick, D. F. and Telford, J. M. (1965). The histamine content and histidine decarboxylase activity of some marine and terrestrial animals from the West Indies. *Comparative Biochemistry and Physiology* **16,** 547–559.

Miller, C. W. and Roberts, R. N. (1964). Alanine synthesis in the nematode *Aphelenchoides ritzemabosi. Phytopathology* **54,** 1177.

Modavi, S. and Isseroff, H. (1984). *Fasciola hepatica*: Collagen deposition and other histopathology in the rat host's bile duct caused by the parasite and by proline infusion. *Experimental Parasitology* **58,** 239–244.

Monteoliva, M. (1963). Los constituyentes nitrogenados de *Ascaridia galli* Schrank, 1788. IV. Aminoacidos conjugadas. *Revista Iberica de Parasitologia* **23,** 75–81.

Monteoliva, M., Rasero, F. S., Gorge, J. L. and Mayor, F. (1965). L-Glutamate-1-carboxylase in intestinal parasites. *Nature (London)* **205,** 1111–1112.

Monteoliva Hernandez, M. (1962). Los constituyentes nitrogenados de *Ascaridia galli* Schrank 1788. III. Aminoacidos libres. *Revista Iberica de Parasitologia* **22,** 255–261.

Monteoliva Hernandez, M., Escobar Bueno, C. and Guevara Pozo, D. (1962). Los constituyentes nitrogenados de *Ascaridia galli* Schrank 1788. II. Aminoacidos totales. *Revista Iberica de Parasitologia* **22,** 49–54.

Morseth, D. J. (1966). Chemical composition of embryophoric blocks of *Taenia hydatigena*, *Taenia ovis* and *Taenia pisiformis* eggs. *Experimental Parasitology* **18**, 347–354.

Moss, G. D. (1970). The excretory metabolism of the endoparasitic digenean *Fasciola hepatica* and its relationship to its respiratory metabolism. *Parasitology* **60**, 1–19.

Munir, W. A. and Barrett, J. (1985). The metabolism of xenobiotic compounds by *Hymenolepis diminuta* (Cestoda:Cyclophyllidea). *Parasitology* **91**, 145–156.

Mustafa, T., Komuniecki, R. and Mettrick, D. F. (1978). Cytosolic glutamate dehydrogenase in *Hymenolepis diminuta* (Cestoda). *Comparative Biochemistry and Physiology* **61B**, 219–222.

Muthukrishnan, S. (1975). Studies on the integument of cestodes II. The nature of protein component in the integument of gravid proglottides of *Taenia hydatigena*. *Acta Histochemica* **53**, 174–181.

Myers, R. F. and Balasubramanian, M. (1973). Nutrient media for plant-parasitic nematodes. V. Amino acid nutrition of *Aphelenchoides rutgersi*. *Experimental Parasitology* **34**, 123–131.

Myers, R. F. and Krusberg, L. R. (1965). Organic substances discharged by plant-parasitic nematodes. *Phytopathology* **55**, 429–437.

Nakamura, T., Yamaguchi, M. and Yanagisawa, T. (1979). Comparative studies on actins from various sources. *Journal of Biochemistry* **85**, 627–631.

Nanda, S., Bhalya, A., Gairola, D., Malhotra, S. K. and Capoor, V. N. (1987). Comparative analysis of amino acids of three species of *Gangesia* (Cestoda: Proteocephalata). *Journal of Helminthology* **61**, 233–239.

Nations, C., Hicks, T. C. and Ubelaker, J. E. (1973). CO_2 production by extracts of *Hymenolepis diminuta* (Cestoda:Hymenolepididae) with aspartate and α-ketoglutarate as substrates. *Journal of Parasitology* **59**, 112–116.

Negus, M. R. S. (1968). The nutrition of sporocysts of the trematode *Cercaria doricha* Rothschild, 1935 in the molluscan host *Turritella communis* Risso. *Parasitology* **58**, 355–366.

Nellaiappan, K. and Ramalingam, K. (1980). Stabilization of egg-shell in *Paraplerurus sauridae* (Digenea:Hemiuridae). *Parasitology* **80**, 1–7.

Nicholas, W. L., Dougherty, E., Hansen E. L., Holm-Hansen, O. and Moses, V. (1960). The incorporation of ^{14}C from sodium acetate 2-^{14}C into the amino acids of the soil inhabiting nematode, *Caenorhabditis briggsae*. *Journal of Experimental Biology* **37**, 435–443.

Nigam, S. C. (1978). Free amino acids from *Ascaridia galli*. *Indian Journal of Parasitology* **2**, 157–158.

Nigam, S. C. (1979). Amino acid composition of nematode parasites. *Indian Journal of Helminthology* **31**, 69–71.

Niyogi, A. and Agarwal, S. M. (1983). Free and protein amino acids in *Lytocestus indicus*, *Introvertus raipurensis* and *Lucknowia indica* parasitizing *Clarias batrachus* (Linn.) (Cestoda:Caryophillidae). *Japanese Journal of Parasitology* **32**, 341–345.

Nordwig, A. and Hayduk, U. (1969). Invertebrate collagens: Isolation, characterization and phylogenetic aspects. *Journal of Molecular Biology* **44**, 161–172.

Okazaki, T., Wittenberg, B. A., Briehl, R. W. and Wittenberg, J. B. (1967). The hemoglobin of *Ascaris* body walls. *Biochimica et Biophysica Acta* **140**, 258–265.

Okuno, Y. (1968). Amino acids of hydrolysed tissues of hog *Ascaris* and *Anisakis* sp. larvae. I. Analysis by thin layer chromatography. *Japanese Journal of Parasitology* **17**, 199–207.

Okuno, Y. (1969). Amino acids in *Ascaris suum* and *Anisakis* sp. larvae. Analysis by thin layer chromatography and amino acid autoanalyser. *Japanese Journal of Parasitology* **18,** 77–86.

Ossikovski, E. (1983). Comparative studies of the proteins and amino acids, and metabolism of helminths. IV. Free amino acids in different organs and developmental stages of *Ascaridia galli* (Schrank, 1788) Freeborn, 1923. *Khelmintologiya* **16,** 54–63.

Ossikovski, E. (1984). Comparative studies of the protein and amino acid content of helminths and of their metabolism. V. Influence of osmotic pressure on the free amino acid content of adult *Ascaridia galli* (Schrank, 1788) Freeborn, 1923. *Khelmintologiya* **17,** 41–51.

Ossikovski, E. and Khlebarov, Z. (1983). Comparative studies of the proteins and amino acids, and metabolism of helminths. III. Amino acid composition of protein hydrolysates of different developmental stages of *Ascaridia dissimilis* Vigueras, 1931. *Khelmintologiya* **16,** 46–53.

Ouazana, R. (1981). Etude du collagène cuticulaire au cours du développement postembryonaire du Nématode *Caenorhabditis elegans*: Comparaison entre larves de premier stade et adultes. *Comptes Rendus des Séances de l'Académie des Sciences* **293,** 467–470.

Ouazana, R. and Gilbert, M. A. (1979). Composition du collagene cuticulaire du Nématode *Caenorhabditis elegans*, lignée sauvage Bergerac. *Comptes Rendus des Séances de l'Académie des Sciences* **288,** 911–914.

Ouazana, R. and Herbage, D. (1981). Biochemical characterization of the cuticle collagen of the nematode *Caenorhabditis elegans. Biochimica et Biophysica Acta* **669,** 236–243.

Ouazana, R., Herbage, D. and Godet, J. (1984). Some biochemical aspects of the cuticle collagen of the nematode *Caenorhabditis elegans. Comparative Biochemistry and Physiology* **77B,** 51–56.

Page, C. R., Defraites, R. F., Newport, G. R. and Velez, R. J. (1978). Diurnal variation of free pool amino acids in *Hymenolepis diminuta. Fourth International Congress of Parasitology* (*Warsaw*) F, 67.

Paltridge, R. W. and Janssens, P. A. (1971). A reinvestigation of the status of the ornithine urea-cycle in adult *Ascaris lumbricoides. Comparative Biochemistry and Physiology* **40B,** 503–513.

Pappas, P. W. (1978). The inability of a trypanorhynchid cestode to utilize CO_2 produced during urea catabolism. *Ohio Journal of Science* **78,** 152–153.

Park, S. H., Kwon, N. S., Lee, H. S. and Song, C. Y. (1983). Aspartate and alanine aminotransferase in *Fasciola hepatica. Korean Journal of Parasitology* **21,** 41–48.

Pascoe, D. (1970). Dehydrogenases in the daughter sporocysts of *Microphallus pygmaeus* (Levinsen, 1881) (Trematoda: Microphallidae). *Zeitschrift für Parasitenkunde* **35,** 7–15.

Pascoe, D. and Richards, R. J. (1970). Variations in the respiratory quotient of the daughter sporocysts of *Cercaria dichotoma* Lebour, 1911 during starvation. *Acta Parasitologica Polonica* **18,** 107–114.

Pathak, K. M. L., Gaur, S. N. S. and Verma, H. C. (1980). Quantitative estimation of amino acids in cysticercus of *Taenia hydatigena. Veterinary Parasitology* **7,** 375–378.

Peanasky, R. J., Bentz, Y., Paulson, B., Graham, D. L. and Babin, D. R. (1984). The isoinhibitors of chymotrypsin/elastase from *Ascaris lumbricoides*: Isolation by affinity chromatography and association with the enzymes. *Archives of Biochemistry and Biophysics* **232,** 127–134.

Pearson, A. G. M., Fincham, A. G., Waters, H. and Bundy, D. A. P. (1985). Differences in composition between *Fasciola hepatica* spines and cestode hooks. *Comparative Biochemistry and Physiology* **81B,** 373–376.

Peczon, B. D., Venable, J. H., Beams, C. G. and Hudson, B. G. (1975). Intestinal basement membrane of *Ascaris suum*: Preparation, morphology and composition. *Biochemistry* **14,** 4069–4075.

Peczon, B. D., Wegener, L. J., Hung, C.-H. and Hudson, B. G. (1977). Intestinal basement membrane of *Ascaris suum*: Characterization of carbohydrate units. *Journal of Biological Chemistry* **252,** 4002–4006.

Perez-Gimenez, M. E., Gimenez, A. and Gaede, K. (1967). Metabolic transformation of ^{14}C-glucose into tissue proteins of *Ancylostoma caninum*. *Experimental Parasitology* **21,** 215–223.

Pollack, J. K. (1957a). The uptake and utilization of ammonium ions by the parasitic roundworm *Ascaris lumbricoides*. *Australian Journal of Science* **19,** 208–209.

Pollack, J. K. (1957b). The metabolism of *Ascaris lumbricoides* ovaries. III. The synthesis of alanine from pyruvate and ammonia. *Australian Journal of Biological Sciences* **10,** 465–474.

Pollack, J. K. and Fairbairn, D. (1955a). The metabolism of *Ascaris lumbricoides* ovaries. I. Nitrogen distribution. *Canadian Journal of Biochemistry and Physiology* **22,** 297–306.

Pollack, J. K. and Fairbairn, D. (1955b). The metabolism of *Ascaris lumbricoides* ovaries. II. Amino acid metabolism. *Canadian Journal of Biochemistry and Physiology* **33,** 307–316.

Polyakova, O. I. (1962). Transamination and reductive amination in *Dictyocaulus filaria*. *Biokhimiya Moscow* **27,** 430–436.

Popiel, I. and Basch, P. F. (1984). Putative polypeptide transfer from male to female *Schistosoma mansoni*. *Molecular and Biochemical Parasitology* **11,** 179–188.

Porter, C. A. and Gamble, W. (1971). Free amino acids content of the healthy and parasitized digestive gland of *Oxytrema silicata* and the rediae of *Nanophyetus salmincola*. *Comparative Biochemistry and Physiology* **40B,** 335–340.

Prescott, L. M. and Campbell, J. W. (1965). Phosphoenolpyruvate carboxylase activity and glycogenesis in the flatworm *Hymenolepis diminuta*. *Comparative Biochemistry and Physiology* **14,** 491–511.

Preston, R. L. (1987). Occurrence of D-amino acids in higher organisms: A survey of the distribution of D-amino acids in marine invertebrates. *Comparative Biochemistry and Physiology* **87B,** 55–62.

Pudles, J., Rola, F. H. and Matida, A. K. (1967). Studies on the proteolytic inhibitors from *Ascaris lumbricoides* var. *suum*. II. Purification, properties and chemical modification of the trypsin inhibitor. *Archives of Biochemistry and Biophysics* **120,** 594–601.

Rahman, M. S., Mettrick, D. F. and Podesta, R. B. (1985). Serotonin and distribution of radiocarbon from D-[^{14}C-U]-glucose in *Schistosoma mansoni*. *Journal of Parasitology* **71,** 403–408.

Rainsford, K. D. (1972). The chemistry of egg-shell formation in *Fasciola hepatica*. *Comparative Biochemistry and Physiology* **43B,** 983–989.

Ramalingam, K. (1973a). Some aspects of the biochemistry of monogenean parasites of marine fishes. *Special Publications of the Marine Biological Association of India*, pp. 208–226.

Ramalingam, K. (1973b). The chemical nature of the egg-shell of helminths—1. Absence of quinone tanning in the egg-shell of the liver fluke, *Fasciola hepatica*. *International Journal for Parasitology* **3,** 67–75.

Ramalingam, K. (1973c). Chemical nature of the egg shell in helminths—II. Mode of stabilization of egg shells of monogenetic trematodes. *Experimental Parasitology* **34,** 115–122.

Rasero, F. S., Monteoliva, M. and Mayor, F. (1968). Enzymes related to 4-aminobutyrate metabolism in intestinal parasites. *Comparative Biochemistry and Physiology* **25,** 693–701.

Rasheed, U. (1981). Transaminase activity in *Lytocestus indicus* and its host. *Proceedings of the Indian Academy of Parasitology* **2,** 115–116.

Rathaur, S., Wittich, R. M. and Walter, R. D. (1988). *Ascaris suum* and *Onchocerca volvulus*: *S*-adenosylmethionine decarboxylase. *Experimental Parasitology* **65,** 277–281.

Read, C. P. (1953). Contributions to cestode enzymology. II. Some anaerobic dehydrogenases in *Hymenolepis diminuta. Experimental Parasitology* **2,** 341–347.

Read, C. P. (1968). Intermediary metabolism of flatworms. *In* "Chemical Zoology" (M. Florkin and B. T. Scheer, eds), Vol. 2, pp. 327–357. Academic Press, London and San Diego.

Rhodes, M. B. and Ferguson, D. L. (1973). *Haemonchus contortus*: Enzymes III. Glutamate dehydrogenase. *Experimental Parasitology* **34,** 100–110.

Ribeiro, P. and Webb, R. A. (1983). The occurrence and synthesis of octopamine and catecholamines in the cestode *Hymenolepis diminuta. Molecular and Biochemical Parasitology* **7,** 53–62.

Richards, R. J. (1969). Qualitative and quantitative estimations of the free amino acids in the healthy and parasitized digestive gland and gonad of *Littorina saxatilis tenebrosa* (Mont.) and in the daughter sporocysts of *Microphallus pygmaeus* (Levinsen, 1881) and *Microphallus similis* (Jagerskiold, 1900) (Trematoda: Microphallidae). *Comparative Biochemistry and Physiology* **31,** 655–665.

Richards, R. J. (1970a). The leakage and transamination of amino acids *in vitro* by the germinal sacs of marine digenea. *Journal of Helminthology* **44,** 231–241.

Richards, R. J. (1970b). The effects of starvation *in vitro* on the free amino acids and sugars in the daughter sporocysts of *Microphallus pygmaeus* (Levinsen, 1881) (Trematoda: Microphallidae). *Zeitschrift für Parasitenkunde* **35,** 31–39.

Rijavec, M. (1965). Ornithinische Transcarbamylase und Arginase bei manchen parasitischen Wurmern der Rinder. *Zeitschrift für Parasitenkunde* **26,** 163–167.

Rijavec, M. (1974). Activity of enzymes of the metabolic path of arginine–proline in the liver fluke. *Third International Congress of Parasitology Abstracts (Munich)* **3,** 1491.

Rijavec, M. and Kurelec, B. (1965). Harnstoffzyklus bei einigen Rinderparasiten (Helminthen). *Zeitschrift für Parasitenkunde* **26,** 168–172.

Rijavec, M. and Kurelec, B. (1966). Die Activitat der ornithinischen Transcarbamylase und Arginase im Gewebe verschiedener Entwicklungsstadien des grossen Leberegels (*Fasciola hepatica* L.). *Zeitschrift für Parasitenkunde* **27,** 99–105.

Robbins, S. H., Hammett, M. and Fried, B. (1979). Light and transmission electron microscopical studies and amino acid analysis of the metacercarial cyst of *Zygocotyle lunata* (Trematoda). *International Journal for Parasitology* **9,** 257–260.

Robinson, D. L. H. (1961). Amino acids of *Schistosoma mansoni. Annals of Tropical Medicine and Parasitology* **55,** 403–406.

Rogers, W. P. (1952). Nitrogen catabolism in nematode parasites. *Australian Journal of Scientific Research* **5B,** 210–222.

Rogers, W. P. (1955). Amino acids and peptides excreted by nematode parasites. *Experimental Parasitology* **4,** 21–28.

Roth, A. A. and Hare, L. V. (1966). Effect of *Schistosoma mansoni* on amino acid levels in chemically defined medium. Cited by Senft, A. W. in *Comparative Biochemistry and Physiology* (1966) **18,** 209–216.

Rothstein, M. (1963). Nematode biochemistry. III. Excretion products. *Comparative Biochemistry and Physiology* **9,** 51–59.

Rothstein, M. (1965). Nematode biochemistry. V. Intermediary metabolism and amino acid interconversions in *Caenorhabditis briggsae. Comparative Biochemistry and Physiology* **14,** 541–552.

Rothstein, M. and Mayo, H. (1964a). Nematode biochemistry. IV. On isocitrate lyase in *Caenorhabditis briggsae. Archives of Biochemistry and Biophysics* **108,** 134–142.

Rothstein, M. and Mayo, H. (1964b). Glycine synthesis and isocitrate lyase in the nematode *Caenorhabditis briggsae. Biochemical and Biophysical Research Communications* **14,** 43–47.

Rothstein, M. and Tomlinson, G. A. (1961). Biosynthesis of amino acids by the nematode *Caenorhabditis briggsae. Biochimica et Biophysica Acta* **49,** 625–627.

Rothstein, M. and Tomlinson, G. (1962). Nematode Biochemistry. II. Biosynthesis of amino acids. *Biochimica et Biophysica Acta* **63,** 471–480.

Roy, T. K. (1975). Effect of amino acids and carbohydrates on the population growth of *Chiloplacus lentus* in axenic culture. *Nematologica* **21,** 12–18.

Roy, T. K. and Srivastava, V. M. L. (1981). Effects of adenosine 3′,5′-cyclic monophosphate, hydrocortisone and epinephrine on certain aspects of metabolism of *Ascaridia galli. Rivista de Parassitologia* **42,** 219–239.

Saber, M. A. and Wu, G. (1985). *In vitro* catabolism of L-histidine by adult *Schistosoma mansoni. Egyptian Journal of Parasitology* **15,** 29–39.

Sahu, R., Gupta, A. K., Gaur, A. S. and Agarwal, S. M. (1987). Studies on certain aspects of protein metabolism during the developmental physiology of *Clinostomum complanatum. Indian Journal of Parasitology* **11,** 37–42.

Salmenkova, E. A. (1962). The composition of free amino-acids in the body fluid of *Ascaris suum* and its changes during cultivation under conditions of protein deficiency. *Meditsinskaya Parazitologiya i Parazitarnie Bolezni* **31,** 664–668.

Savel, J. (1955). Etudes sur la constitution et le metabolisme protéiques d' *Ascaris lumbricoides* Linné 1758 Chapitre III: Les produits d'excretion du catabolism azote de l'*Ascaris* du proc. *Revue de Pathologie Générale et Comparée* **55,** 213–279.

Sawma, J. T., Isseroff, H. and Reino, D. (1978). Proline in fascioliasis: IV. Induction of bile duct hyperplasia. *Comparative Biochemistry and Physiology* **61A,** 239–243.

Schaeffer, J. M. and Bergstrom, A. R. (1988). Identification of gamma-aminobutyric acid and its binding sites in *Caenorhabditis elegans. Life Sciences* **43,** 1701–1706.

Scott, H. L. and Whittaker, F. H. (1970). *Pelodera strongyloides* Schneider 1866: A potential research tool. *Journal of Nematology* **2,** 193–203.

Seed, J. L., Kilts, C. D. and Bennett, J. L. (1980). *Schistosoma mansoni*: Tyrosine, a putative *in vivo* substrate of phenol oxidase. *Experimental Parasitology* **50,** 33–44.

Senft, A. W. (1963) Observations on the amino acid metabolism of *Schistosoma mansoni* in a chemically defined medium. *Annals of the New York Academy of Sciences* **113,** 272–288.

Senft, A. W. (1965). Recent developments in the understanding of amino-acid and protein metabolism by *Schistosoma mansoni in vitro. Annals of Tropical Medicine and Parasitology* **59,** 164–168.

Senft, A. W. (1966). Studies in arginine metabolism by schistosomes—I. Arginine uptake and lysis by *Schistosoma mansoni. Comparative Biochemistry and Physiology* **18,** 209–216.

Senft, A. W., Miech, R. P., Brown, P. R. and Senft, D. G. (1972). Purine metabolism in *Schistosoma mansoni. International Journal for Parasitology* **2,** 249–260.

Sharma, R. K. and Singh, K. (1977). Studies on glutamic-oxaloacetic (GOT) and glutamic-pyruvate (GPT) transaminases of swine kidney worm *Stephanurus dentatus* (Diesing, 1839) I. Assay and general properties. *Zeitschrift für Parasitenkunde* **54,** 251–256.

Sheers, M., Edwards, S. R., Moore, R. J., Montague, P. E. and Campbell, A. J. (1980). Proline production by the common liver fluke *Fasciola hepatica in vitro* and *in vivo. Proceedings of the Australian Biochemical Society* **13,** 74.

Sheers, M., Campbell, J., Beames, D. J., Edwards, S. R., Moore, R. J. and Montague, P. E. (1982). Fasciolicidal potential of proline analogues and proline biosynthesis inhibitors. *International Journal for Parasitology* **12,** 47–52.

Shishkova-Kasatochkina, O. A., Koloskova, T. G. and Sokhina, L. I. (1976). Action of the enzymes arginase and urease in a possible regulatory mechanism at tissue level of osmotic and temperature homeostasis in fish nematodes. In "Kratkie tezisy doklodov II Vsesoyuznogo simpoziuma po parazitam i boleznyam morskikh zhivotnykh", pp. 71–72, Kalinigrad.

Shishkova-Kasatochkina, O. A., Dubovskaya, A. Ya. and Koloskova, T. G. (1986). Activity of the enzymes arginase and urease and the final products of protein metabolism in certain nematodes of different locations. *Trudy Gelmintologicheskoi Laboratorii* **34,** 138–141.

Siddiqui, S. S. and Babu, P. (1980). Kynurenine hydroxylase mutants of the nematode *Caenorhabditis elegans. Molecular and General Genetics* **179,** 21–24.

Siddiqui, S. S. and von Ehrenstein, G. (1980). Biochemical genetics of *Caenorhabditis elegans. In* "Nematodes as Biological Models" (B. M. Zuckerman, ed.), Vol. 1, pp. 285–303. Academic Press, London and San Diego.

Simmonds, R. A. (1958). Studies on the sheath of fourth stage larvae of the nematode parasite *Nippostrongylus muris. Experimental Parasitology* **7,** 14–22.

Simmons, J. E. (1961). Urease activity in trypanorhynch cestodes. *Biological Bulletin* **121,** 535–546.

Simmons, J. E. (1969). Composition of the amino acid pools of some cestodes of elasmobranch fishes of the Woods Hole area. *Experimental Parasitology* **26,** 264–271.

Singh, G. and Srivastava, V. M. L. (1983). Metabolism of amino acids in *Ascaridia galli*: Transamination. *Zeitschrift für Parasitenkunde* **69,** 783–788.

Singh, G. and Srivastava, V. M. L. (1984). A general microsensitive assay method for transaminases. *Indian Journal of Experimental Biology* **22,** 91–93.

Singh, G., Pampori, N. A. and Srivastava, V. M. L. (1983a). Metabolism of amino acids in *Ascaridia galli*: Decarboxylation reactions. *International Journal for Parasitology* **13,** 305–307.

Singh, G., Pampori, N. A. and Srivastava, V. M. L. (1983b). Amino acid metabolizing enzymes of *Ascaridia galli. Indian Journal of Parasitology* **7,** 75–77.

Singh, G., Pampori, N. A. and Srivastava, V. M. L. (1984). ATP production in parasitic nematodes and effect of anthelmintics. *Indian Journal of Experimental Biology* **22,** 50–53.

Singh, G., Pampori, N. A. and Srivastava, V. M. L. (1985). Metabolic fate of alanine in cuticle hypodermis muscle system of the fowl nematode *Ascaridia galli. Indian Journal of Experimental Biology* **23,** 708–710.

Singh, G., Gupta, S., Katiyar, J. C. and Srivastava, V. M. L. (1987). Amino acid metabolism in *Ancylostoma ceylanicum* and *Nippostrongylus brasiliensis. Journal of Helminthology* **61,** 84–88.

Slonka, G. F., Ridley, R. K. and Leland, S. E. (1973). The use of *in vitro*-grown *Cooperia punctata* (Nematoda: Trichostrongyloidea) to study incorporation of carbon from D-glucose-U-^{14}C into major chemical fractions. *Journal of Parasitology* **59,** 282–288.

Smart, D. (1987). Biogenic amine metabolism in parasitic nematodes. Ph.D. thesis, University College of Wales, Aberystwyth.

Smart, D. (1988). Catecholamine synthesis in *Ascaridia galli* (Nematoda). *International Journal for Parasitology* **18,** 485–492.

Sokhina, L. I. (1975). Study of arginase in *Fasciola hepatica*. *In* "Problemy parazitologii Materialy VIII nauchnoi konferentsii parazitologov UkSSR Chast'2, pp. 186–187, Kiev.

Sokhina, L. I. (1976). Localization of the enzymes arginase and urease in the organelles of the nematode *Contracaecum aduncum*. *In* "Kratkie tezisy dokladov II Vsesoyuznogo simpoziuma po parazitam i boleznyam morskikh zhivotnykh", pp. 59–60, Kaliningrad.

Sokhina, L. I. and Koloskova, T. G. (1978). The activity of the enzymes arginase and urease as a factor in the adaptation of nematodes to parasitism. *Trudy Gelmintologicheskoi Laboratorii* **28,** 104–108.

Sokhina, L. I. and Shishkova-Kasatochkina, O. A. (1979). Studies on the arginase activity in the ontogenesis of *Ascaridia galli* and effect of thyroxine hormone on the synthesis of the given enzyme. *Helminthologia* **16,** 287–292.

Soutter, A. M., Walkey, M. and Arme, C. (1980). Amino acids in the plerocercoid of *Ligula intestinalis* (Cestoda:Pseudophyllidea) and its fish host *Rutilus rutilus*. *Zeitschrift für Parasitenkunde* **63,** 151–158.

Srivastava, V. M. L., Misra, S. and Chatterjee, R. K. (1987). Amino acid metabolism in *Dipetalonema viteae*. *Indian Journal of Experimental Biology* **25,** 416–418.

Srivastava, V. M. L., Saz, H. J. and deBruyn, B. (1988). Comparisons of glucose and amino acid use in adults and microfilariae of *Brugia pahangi*. *Parasitology Research* **75,** 1–6.

Stjernholm, R. L. and Warren, K. S. (1974). *Schistosoma mansoni:* Utilization of exogenous metabolites by eggs *in vitro*. *Experimental Parasitology* **36,** 222–232.

Sulston, J., Dew, M. and Brenner, S. (1975). Dopaminergic neurons in the nematode *Caenorhabditis elegans*. *Journal of Comparative Neurology* **163,** 215–226.

Tanaka, R. D. and MacInnis, A. J. (1980). Analyses of the pseudocoelomic fluid from *Moniliformis dubius*. *Journal of Parasitology* **66,** 354–355.

Tandon, R. S. (1968). Amino acid composition of the Indian liver fluke *Fasciola indica*. *Zeitschrift für Parasitenkunde* **30,** 149–151.

Tandon, R. S. and Misra, K. C. (1980). Threonine and serine dehydratase activity in the buffalo liver-fluke *Fasciola indica*. *Journal of Helminthology* **54,** 259–262.

Tandon, R. S. and Misra, K. C. (1984). Aminotransferases in trematode parasites of fish and mammalian hosts. *Indian Journal of Parasitology* **8,** 93–96.

Tao, I. H. and Huang, T. Y. (1965). Studies on arginase of *Schistosoma japonica*. *Scientia sinensis* **14,** 417–422.

Taylor, A. E. R. and Haynes, W. D. G. (1966). Studies on the metabolism of larval tapeworms (Cyclophyllidea: *Taenia crassiceps*). I. Amino acid composition before and after *in vitro* culture. *Experimental Parasitology* **18,** 327–331.

Taylor, J. B., Vidal, A., Torpier, G., Meyer, D. J., Roitsch, C., Balloul, J.-M., Southan, C., Sondermeyer, P., Pemble, S., Lecocq, J.-P., Capron A. and Ketterer, B. (1988). The glutathione transferase activity and tissue distribution of a cloned Mr 28K protective antigen of *Schistosoma mansoni*. *EMBO Journal* **7,** 465–472.

Thong, K.-W. and Coombs, G. H. (1985). L-Serine sulphydrase activity in trichomonads. *IRCS Medical Science* **13,** 495–496.

Titanji, V. P. K., Mbacham, W. F. and Mbakop, A. (1988). Isolation and biochemical composition of the cuticle of *Onchocerca volvulus*. *Tropical Medicine and Parasitology* **39,** 100–104.

Toro-Goyco, E. and del Valle, M. R. (1970). *Schistosoma mansoni*. I. Chemical composition of eggs. *Experimental Parasitology* **27,** 265–272.

Torre-Blanco, A. and Alvizouri, A. M. (1987). *In vitro* hydroxylation of proline in the collagen of the cysticercus of *Taenia solium*. *Comparative Biochemistry and Physiology* **88B,** 1213–1217.

Torre-Blanco, A. and Toledo, I. (1981). The isolation, purification and characterization of the collagen of *Cysticercus cellulosae*. *Journal of Biological Chemistry* **256,** 5926–5930.

Trivedi, K. K. and Gupta, S. P. (1987). Biochemical studies on kinetic properties of glutamic-oxaloacetic (GOT) and glutamic-pyruvic (GPT) transaminases of some nematodes parasitic in ruminants. *Acta Parasitologica Polonica* **31,** 283–290.

Turner, A. C., Lushbaugh, W. B. and Hutchison, W. F. (1986). *Dirofilaria immitis*: Comparison of cytosolic and mitochondrial glutamate dehydrogenases. *Experimental Parasitology* **61,** 176–183.

Ueno, Y. (1960). Quantitative studies on the metabolic intermediates of *Ascaris lumbricoides* var. *suis* (I). Fatty acids, ammonia and amino acids. *Journal of the Japanese Biochemical Society* **32,** 142–147.

Vanfleteren, J. R. (1973). Amino acid requirements of the free-living nematode *Caenorhabditis briggsae*. *Nematologica* **19,** 93–99.

Vanfleteren, J. R., Van Bun, S. M., Delcambe, L. L. and Van Beeumen, J. J. (1986). Multiple forms of histone H2B from the nematode *Caenorhabditis elegans*. *Biochemical Journal* **235,** 769–773.

Vanfleteren, J. R., Van Bun, S. M. and Van Beeumen, J. J. (1987a). The primary structure of histone H3 from the nematode *Caenorhabditis elegans*. *FEBS Letters* **211,** 59–63.

Vanfleteren, J. R., Van Bun, S. M. and Van Beeumen, J. J. (1987b). The primary structure of histone H2A from the nematode *Caenorhabditis elegans*. *Biochemical Journal* **243,** 297–300.

Vanfleteren, J. R., Van Bun, S. M. and Van Beeumen, J. J. (1988). The primary structure of the major isoform (H1.1) of histone H1 from the nematode *Caenorhabditis elegans*. *Biochemical Journal* **255,** 647–652.

Van Grembergen, G. (1944). Le métabolisme respiratoire du cestode *Moniezia benedeni* (Moniez, 1879). *Enzymologia* **11,** 268–281.

Van Grembergen, G. and Pennoit de Cooman, E. (1944). Experimenteele Gegevens over het Stikstofmetabolisme der Platyhelminthen. *Natuurwetenschappelijk Tijdschrift* **26,** 91–97.

Van Gundy, S. D., Bird, A. F. and Wallace, H. R. (1967). Aging and starvation in larvae of *Meloidogyne javanica* and *Tylenchulus semipenetrans*. *Phytopathology* **57,** 559–571.

Varma, A. and Sharma, P. N. (1984). Chromatographic study of amino acids in *Paramphistomum cervi*. *Indian Journal of Parasitology* **8,** 81.

Vernberg, W. B. and Hunter, W. S. (1963). Utilization of certain substrates by larval and adult stages of *Himasthla quissetensis*. *Experimental Parasitology* **14,** 311–315.

Vigliershio, D. R. and Görtz, J. H. (1972). *Anisakis physeteris*: Amino acids in body tissues. *Experimental Parasitology* **32,** 140–148.

Wack, M., Komuniecki, R. and Roberts, L. S. (1983). Amino acid metabolism in the rat tapeworm *Hymenolepis diminuta. Comparative Biochemistry and Physiology* **74B,** 399–402.

Waite, J. H. and Rice-Ficht, A. C. (1987). Presclerotized eggshell protein from the liver fluke *Fasciola hepatica. Biochemistry* **26,** 7819–7825.

Waitz, J. A. (1963). Glycolytic enzymes of the cestode *Hydatigera taeniaeformis. Journal of Parasitology* **49,** 285–293.

Walter, R. D. (1988). Polyamine metabolism of filaria and allied parasites. *Parasitology Today* **4,** 18–20.

Ward, P. F. V. and Crompton, D. W. T. (1986). Linked metabolism of L-serine and L-alanine by *Moniliformis moniliformis* (Acanthocephala) *in vitro. Parasitology* **93,** 333–340.

Ward, P. F. V. and Crompton, D. W. T. (1987). Features of amino acid metabolism in *Moniliformis moniliformis* (Acanthocephala) *in vitro. Parasitology* **94,** 533–541.

Warren, L. G. and Poole, W. J. (1970). Biochemistry of the dog hookworm II. Nature and origin of the excreted fatty acids. *Experimental Parasitology* **27,** 408–416.

Watson, M. R. and Silvester, N. R. (1959). Studies of invertebrate collagen preparations. *Biochemical Journal* **71,** 578–584.

Watts, S. D. M. (1970a). The amino acid requirements of the rediae of *Cryptocotyle lingua* and *Himasthla leptosoma* and of the sporocysts of *Cercaria emasculans* Pelseneer, 1900. *Parasitology* **61,** 491–497.

Watts, S. D. M. (1970b). Transamination in homogenates of rediae of *Cryptocotyle lingua* and of sporocysts of *Cercaria emasculans* Pelseneer, 1900. *Parasitology* **61,** 499–504.

Watts, S. D. M. and Atkins, A. M. (1983). Application of a quick, simple and direct radiometric assay for 4-aminobutyrate : 2-oxoglutarate aminotransferase to studies of the parasitic nematode *Nippostrongylus brasiliensis. Comparative Biochemistry and Physiology* **76B,** 899–906.

Watts, S. D. M. and Atkins, A. M. (1984). Kinetics of 4-aminobutyrate : 2-oxoglutarate aminotransferase from *Nippostrongylus brasiliensis. Molecular and Biochemical Parasitology* **12,** 207–216.

Watts, S. D. M. and Atkins, A. M. (1986). Mechanism based inactivation of GABA-transferase from a nematode parasite. *Biochemical Society Transactions* **14,** 452–453.

Webb, R. A. (1986). The uptake and metabolism of L-glutamate by tissue slices of the cestode *Hymenolepis diminuta. Comparative Biochemistry and Physiology* **85C,** 151–162.

Webster, L. A. and Wilson, R. A. (1970). The chemical composition of protonephridial canal fluid from the cestode *Hymenolepis diminuta. Comparative Biochemistry and Physiology* **35,** 201–209.

Weinstein, P. P. and Haskins, W. T. (1955). Chemical evidence of an excretory function for the so-called excretory system of the filariform larva of *Nippostrongylus muris. Experimental Parasitology* **4,** 226–243.

Wertheim, G., Zeledon, R. and Read, C. P. (1960). Transaminases of tapeworms. *Journal of Parasitology* **46,** 497–499.

Wieser, W. and Platzer, U. (1983). A novel metabolic response to anoxia in an invertebrate, the snail *Helix pomatia*, involving the reaction catalysed by arginase. *Molecular Physiology* **4,** 155–164.

Winkelman, L. (1976). Comparative studies of paramyosins. *Comparative Biochemistry and Physiology* **55B,** 391–397.

Wittenberg, B. A., Okazaki, T. and Wittenberg, J. B. (1965). The hemoglobin of *Ascaris* perienteric fluid 1. Purification and spectra. *Biochimica et Biophysica Acta* **111,** 485–495.

Wittich, R.-M., Kilian, H.-D. and Walter, R. D. (1987). Polyamine metabolism in filarial worms. *Molecular and Biochemical Parasitology* **24,** 155–162.

Wolf-Spengler, M. L. and Isseroff, H. (1983). Fascioliasis: Bile duct collagen induced by proline from the worm. *Journal of Parasitology* **69,** 290–294.

Wright, D. J. (1975). Studies on nitrogen catabolism in *Panagrellus redivivus,* Goodey, 1945 (Nematoda:Cephalobidae). *Comparative Biochemistry and Physiology* **52B,** 255–260.

Yancy, P. H., Clark, M. E., Hand, S. C., Bowlus, R. D. and Somero, G. N. (1982). Living with water stress: Evolution of osmolyte systems. *Science* (New York) **217,** 1214–1222.

Yarygina, G. V., Vykhrestyuk, N. P. and Klochkova, V. I. (1982). The amino acid composition of collagenous proteins of the trematodes *Calicophoron erschowi, Eurytrema pancreaticum* and the cestodes *Bothriocephalus scorpii* and *Nybelinia* sp. larvae. *Zhurnal Evolyutsionnoi Biokhimii i Fiziologii* **18,** 564–567.

Zavras, E. T. and Roberts, L. S. (1984). Developmental physiology of cestodes: Characterization of putative crowding factors in *Hymenolepis diminuta. Journal of Parasitology* **70,** 937–944.

Zenka, J. and Prokopic, J. (1983). Activities of some enzymes in the perienteric fluid of *Ascaris suum. Folia Parasitologica* **30,** 373–376.

Zillman, U., Sachs, R. and Ebert, F. (1987). Isoenzymes of the lung fluke *Paragonimus uterobilateralis* from Liberia. *Tropical Medicine and Parasitology* **38,** 320–322.

Zurita, M., Bieber, D., Ringold, G. and Mansour, T. E. (1987). Cloning and characterization of a female genital complex cDNA from the liver fluke *Fasciola hepatica. Proceedings of the National Academy of Sciences (USA)* **84,** 2340–2344.

Cultivation of Helminths in Chick Embryos

BERNARD FRIED AND LOUIS T. STABLEFORD

Department of Biology, Lafayette College, Easton, Pennsylvania 18042, USA

ADVANCES IN PARASITOLOGY VOL. 30
ISBN 0-12-031730-3

I. Introduction

Helminths can be obtained from naturally infected hosts or from laboratory infections which use larval stages to infect definitive hosts (*in vivo* studies). Helminths can also be maintained or cultured *in vitro* for various periods. However, development *in vitro* is often abnormal and worms may show a decline in their physiological status during long-term culture. There has been relatively limited success in the *in vitro* culture of helminths from the larval stage to the ovigerous adult; helminth cultivation methods lag behind those of viral, bacterial, fungal and protozoan culture.

Ectopic sites (e.g. anterior chamber of the eye, coelomic cavity of vertebrates, or chick embryos) provide sites that may be intermediate between the normal *in vivo* one and the *in vitro* culture environment. Relatively little information is available on the development of helminths in ectopic sites. Of the available ectopic sites, the chick embryo has been explored more intensely, particularly the chick chorioallantoic membrane (or chorioallantois, hereafter referred to as CAM), than any of the others. Less use has been made of other chick embryo sites such as the yolk sac, allantois, albumen, blood vessels of the embryo, and the subchorioallantoic space. Of all helminths studied in chick embryos, most work has been done on the hermaphroditic digenetic trematodes.

This chapter examines the literature on the biology, physiology and development of helminths cultivated on the CAM and in other chick extraembryonic sites. The significant literature from the field of developmental biology on the structure, physiology and biochemistry of the chick embryo and its extraembryonic membranes is examined and related to the cultivation of helminths in chick embryos wherever possible.

II. Use of Chick Embryos in Biological and Biomedical Sciences

A. General

The chick embryo has been used extensively for biological and biomedical studies in the twentieth century. It has been used in microbiology and virology to cultivate bacteria, rickettsiae and viruses (Buddingh, 1952; Blaškovič and Styk, 1967); in protozoology for developmental and culture studies on intra- and extracellular protozoans (Pipkin and Jensen, 1958; Long, 1978); in developmental biology for studies on angiogenesis, tumorgenesis, organ and tissue transplantation (Hamburger, 1960; Rugh, 1962; Romanoff and Romanoff, 1949; Paul, 1960); and in toxicology as a site for drug testing (Leighton *et al.*, 1985). Less extensive use has been made of the chick embryo in helminthology, and most of the studies have been limited to basic biological research (Fried, 1969, 1989). The techniques and precautions involved in avian embryo studies have been discussed in numerous works (Rugh, 1962; Buddingh, 1952; Pipkin and Jensen, 1958; Paul, 1960; Blaškovič and Styk, 1967).

The reasons for using chick embryos in biological and biomedical research are as follows: fertile chick eggs are inexpensive, easy to obtain and handle, and technologically easy to manipulate in experiments. The only major laboratory equipment needed is an incubator that allows for controlled temperature and humidity. Problems of animal room space, animal care and increasingly restrictive governmental regulation of the same are avoided and the chance of the eggs becoming cross-contaminated with infectious agents is minimized. The embryo, protected by a shell and membranes, provides a sterile environment for experimentation. If sterility is compromised by the absence of asepsis during manipulations on the embryo, the results are obvious because of contamination to the egg.

The chick embryo and its extraembryonic membranes provide various tissues in different stages of development for experimental purposes. The unspecialized nature of the chick embryo makes it a compatible site for transplantation studies. The avian egg, once thought to be immunologically incompetent, but now known to produce immunogenic substances, is still easier to infect with microorganisms and metazoans than are many adult animals. The embryo also responds to injury with typical inflammatory reactions and is useful in drug testing. The CAM has recently been used in toxicology as an alternative test site for the controversial Draize rabbit eye bioassay (Leighton *et al.*, 1985).

The most widely used site in the chick embryo is the CAM, which consists of an outer chorionic ectoderm, a middle layer of connective tissue or mesenchyme permeated with arteries, veins and capillaries and an inner layer of allantoic endoderm. Access to the embryo is easy with numerous routes available, of which the following are frequently used: (1) inoculation onto or under the CAM; (2) inoculation into a large terminal venous sinus over the CAM; (3) intraembryonic inoculation, usually into the musculature or the cerebrum; (4) inoculation into the yolk sac; (5) inoculation into other embryonic membranes and sites, e.g. allantois, albumen.

B. PRE-1960s STUDIES ON HELMINTHS IN CHICK EMBRYOS

Prior to the early 1960s, the chick embryo had not been well-explored in helminthology except as a source of tissue extract. McCoy (1936) inoculated *Trichinella spiralis* (Nematoda) larvae into chick embryos and obtained partial worm development. Smyth (1958, 1959) observed some development of *Diphyllobothrium dendriticum* (Cestoda) plerocercoids injected into the yolk sac and amniotic cavity of 3- to 9-day-old embryos. Only two pre-1960s studies are available on trematodes. Ferguson (1949) placed metacercariae of *Posthodiplostomum minimum* on the CAM of eggs but the metacercariae died within 2 days and showed no signs of development. Brackett and Beckman (1942) could not infect chick embryos with avian schistosome cercariae of the genus *Trichobilharzia* inoculated on the CAM. The membrane was first used successfully as a development site for trematodes when Fried (1962a) transplanted mechanically excysted metacercariae of *Philophthalmus hegeneri* to the CAM of chick embryos.

C. POST-1960s STUDIES ON HELMINTHS IN CHICK EMBRYOS

Earlier reviews discussed the uses of chick embryos in trematode studies (Fried, 1969, 1989; Long, 1978; Clegg and Smith, 1987; Smyth and Halton, 1983). Information on helminths other than trematodes is scant and not discussed in any review. In this chapter, details of the worm embryo studies are discussed under the individual species of helminths considered (see Secion III).

Chick embryos have been used mainly to study the survival, growth and development of trematodes, particularly digenetic trematodes. In trematode work, the chick embryo has been used for the following purposes: to grow and identify unknown species of trematodes where definitive hosts were not known or not available (Shimazu, 1974); to study the extent of intraspecific variation of trematode species (Fried, 1962a; Fried and Pentz, 1983); to determine the extent of temperature tolerance of trematodes (Fried, 1965; Fried and Foley, 1969); to determine the tolerance of trematodes to antibiotics and X-irradiation (Austin and Fried 1972a; Fried and Davis, 1972); to investigate worm wound healing and repair of damage following experimental damage or removal of parts (Austin and Fried, 1972b); to investigate worm pairing and, self *vs* cross-fertilization along with other aspects of reproductive behavior (Fried and Roberts, 1972; Fried and Holmes, 1979; Fried *et al.*, 1980); to examine the effects of added serum to the chick embryo on the subsequent growth and development of worms (Irwin and Saville, 1988a,b); to determine the ability of cercariae to develop into adults without passing through the metacercarial state (Fried and Groman, 1985); to examine the feeding of trematodes in chick embryos (Fried and Caruso, 1970; Wisnewski *et al.*, 1986); to examine histopathological effects of trematodes on the CAM (Fishbein *et al.*, 1985; Huffman *et al.*, 1984); to examine histochemical and ultrastructural changes of trematodes grown in chick embryos (Fried *et al.*, 1985; Fried and Fujino, 1984; Fried and Mishkind, 1985); to examine the site-finding behavior of trematodes (Fried and Diaz, 1987); to examine the pathobiochemical effects of trematodes on the CAM (Fried and Bradford, 1984); to examine the development of cestode larvae in chick embryos (Parmeter and Gemmell, 1974; Mueller, 1966); and to examine the development of nematode larvae in chick embryos (Winward, 1976).

D. TECHNIQUES USED TO CULTURE HELMINTHS IN CHICK EMBRYOS

The procedures described below were designed for studies on trematodes in chick embryos. Relatively few studies arc available on cestodes and nematodes in chick embryos and there are no studies on acanthocephalan cultivation in chick embryos. The techniques described below are also applicable to studies on helminths other than trematodes.

Various techniques have been used to cultivate trematodes in chick embryos. The first successful study described by Fried (1962a) used a modified Woodruff and Goodpasture (1931) procedure. In this procedure, 8- to 10-day-old fertile eggs are used and a window about 1 × 1 × 1 cm is cut in the upper surface of the shell. The shell membrane under the window is moistened with a drop of sterile saline and removed with a fine forceps. Using aseptic procedures, worms are placed on the CAM in a minimal amount of sterile saline with a pipette. The window is closed with transparent tape or cellophane and the egg placed in a humidified incubator at 38 ± 1°C. The egg is usually examined for parasites from 1 to 7 days later. The CAM begins to dry out on about day 17 or 18; if worms are to be maintained on the CAM for relatively long periods of time, they must be transferred (serial transfer) to the CAM of new 8- to 10-day-old embryos (Fried, 1962a). To examine the egg, the tape is removed from the window and the shell is chipped away with a forceps revealing the surface of the CAM. There may be some drying-up of the CAM immediately under the window and most parasites are usually recovered from the moist peripheral areas of the CAM. Parasites can be removed from the surface of the CAM with a fine forceps or by pipet, examined alive and, if aseptic procedures are used, transferred serially to new chorioallantoic membranes. Usually, parasites removed from the CAM are fixed for light or scanning electron microscopy (Fried and Pentz, 1983; Fried and Fujino, 1984; Fried and Mishkind, 1985). Pieces of CAM with parasites attached can be removed for *in situ* histopathologic or pathobiochemical studies (Fried and Bradford, 1984; Fishbein *et al.*, 1985; Huffman *et al.*, 1984).

The technique used most frequently for *in ovo* studies on trematodes is that of Zwilling (1959), as modified by Fried (1973). This technique is simpler to use than the modified Woodruff and Goodpasture (1931) procedure. In the modified Zwilling procedure, 3-day-old fertile hens' eggs are used and a hole is punched in the shell at the albumen end of the egg, 2 ml of which is removed with the aid of a hypodermic syringe. The removal of the albumen allows the developing embryo to move away from the shell membrane. A window is then made in the upper surface of the shell as described previously. The shell and shell membrane under the window are removed with a fine forceps and the window is covered with tape. The egg is placed in a humidifed incubator at 38 ± 1°C for another 3–5 days. On days 6–8, at which time the CAM is well-developed, the egg is suitable for worm implantation. The tape is lifted off the shell and worms placed on the CAM; the egg is resealed with tape, returned to the incubator and examined for worms 1–8 days later as described previously.

Another procedure developed by Auerbach *et al.* (1974) and modified by Fried *et al.* (1980) uses chick embryos cultured *in vitro*. This technique uses a double petri dish design. The lower lid of a 10-cm petri dish serves as the culture chamber and is placed within the lower lid of a 15-cm petri dish which serves as a moist chamber. Using aseptic precautions, a 3-day-old fertile egg is cracked open and its contents placed into the inner petri dish. The outer dish is filled with sterile distilled water and a lid is then placed over the larger petri dish. The chamber is then placed in a humidified incubator at 38°C for 2–3 days. For reasons that are not clear, about 50% of the embryos that are prepared die between days 3 and 6 regardless of the care taken to prepare the cultures. Beyond day 6, losses are minimal. The CAM develops extensively over the chick embryo and is ideally suited to receive worms by days 6–8. This technique has the obvious advantage of allowing continuous observation of worm survival, growth, development and behavior on a relatively flat surface.

A recent technique developed by Irwin and Saville (1988b) makes use of the subchorioallantoic space of 7- to 10-day-old chick embryos. In this procedure, worms are inoculated by hypodermic syringe directly into the subchorioallantoic space and develop on the undersurface of the CAM rather than the upper surface as in the aforementioned procedures. Irwin and Saville (1988b) also noted that the addition of chick serum inoculated into the subchorioallantoic space enhanced the development of microphallid trematodes in chick embryos.

E. SOURCES OF TREMATODES USED IN CHICK EMBRYO STUDIES

Most studies have used free (unencysted) or excysted metacercariae, but some have used adult trematodes and even cercariae. Cercariae have been successfully used in only one study (Fried and Groman, 1985) involving the marine avian echinostome, *Himasthla quissetensis*. The cercariae released cystogenous material on the CAM, bypassed the metacercarial stage and developed directly into adults.

Free (unencysted) metacercariae, such as those found in brachylaimids, microphallids and some strigeids, are excellent subjects for CAM transplant studies. However, most trematode metacercariae occur as cysts (encysted metacercariae) and must be excysted first, usually by chemical means. For at least six species of trematodes (Fried and Ramundo, 1987), the chemical excystation of encysted metacercariae is easily achieved in 0.5% bile salts + 0.5% trypsin in Earle's balanced salt solution adjusted to a pH of 8.0 ± 0.2 with 7.5% $NaHCO_3$. Excysted metacercariae are then rinsed in several changes of sterile Locke's solution supplemented with 200 units penicillin ml^{-1} and 200 µg streptomycin ml^{-1} prior to worm inoculation onto the CAM. During transfer and inoculation procedures, aseptic or oligoseptic procedures associated with work in microbiology and virology laboratories should be followed.

Some encysted metacercariae can be excysted by mechanical means (the larvae of *Philophthalmus hegeneri* or *Clinostomum marginatum*: Fried, 1962a; Fried and Foley, 1970). Philophthalmid metacercariae can also be excysted by elevating the temperature of the saline containing the cysts to 35–45°C; such excysted metacercariae provide good material for transplantation studies (Fried, 1981).

When encysted metacercariae cannot be excysted *in vitro*, the cysts may be grown in hosts, and juveniles or adult worms obtained after necropsy can then be used for transplant studies. Worms removed from definitive host sites, particularly the gut, usually need extensive treatment with antibiotics prior to transfer to membranes.

The trematode larvae most suitable for CAM studies are the progenetic types, i.e. those that reach sexual maturity rapidly in vertebrate hosts (Smyth and Halton, 1983). Many strigeids, brachylaimids and microphallids are of this type. Unfortunately, the non-progenetic digeneans such as fasciolids and echinostomes are better known to most parasitologists. However, neither fasciolids nor echinostomes have been cultured to ovigerous adults from excysted metacercariae, either *in vitro* or *in ovo* (Clegg and Smith, 1987; Fried, 1978; Fried and Pentz, 1983). The 37-collar-spined echinostome, *Echinostoma trivolvis* (synonym of *E. revolutum*: see Huffman and Fried, 1990), does show considerable post-metacercarial development on the CAM up to the uterine coiling stage, which is about equivalent to worms grown for 7 days in domestic chicks (Fried and Pentz, 1983). This echinostome may prove suitable as a model for additional studies on the growth and development of trematodes on the CAM.

III. Helminths Cultivated in Chick Embryos

A. General

The following section considers studies on helminths cultivated in chick embryos since 1960. Most of the studies have been on digenetic trematodes. The arrangement of the section is based on the list of families of the Digenea provided by La Rue (1957). For helminths other than the Digenea, each species for which there is information is listed under Monogenea, Cestoda or Nematoda. As mentioned previously, there are no published studies on Acanthocephala in chick embryos.

Table 1 lists the studies (arranged in alphabetic order by species) on digenetic trematodes grown in chick embryos from the 1960s to the present.

B. Strigeidae

1. Cotylurus strigeoides

Excysted tetracotyles of *Cotylurus strigeoides* showed post-metacercarial growth and development on the CAM, but did not become ovigerous there (Fried *et al.*, 1978). *In vitro* cultivation with media supplemented with chicken serum or chick mucosal extract allowed for better growth and development of this species than did the CAM (Fried *et al.*, 1978).

C. Diplostomatidae

1. Posthodiplostomum minimum minimum

Fried (1970) excysted metacercariae of *Posthodiplostomum minimum minimum* obtained from various cyprinid fishes in 0.1% pepsin-Ringer's-HCl-pH

2.5. Using the transplant procedure of Fried (1962a), excysted metacercariae became sexually mature on the CAM within 3 days at 41°C and eggs from CAM-worms embryonated. When the CAM was maintained at 37.5°C, the worms survived for up to 10 days, but showed only minimal post-metacercarial development.

2. Alaria *spp.*

Madsen and Johnson (1974) used the chorioallantoic technique of Fried (1962a) to study the post-larval development of mesocercariae of *Alaria arisaemoides*, *A. marcianae* and *A. mustelae*. Although mesocercariae survived for up to 10 days in chick embryos and often penetrated the surface of the CAM (becoming lodged in the mesenchyme of the membrane and other sites in the egg), post-larval development did not occur. Penetration of the chorionic ectoderm by a mesocercaria and location of this larva in the mesenchyme is shown in Fig. 1.

3. Diplostomum *spp.*

Kannangara and Smyth (1974) contributed a significant study on the *in vitro* cultivation of *Diplostomum spathaceum* and *D. phoxini* metacercariae. These metacercariae lack genital rudiments and are more difficult to culture than trematodes with larvae containing well-developed genital anlagen. Kannangara and Smyth (1974) explored various techniques, i.e. strict *in vitro* cultivation procedures as well as *in ovo* procedures, and used numerous media including liquid, semi-solid and diphasic. Although they had considerable success with various *in vitro* culture procedures, they reported no significant growth or development of their diplostomatids in chick embryos. They also found that their diplostomatid larvae could be stored in the defined medium NCTC 135 at 4°C for at least 2 months and still retain viability. Interestingly, at the end of their study, they commented on the impossibility of developing a "universal" medium for the cultivation of digenetic trematodes.

4. Diplostomum spathaceum

Free metacercariae were obtained from the lenses of naturally infected *Splodinotus grunniens* fishes in the USA and transplanted to the CAMs of 5- to 12-day-old chick embryos (Leno and Holloway, 1986). Worms were removed from the membranes 2–8 days later and transferred to new CAMs so that some worms survived serial transfer on both CAMs for 7–14 days. Some worms went through the typical trematode development stages (i.e.

Table 1 *Cultivation of trematodes in chick embryos*

Species	*Initial stage*	*Maximum survival (days)*	*Significance of study*	*Reference*
Alaria arisaemoides	Me	10	CAM, SWD	Madsen and Johnson (1974)
Alaria marcianae	Me	5	CAM, SWD	Madsen and Johnson (1974)
Alaria mustelae	Me	10	CAM, SWD	Madsen and Johnson (1974)
Amblosoma suwaense	M	7	CAM, Alb, PLD*	Fried *et al.* (1981)
Amblosoma suwaense	M	15	CAM, Alb, PLD*	Shimazu (1974)
Clinostomum marginatum	M-Fr	10	CAM, PLD*	Fried and Foley (1970)
Clinostomum marginatum	M-Fi	6	CAM, Alb*	Larson and Uglem (1990)
Cotylurus strigeoides	M	8	CAM, PLD	Fried *et al.* (1978)
Cyathocotyle bushiensis	M	8	CAM, PLD	Fried and Ramundo (1987)
Diplostomum spathaceum	M	14	CAM, PLD*	Leno and Holloway (1986)
Diplostomum spathaceum	M	15	Al-Cav, PLD*, Ser	Irwin and Saville (1988a)
Echinostoma trivolvis	J&A	22	CAM, PLD	Fried *et al.* (1968)
Echinostoma trivolvis	A	3	CAM, Histoglyc	Fried and Kramer (1968)
Echinostoma trivolvis	A	7	CAM, Temp-tol	Fried and Foley (1969)
Echinostoma trivolvis	A	4	Feed	Fried and Caruso (1970)
Echinostoma trivolvis	M	14	CAM, Ex, Zwill	Fried and Butler (1978)
Echinostoma trivolvis	M	10	*In vivo vs* CAM studies	Fried and Pentz (1983)
Echinostoma trivolvis	M	7	SEM of CAM worms	Fried and Fujino (1984)
Echinostoma trivolvis	M	7	TEM on CAM, Feed	Wisnewski *et al.* (1986)
Echinostoma trivolvis	A	3	Pairing and site finding on CAM	Fried and Diaz (1987)
Fasciola hepatica	M	5	CAM, PLD, Ex	Fried and Butler (1979)
Haematoloechus sp.	J&A	7	PLD*, SWD	Fried and Weaver (1969)
Himasthla quissetensis	C	7	EN, PLD	Fried and Groman (1985)
Leucochloridiomorpha constantiae	M	6	SWD, CAM antibiotic studies	Austin and Fried (1972a)
Leucochloridiomorpha constantiae	M&A	6	Wound healing on CAM	Austin and Fried (1972b)
Leucochloridiomorpha constantiae	M	8	CAM, PLD*	Harris *et al.* (1972)

Leucochloridiomorpha constantiae	M&A	1	SWD, Pairing on CAM	Fried and Roberts (1972)
Leucochloridiomorpha constantiae	M	7	CAM, X-ray tolerance	Fried and Davis (1972)
Leucochloridiomorpha constantiae	M	7	CAM, Fert, Histopath	Fried and Holmes (1979)
Leucochloridiomorpha constantiae	M	7	*In vitro vs in ovo* CAM studies	Fried *et al.* (1980)
Leucochloridiomorpha constantiae	M	7	Path-chem on CAM	Fried and Bradford (1984)
Leucochloridiomorpha constantiae	M	7	Histopath on CAM	Fishbein *et al.* (1985)
Leucochloridiomorpha constantiae	M	5	SEM on CAM	Fried and Mishkind (1985)
Leucochloridiomorpha constantiae	M	7	AIP on CAM	Fried *et al.* (1985)
Leucochloridium variae	M	14	CAM, PLD*	Fried (1973)
Microphallus pygmaeus	M	6	sub-CAM*, PLD*, Ser	Irwin and Saville (1988b)
Philophthalmus hegeneri	M	21	PLD, CAM*	Fried (1962a)
Polystomoides sp.	A	7	CAM, Temp-tol	Fried (1965)
Posthodiplostomum minimum minimum	M	10	CAM, PLD*	Fried (1970)
Schistosoma mansoni	A	2	CAM, SWD, Pairing	Fried *et al.* (1982)
Sphaeridiotrema globulus	M	9	CAM, PLD*	Fried and Huffman (1982)
Sphaeridiotrema globulus	M	7	Path-CAM	Huffman *et al.* (1984)
Spirorchis elegans	A	3	CAM, ELC	Fried and Tornwall (1969)
Spirorchis scripta	A	3	CAM, ELC	Fried and Tornwall (1969)
Zygocotyle lunata	M	5	CAM, PLD	Fried and Nelson (1978)

Key to abbreviations: A, adult; Alb, albumen; Alb*, ovigerous in albumen; Al-Cav, allantoic cavity; AIP, alkaline phosphatase activity; C, cercaria; CAM, chick chorioallantois; E, egg; ELC, eggs laid on CAM; Ex, excystation study; Feed, feeding studies on CAM; Fert, fertilization studies; Histo-glyc, used for histochemical glycogen studies; J, juvenile; L, larva; M, metacercaria; Me, mesocercaria; M-Fi, metacercariae from fishes; M-Fr, metacercariae from frogs; Path-CAM, pathology studies on CAM; PLD, post-larval development; Path-chem, pathobiochemistry studies on the CAM; PLD*, post-larval development to the ovigerous stage; SEM, scanning electron microscopy; Ser, serum added to egg; Sub-CAM, subchorioallantoic space; SWD, survival without development; TEM, transmission electron microscopy; Temp-tol, used for temperature tolerance studies; Zwill, use of Zwilling procedure.

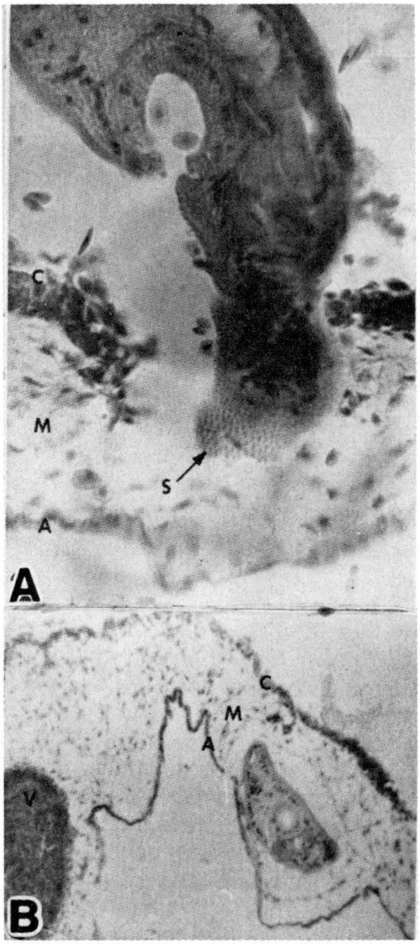

FIG. 1. (A) Section of *Alaria* mesocercaria penetrating CAM, 10 min post-infection. Hematoxylin and eosin. (B) Section of *Alaria* mesocercaria in mesenchyme. Hematoxylin and eosin. *Abbreviations*: A, allantoic endoderm; C, chorionic ectoderm; M, mesenchyme; S, anterior worm spination. Scale bars not given in the original publication from Madsen and Johnson, 1974. Reproduced with the permission of the American Microscopical Society.

immature, genital rudiment, testes, follicular ovary, vitellaria and ovigerous stages); they fed on blood from the CAM and oviposited there. Eggs removed from CAMs and embryonated in tap water produced viable miracidia, but infectivity to the first intermediate snail host was not tested. The results of this study suggest that the CAM shares characteristics with the intestine of the definitive host, a piscivorous bird.

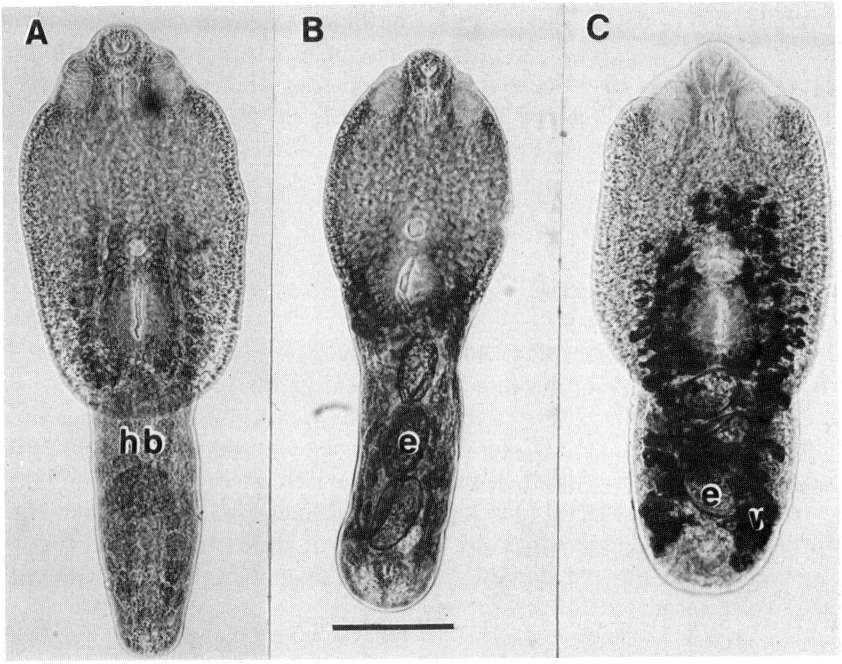

FIG. 2. Fixed and stained whole mounts of *Diplostomum* sp. grown in chick embryos for 14 days. Scale bar = 100 μm and is applicable for all three worms. Metacercariae of *Diplostomum* sp. are undeveloped and lack a hindbody (photograph of metacercaria is not available). (A) Metacercaria from the retina of *Perca fluviatilis* grown on the CAM; shelled eggs not present in the hindbody (hb). (B) Metacercaria from the retina of *Gymnocephalus cernua* grown on the CAM. Note presence of three shelled-eggs (e) in the hindbody. (C) Metacercaria from the retina of *Perca fluviatilis*. Note presence of eggs (e) in hindbody and extensive vitelline (v) development. Photographs courtesy of S. W. B. Irwin and J. C. Chubb.

Irwin and Saville (1988a) injected the metacercariae of *D. spathaceum* into the allantoic cavity of chick embryos and, using serial transfer procedures, maintained worms there for 15 days. Some eggs received daily injections of 0.2 ml chicken serum added to the allantoic cavity. Considerable variation in development was seen in the worms, but the worms from embryos that had serum added had larger hind bodies than the worms from eggs without added serum. There was no evidence that worms ingested blood and only one digenean, from the serum supplemented group, became ovigerous. Irwin *et al.* (1989) collected the metacercariae of *Diplostomum* sp. from the eyes of the freshwater fishes *Gymnocephalus cernua*, *Perca fluviatilis* and *Rutilis rutilis* from North Wales and *Salmo gairdneri* from Northern Ireland. The metacercariae were inoculated into the CAM of fertile eggs which had been incubated for 7 days. After 1 week, the digeneans were transferred to a second batch of eggs for an additional 7 days. Chick serum was injected into the CAM on days 10–14 of each egg's incubation. Under these conditions, the metacercariae developed to varying levels of maturity including the formation of shelled eggs (Fig. 2).

D. CYATHOCOTYLIDAE

1. Cyathocotyle bushiensis

Encysted metacercariae of *Cyathocotyle bushiensis* were excysted in an alkaline bile salt-trypsin medium without acid pepsin pre-treatment (Fried and Ramundo, 1987). Excysted metacercariae were cultivated to ovigerous adults *in vitro* in NCTC 135 plus 20% hen's egg yolk or in NCTC 135 plus 50% chicken serum. Although excysted metacercariae implanted on the CAM showed post-larval development and increased somatic growth, worms did not become ovigerous by day 8, at which time cultures were terminated. One-day-old domestic chicks could not be infected with the encysted metacercariae of this species.

E. CLINOSTOMATIDAE

1. Clinostomum marginatum

Fried *et al.* (1970) studied the chemical excystation of the yellow grub, *Clinostomum marginatum*, obtained from naturally infected *Rana pipiens* frogs and obtained 100% excystation within 30 min when metacercariae were placed in pepsin-Ringer's-HCl-pH 2.3 at 40°C. In a related study, Fried and Foley (1970) grew *C. marginatum* in the domestic chick and on the CAM. Chicks fed cysts usually lost the infection from the mouth cavity within 3 days post-exposure. Interestingly, excysted metacercariae of this species implanted on the CAM became ovigerous by day 4 when membranes were maintained at either 35 or 37.5°C. Both mono- and multiple metacercarial infections on the CAM produced worms with viable eggs within 10 days post-inoculation. *C. marginatum* is apparently capable of self-fertilization on the CAM. Although miracidia were obtained from CAM-cultured worms, attempts to infect snails were not made. Worms grown on the CAM were larger and contained more eggs than any worm obtained from the domestic chick. The body surface of the metacercaria changed from spinose to aspinose during development on the CAM. Several photomicrographs of *C. marginatum* excysted metacercariae and adults grown in chicks and on the CAM are shown in Fig. 3. Larson and Uglem (1990) recently cultivated mechanically excysted the metacercariae of *C. marginatum* from yellow perch (*Perca flavescens*) on the CAM. The techniques used in their

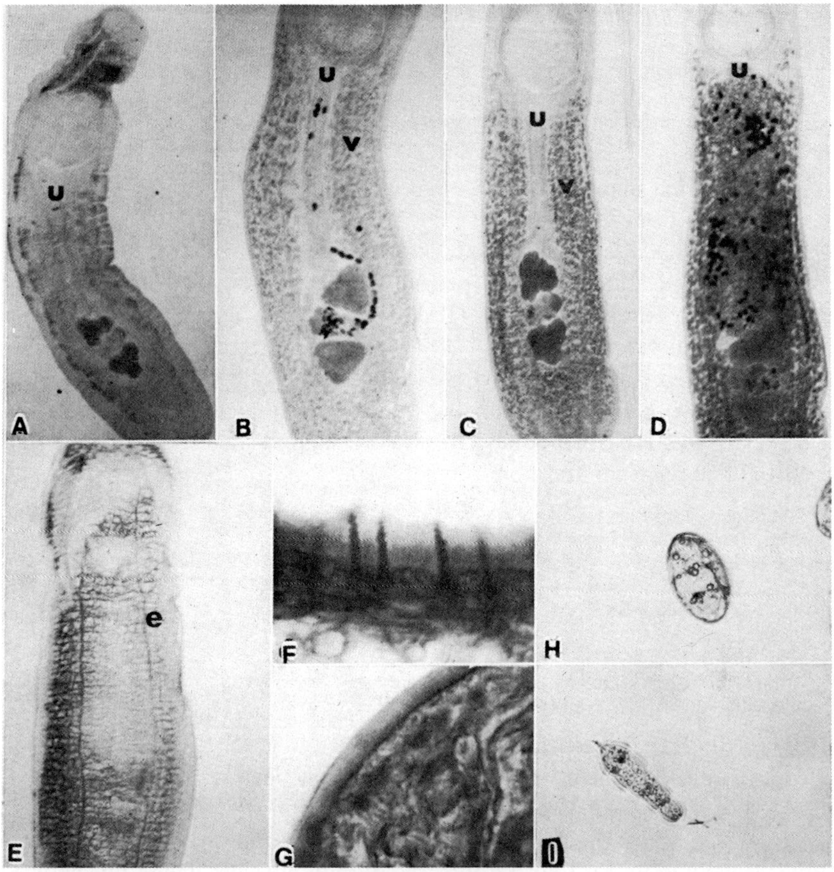

FIG. 3. (A) One-day-old *Clinostomum marginatum* from the chick. (B) Six-day-old ovigerous *C. marginatum* from CAM. (C) Three-day-old ovigerous *C. marginatum* from CAM. (D) Seven-day-old ovigerous *C. marginatum* from CAM. (E) Excysted metacercaria of *C. marginatum*. (F) Tegumentary spines from excysted metacercaria of *C. marginatum*. (G) Aspinose tegument from 3-day-old *C. marginatum* grown in chick (CAM-worms also show aspinose tegument). (H) Fully-embryonated egg from CAM-worm. (I) Miracidium hatched from egg produced by CAM-worm. Abbreviations: e, excretory system; u, uterus; v, vitellaria. (A)–(E) are whole mounts stained in Gower's carmine; (F) and (G) are paraffin sections stained in hematoxylin and eosin; (H) and (I) are fresh preparations. Scale bars not given in the original publication (from Fried and Foley, 1970). Reproduced with the permission of the American Society of Parasitologists.

studies were modeled after Fried (1962a). They recovered relatively few worms on the CAM by day 5, and the majority of worms were in the albumen. Only worms from the albumen were ovigerous, but worm-egg viability was not tested. The results from Larson and Uglem's (1990) study suggest that *Clinostomum* from fishes is a different species from the one in frogs.

F. SCHISTOSOMATIDAE

1. Schistosoma mansoni

Adult *Schistosoma mansoni*, 6–8 weeks old, were removed from the veins of mice, treated in antibiotics and placed on the CAMs of 10-day-old chick embryos either singly or as pairs (Fried *et al.*, 1982). Single worms or pairs were able to survive at this site at 37°C for up to 2 days but oviposition did not occur. Heterosexual pairing as well as male and female homosexual pairing were observed. Some pairs that did not uncouple on the CAM became attracted to other pairs. The CAM is a good site for observations on both the heterosexual and homosexual attraction of *Schistosoma mansoni*.

G. SPIRORCHIIDAE

1. Spirorchis *spp.*

Adult turtle blood flukes, *Spirorchis elegans* and *S. scripta*, removed from the blood and tissue sites of painted turtles, *Chrysemys picta bellei*, were maintained on the chick chorioallantois at 30 ± 1°C for up to 3 days (Fried and Tornwall, 1969). One worm recovered 3 days post-implantation and laid about 125 eggs on the CAM, but attempts to embryonate these eggs were unsuccessful. Spirorchids maintained in Ringer's solution at room temperature survived no longer than 6 hours and did not lay eggs. This study showed that blood flukes of a poikilothermic host could survive and oviposit on the CAM.

H. BRACHYLAIMIDAE

Three species of brachylaimids, *Leucochloridiomorpha constantiae*, *Leucochloridium variae* and *Amblosoma suwaense*, have been cultivated from the metacercarial stage to the ovigerous adult on the CAM. As adults, these trematodes live in the lower intestinal tract, or bursa of Fabricius, of a variety of avian hosts and the larval forms used to initiate the cultivation studies are obtained from snails. *L. constantiae* and *A. suwaense* larvae are available as free (unencysted) metacercariae from naturally infected freshwater viviparid snails, *Campeloma decisum*. *L. variae* is available as encysted metacercariae from the broodsacs (modified sporocysts) of terrestrial snails in the genus *Succinea*.

1. Leuchochloridiomorpha constantiae

L. constantiae metacercariae have been cultivated to ovigerous adults on the CAM (Harris *et al.*, 1972) using the transplantation procedure of Fried (1962a). The development of this worm on the membrane lags slightly behind that of worms grown in the domestic chick. Worms obtained from the membrane contain normal eggs and appear almost as robust as worms grown in the chick. Whereas sexual maturation usually occurs in 3 days in the chick, it takes about 4 days on the CAM. *L. constantiae* metacercariae have also been cultivated *in vitro* to ovigerous adults within 4 days in NCTC 135 plus 20% hen's egg yolk at 37.5–41°C, pH 7.2–8.0 in static cultures with a gas phase of air (Fried and Contos, 1973). Some worm eggs were shelled, but their viability was not determined. The procedures and media preparation in the aforementioned study followed those of Berntzen and Macy (1969) on *Sphaeridiotrema globulus*.

L. constantiae metacercariae have been used to determine the effects of antibiotic treatment on the subsequent development on the CAM (Austin and Fried, 1972a). Metacercariae maintained for 1 day at 37.5°C in 2 ml of Locke's solution containing 20 000 units penicillin ml^{-1} and then transferred to the CAM, developed in an identical manner to worms not treated with penicillin prior to transfer to the membrane. Streptomycin treatment at the level of 100 000 $\mu g\ ml^{-1}$ streptomycin allowed only 20% of the treated worms to become ovigerous, whereas treatment with 1000 $\mu g\ ml^{-1}$ streptomycin allowed 50% of the worms to become ovigerous on the membrane. In untreated controls, 100% of the worms became ovigerous on the membrane.

L. constantiae metacercariae or adults were maintained *in vitro* for up to 1 day to study worm pairing (Fried and Roberts, 1972). The study used 5-cm sterile plastic petri dishes containing a substrate of 8–10 ml of 0.85% ion agar and an overlay of 3–4 ml sterile Locke's solution. Two metacercariae or two adults were placed on the agar 20 mm apart and the cultures were maintained at 37°C on a slide warmer. Observations on *in vitro* worm pairing were made during a 24-hour period. Fried and Roberts (1972) also showed that metacercariae or adults of *L. constantiae* were capable of pairing on the surface of the CAM in a manner similar to that seen *in vitro* or in the bursa of Fabricius of the domestic chick. The utility of the CAM as a site for behavioral studies was ascertained (Fig. 4). Fried and Holmes (1979)

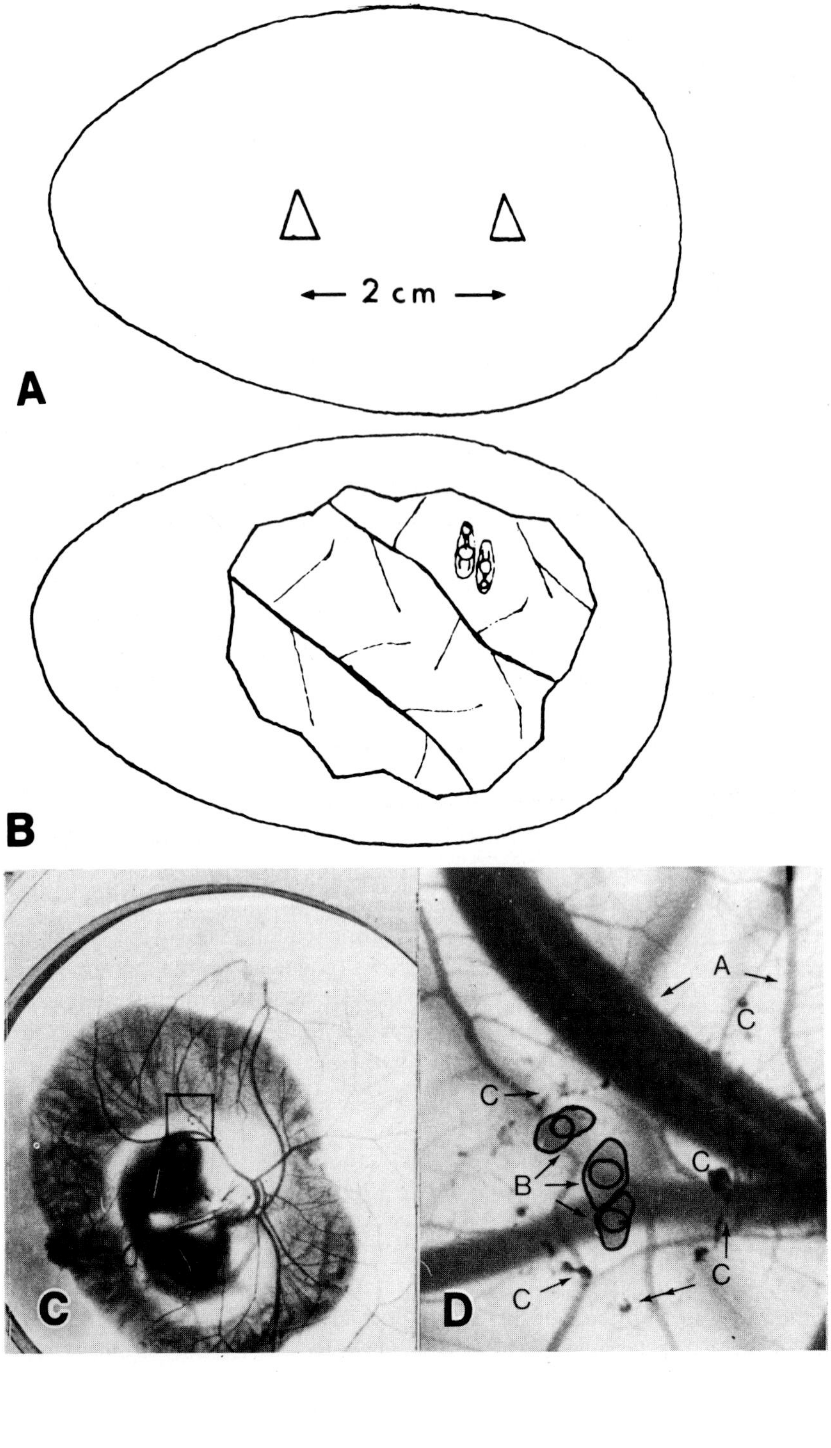
2 cm
A
B
A
C
C
B
C
C
C
C
D

determined that single worms grown on the CAM were capable of self-fertilization and the production of eggs with viable miracidia. They also observed that worm attachment to the surface of the CAM was similar to that previously seen in the bursa of Fabricius of the domestic chick. Fried *et al.* (1980) used chick embryos cultivated *in vitro* according to Auerbach *et al.* (1974) to study the growth, development and pairing of *L. constantiae* metacercariae. Worms grown in this environment showed post-metacercarial development, became ovigerous and exhibited pairing and clustering behavior (Fig. 4). The growth of worms from chick embryos cultivated *in vitro* lagged behind that of worms grown in the bursa of Fabricius of the chick. Histopathologic studies were made on metacercariae of *L. constantiae* grown for 1–12 days in chick extraembryonic membranes maintained *in ovo* or *in vitro* (Fishbein *et al.*, 1985). Studies were made on stained hematoxylin and eosin paraffin sections of membranes that retained worms (Fig. 5). Worms in 4- and 5-day-old embryos attached to the chorion and contained chorionic and yolk-sac tissue plugs in their acetabula and oral suckers. Worms in 6- to 15-day-old embryos attached to the CAM and embraced chorionic ectoderm, mesoderm and host blood vessels within their acetabula and oral sucker tissue plugs. The chorionic ectoderm was hypertrophied in areas underlying the worms.

Thin-layer chromatographic procedures were used to examine the effects of infection by *L. constantiae* metacercariae on the neutral lipid composition of the CAM (Fried and Bradford, 1984). Although there were no alterations in the free fatty acid and triacylglycerol fractions of the infected CAM, there was a significant decrease in the free sterol and steryl ester fractions of infected membranes. Fried and Mishkind (1985) used scanning electron microscopy to study the topography of *L. constantiae* grown on the CAM. The ultrastructure of *L. constantiae* on the CAM was similar to that reported by Font and Wittrock (1980) for this digenean grown in the coelom and bursa of Fabricius of the domestic chick. Ultrastructural evidence from Fried and Mishkind (1985) on *L. constantiae* along with observations by Fried and Fujino (1984) on the topography of *Echinostoma trivolvis* (now a synonym of *E. revolutum*) in chick embryos provide further evidence that the chick embryo is a suitable model to study the post-metacercarial growth and

FIG. 4. (A) *In ovo* CAM technique in which single *Leucochloridiomorpha constantiae* worms were placed in each window, 2 cm from one another. (B) Schematic showing typical worm pairing on the CAM. (C) *L. constantiae* maintained in chick embryos cultivated *in vitro* in a petri dish. Details of the inset are shown in D. (D) Three *L. constantiae* worms clustered on the CAM. The worms have been outlined in India ink for clarity. *Abbreviations*: A, blood vessels; B, worms; C, hematin (from Fried, 1986). Reproduced with the permission of Plenum Publishing Co.

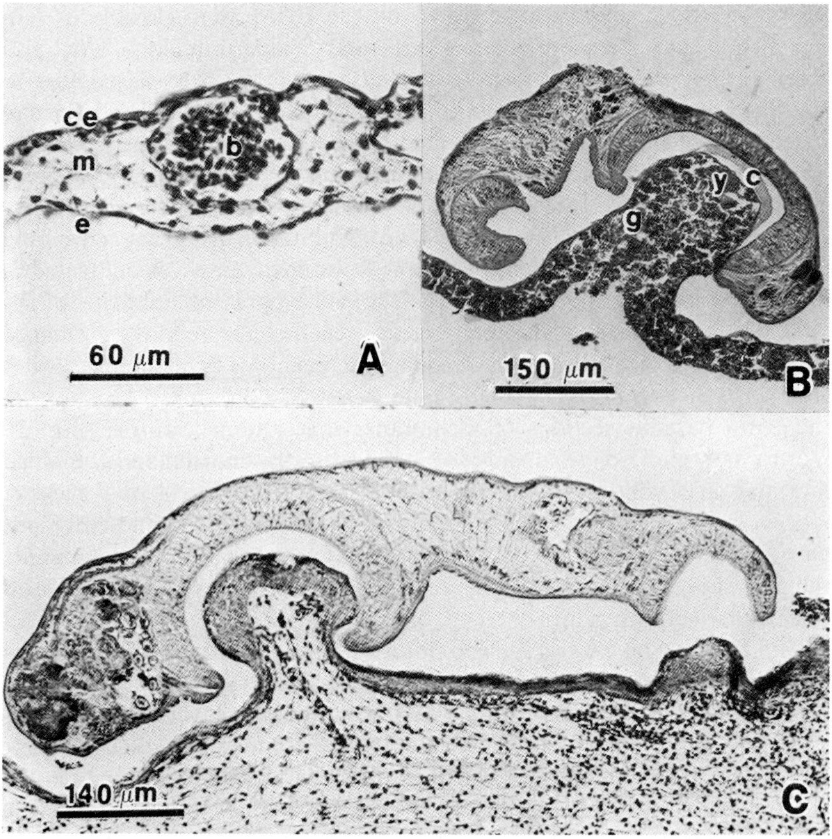

FIG. 5 (A) Uninfected 8-day-old CAM. (B) Two-day-old *Leucochloridiomorpha constantiae* on a 5-day-old chorion; note the acetabular plug containing chorion. (C) Four-day-old *L. constantiae* on a 10-day-old CAM; note ectodermal hypertrophy within the tissue plugs. *Abbreviations*: b, blood vessel; c, chorion; ce, chorionic ectoderm; e, endoderm; g, yolk globules; m, mesoderm; y, yolk sac. Photographs are of paraffin sections stained with hematoxylin and eosin (from Fishbein *et al.*, 1985). Reproduced with the permission of the American Microscopical Society.

development of trematodes. Fried *et al.* (1985) used whole-mount histochemical procedures to demonstrate alkaline phosphatase activity primarily in the excretory and reproductive systems of *L. constantiae* developing on the CAM from the free metacercaria to the ovigerous adult (Fig. 6).

2. Leucochloridium variae

Leucochloridium variae metacercariae inoculated onto 3-day-old domestic chick embryos have developed into ovigerous adults within 2 weeks post-inoculation (Fried, 1973). The larvae of this species were obtained from the broodsac (a modified sporocyst) of the land snail, *Succinea ovalis*, as described by Lewis (1974). Fried (1973) reported the development of ovigerous *L. variae* on the CAM within 14 days post-inoculation of excysted metacercariae. Because the CAM is not present in 3-day-old embryos, worms apparently migrate to this membrane as it develops. This technique (essentially a modified Zwilling, 1959, procedure) is particularly useful for worms that require a long time to reach sexual maturity. When using this technique, a large number of eggs should be employed because it is relatively easy to contaminate 3-day-old embryos, regardless of the aseptic precautions taken in treating the worms prior to implantation.

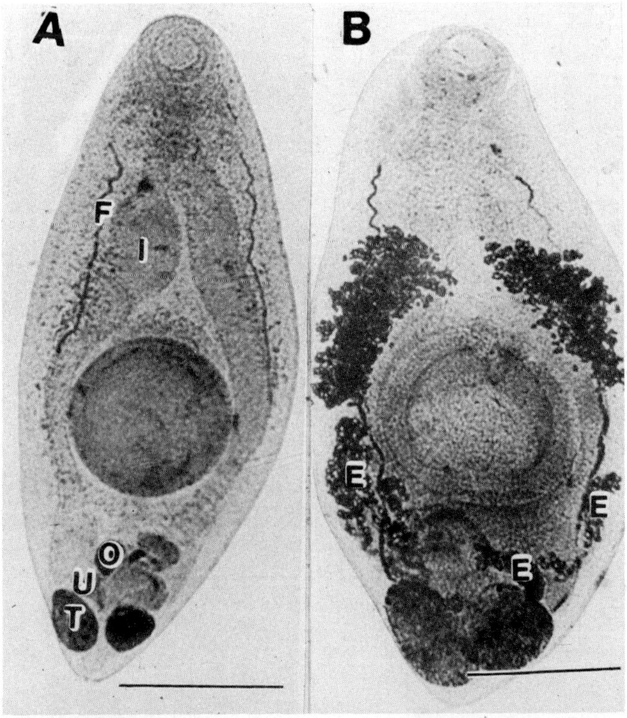

FIG. 6. (A) One-day-old CAM-*Leucochloridiomorpha constantiae* stained for alkaline phosphatase activity (AIP) using the β-glycerophosphate calcium salt method on whole worms fixed in cold 10% neutral buffered formalin. Note reactivity mainly in excretory and reproductive systems. (B) CAM-worm, 4-days old, prepared as described for A. Note intense reactivity mainly in reproductive structures. *Abbreviations*: E, eggs; F, flammiferous tubule; O, ovary; T, testis; U, uterus. Scale bar in A = 430 μm; scale bar in B = 450 μm (from Fried *et al.*, 1985). Reproduced with the permission of the American Society of Parasitologists.

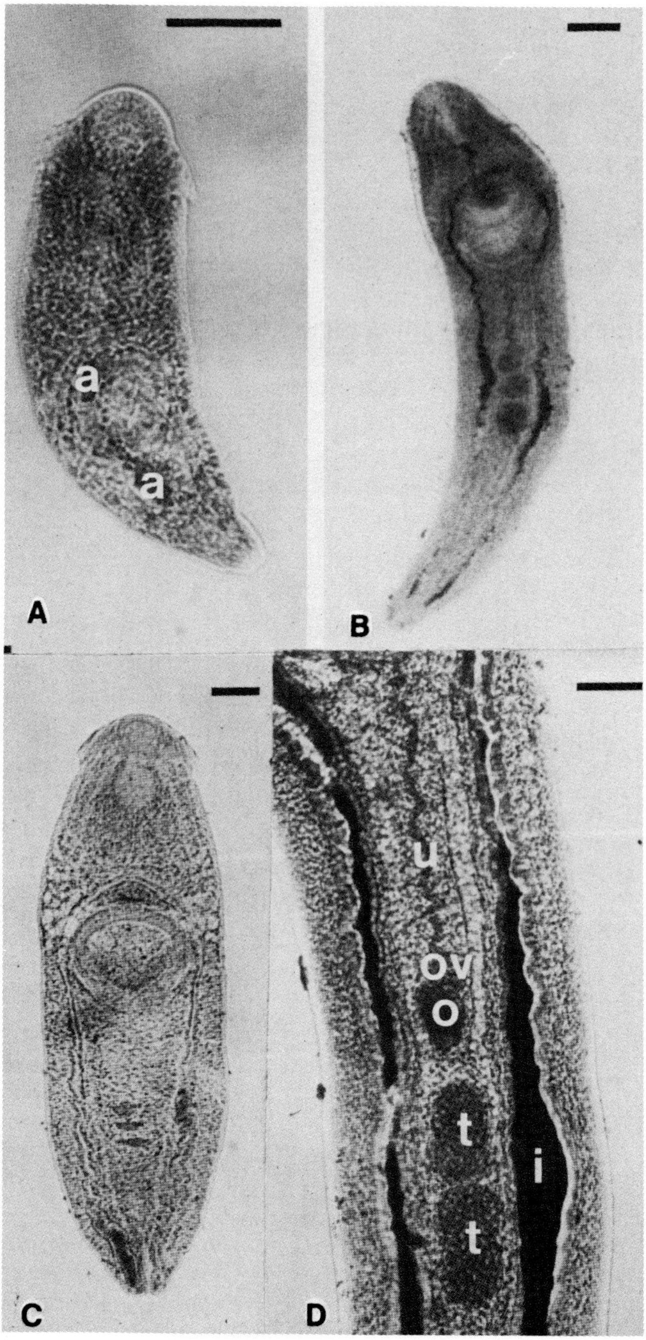
a
a
A
B
u
ov
o
t
t
i
C
D

3. Amblosoma suwaense

Shimazu (1974) described the free metacercaria of this brachylaimid from *Sinotaia quadrata* snails in Japan, and Font (1980) found this species in *Campeloma decisum* snails in the USA. Both workers reported the development of the metacercaria to the ovigerous adult in chick embryos. *A. suwaense*, for which no natural definitive host is known, is the first trematode grown from the metacercaria to the ovigerous adult in chick embryos. Fried (1969) mentioned that one advantage of the chick embryo culture procedure is that it may allow for the maturation of larval trematodes where definitive hosts are not known or not available as experimental hosts.

Fried *et al.* (1981) extended the earlier studies of Shimazu (1974) and Font (1980) on this brachylaimid and grew free metacercariae in chick embryos using the *in ovo* procedure of Zwilling (1959) and chick embryos cultivated *in vitro* according to Auerbach *et al.* (1974). Similar worm development was seen using both techniques, with worms developing mainly in the albumen and becoming ovigerous by day 4. During culture, worms egested their metacercarial pigment (presumably melanin), and transformed their rugose metacercarial tegument into a smooth one. Growth studies on *in ovo* worms showed that the body area of ovigerous worms declined, compared to that of the free metacercariae, but gonadal and vitelline area increased.

J. ECHINOSTOMATIDAE

1. Echinostoma trivolvis (*synonym of* E. revolutum; *see Huffman and Fried 1990*)

Echinostoma trivolvis was first cultivated on the CAM using metacercariae excysted *in vivo* and pre-ovigerous adults (Fried *et al.*, 1968). Fried and Foley (1969) examined 7-day-old pre-ovigerous *E. trivolvis* on the CAM and observed red or black pigment in the intestinal ceca of CAM-grown worms. They also found that this organism could tolerate a wide temperature range on the CAM (30–43°C) and still show growth and development. Optimal

FIG. 7. Photomicrographs of an excysted metacercaria (A) and of 7-day-old CAM-worms (B, C, D) of *Echinostoma trivolvis* fixed in alcohol-formalin-acetic acid and stained in carmine. (A) Excysted metacercaria; scale bar = 50 μm. (B) Seven-day-old worm; scale bar = 100 μm. (C) Seven-day-old worm; scale bar = 100 μm. (D) Seven-day-old worm; scale bar = 50 μm. *Abbreviations*: a, genital anlage; i, intestinal cecum; o, ootype; ov, ovary; t, testis; u, uterus. Note considerable intraspecific variation in body shape and gonads in the 7-day-old CAM-worms (from Fried and Pentz, 1983). Reproduced with the permission of the editor of the International Journal for Parasitology.

growth and development was between 37 and 39°C. Fried and Butler (1978) examined the infectivity and development of *E. trivolvis* on the CAM using chemically excysted metacercariae to infect the embryos. Fried and Pentz (1983) compared three different cultivation procedures for *E. trivolvis* on the CAM, and found that the *in ovo* procedure of Zwilling (1959) was the easiest to use and provided the best worm recovery.

Fried and Kramer (1968) used histochemical methods to determine if *E. revolutum* depletes its glycogen reserves in a non-nutrient medium and resynthesizes glycogen when placed in a nutrient medium or on the CAM. Worms starved for 48 hours and then incubated on the CAM for 48 hours resynthesized lost glycogen, but not to the same extent as those maintained in a Tyrode's-glucose medium after starvation.

In vitro excysted metacercariae of *E. trivolvis* have been cultivated on the CAM to uterine coiling stage, i.e. to a stage about equivalent to the growth and development achieved in the domestic chick after 7 days (Fried and Butler, 1978; Fried and Pentz, 1983). Fried and Pentz (1983) also studied the intraspecific variation of *E. trivolvis* grown from excysted metacercariae to 7-day-old adults on the CAM. The variation in body size and shape and in gonadal structures was very evident in CAM-grown worms (Fig. 7). Scanning electron microscopy (SEM) showed that the topography of worms grown on the CAM was normal and appeared identical to that seen in pre-ovigerous worms grown in domestic chicks (Fried and Fujino, 1984). Single adults of *E. trivolvis* were placed in chick embryos located on the CAM above the embryo. The location of a worm changes in the presence of a second worm suggesting that worm site location may be altered by the release of chemoattractants by another worm (Fried and Diaz, 1987).

Wisnewski *et al.* (1986) examined the feeding of *Echinostoma revolutum* on the CAM *vs* that in worms grown in the domestic chick. Histochemical studies, solubility tests for hematin and X-ray microanalysis of worm cecal contents showed that CAM-grown worms fed on blood from the CAM, whereas chick-worms fed on host intestinal mucosa. A recent TEM study by Fried (unpublished) showed normal ultrastructure of 7-day-old CAM worms; the tegument, subtegument, musculature, gut and excretory system

FIG. 8. Transmission electron micrographs of 7-day-old CAM worms of *Echinostoma trivolvis* fixed in 4% glutaraldehyde, post-fixed in osmium tetroxide and embedded in Epon. Thin sections were stained in uranyl acetate and lead citrate and viewed on a Jeol 100CX transmission electron microscope. (A) Section through tegument (t) and subtegumentary muscle (m). (B) Section through intestinal cecum showing lumen with fragmented erythrocytes (e) and other ingesta obtained from the CAM. (C) Section through the intestinal cecum showing gastrodermal lamellae (la) projecting into the lumen (l). (D) Section through an excretory tubule showing an epithelial cell with lipid (li) inclusions. Photomicrographs courtesy of Gerry Brennan.

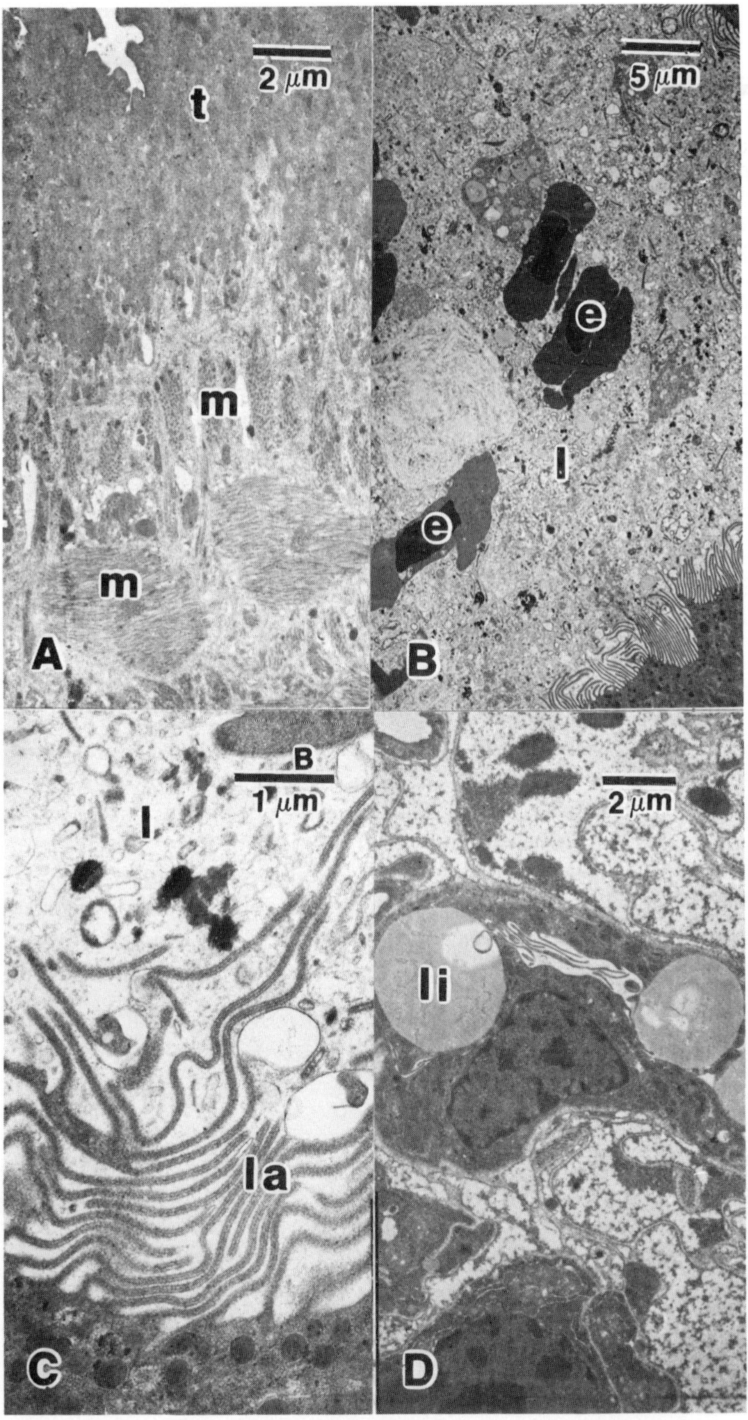

2 µm
t
m
m
A
5 µm
e
l
e
B
B
1 µm
l
la
C
2 µm
li
D

showed the typical ultrastructural characteristics of trematodes grown *in vivo*. The gut of CAM-worms showed lamellae associated with the gastrodemis and the lumen contained nucleated erythrocytes in various stages of fragmentation (Fig. 8).

2. Hismasthla quissetensis

Cercariae of *Himasthla quissetensis* from the marine snail, *Ilyanassa obsoleta*, were cultivated on the CAM, and one worm serially transferred to a new CAM became ovigerous (Fried and Groman, 1985). Most cercariae lost their tails, extruded cystogenous material and showed post-cercarial development without forming a metacercarial cyst. Because many of these cercariae were able to develop to adults without going through a metacercarial stage, a period of encystment for physiological reorganization may not be necessary to utilize nutrients. Attempts to cultivate numerous other trematodes on the CAM from the cercarial stage were unsuccessful (Fried and Groman, 1985).

K. FASCIOLIDAE

1. Fasciola hepatica

Chemically excysted metacercariae of *Fasciola hepatica* were implanted onto the CAMs of 7- to 10-day-old chick embryos maintained at 39°C (Fried and Butler, 1979). Only 4 ($<0.5\%$) of 1250 metacercariae were recovered from 50 inoculated CAMs (only one worm per CAM) from 4 hours to 6 days post-implantation. Some worms ingested and probably utilized blood from the CAM and showed somatic growth and development of reproductive structures. Worms grown for 5–6 days on the CAM were developmentally equivalent to those raised in mice for 4–5 days (Davies and Smyth, 1978).

L. PSILOSTOMATIDAE

1. Sphaeridiotrema globulus

Encysted metacercariae of *Sphaeridiotrema globulus* were obtained from *Goniobasis virginica* snails in the USA and excysted chemically in an alkaline bile salts-trypsin medium at 41°C in the absence of acid pepsin pre-treatment (Fried and Huffman, 1982). Excysted metacercariae transferred to the CAMs of 6- to 10-day-old chick embryos maintained at 41°C developed into ovigerous adults within 4 days. Encysted metacercariae fed to 1-day-old chicks became ovigerous in the lower ileum by day 4. The development times were similar whether worms were grown *in vivo* or *in ovo*. In a later study

(Huffman *et al.*, 1984), histopathological events seen in the gut of the natural host, the mute swan (*Cygnus olor*), closely paralleled that seen in the gut of the domestic chick and the CAM. Worms produced large hemorrhagic zones on the CAM and an intimate association (a host–parasite interface) between worm and CAM was noted (Fig. 9). Thus, the CAM provides an excellent model for histopathological studies on this trematode.

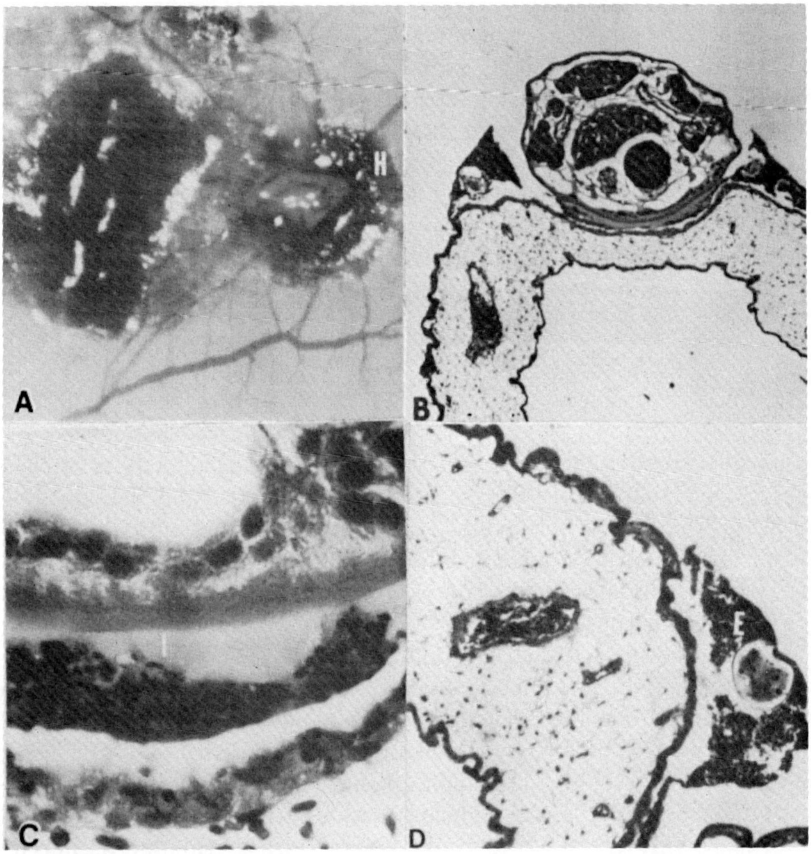

FIG. 9. Photographs showing reaction of *Sphaeridiotrema globulus* on the CAM. (A) Living *S. globulus* photographed *in situ* on the CAM; note the hemorrhagic zone (H) produced by the parasite. (B) A histologic section stained with hematoxylin and eosin showing *S. globulus* on the CAM. (C) A histologic section stained with hematoxylin and eosin showing the interface (I) between the parasite (above) and the CAM (below). The interface (I) zone stains heavily with eosinophil. (D) This hematoxylin- and eosin-stained section shows encapsulation (E) of an egg from the parasite on the CAM. Scale bars were not given in the original publication (from Huffman *et al.*, 1984). Reproduced with the permission of the editor of the *American Journal of Veterinary Research*.

M. PHILOPHTHALMIDAE

1. Philophthalmus hegeneri

Using mechanically excysted metacercariae, Fried (1962a) cultivated this eyefluke to the ovigerous adult on the CAM of eggs prepared by a modified Woodruff and Goodpasture (1931) technique. This was the first report of the successful cultivation of a trematode on the CAM. Worms were serially transferred twice to new CAMs after 7 days survival on the first CAM for a total survival time of 21 days on all membranes. The growth of worms from the CAM lagged behind that of digeneans grown in the eyes of domestic chicks and these worms became ovigerous later than those from chicks (Fried, 1962b). This study suggested that certain physicochemical conditions on the CAM must be similar to those in the normal site of the parasite to allow for the successful *in ovo* cultivation of worms.

N. PARAMPHISTOMIDAE

1. Zygocotyle lunata

Chemically excysted metacercariae were used to study the effects of chick embryo age on parasite development on the CAM (Fried and Nelson, 1978). Larvae implanted into chick embryos on day 3, survived but did not show post-metacercarial development. However, larvae placed on the CAMs of 10-day-old membranes showed comparable development to worms grown in the chick cecum. The age of the developing embryo is an important factor in achieving successful post-metacercarial development of some trematodes. In contrast to the findings in the *Z. lunata* study, excysted metacercariae of the brood sac trematode, *Leucochloridium variae*, implanted onto the CAMs of 3-day-old embryos became ovigerous in that site (Fried, 1973). The age of the CAM used for cultivation studies is more critical for some trematodes than others. Some *Z. lunata* produced a lesion containing cellular debris on top of the CAM, whereas others did not show such a lesion but produced a plug of chorionic ectoderm with the aid of the acetabulum (Fig. 10).

O. HAPLOMETRIDAE

1. Haematoloechus *sp.*

Fried and Weaver (1969) observed the survival and feeding of adult frog lung flukes, *Haematoloechus* sp., for 1 week on the CAM maintained at

30°C. Pre-ovigerous, transparent forms (about 1–2 mm in length and 0.5 mm in width), when implanted on the CAM and maintained there for 5–6 days at 30°C, tripled their length and width and became ovigerous. The intestine of the worm distended with blood and the body developed pigment.

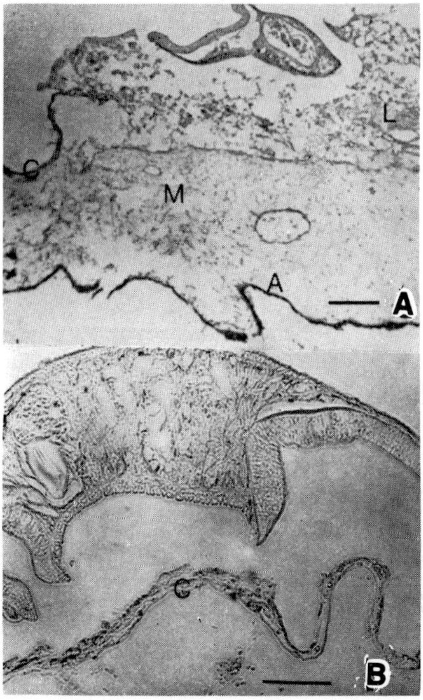

FIG. 10. Photomicrographs of stained sections of *Zygocotyle lunata* on the CAM. (A) An alcian blue cryostat section of a worm grown for 5 days on the CAM. Worm is associated with a lesion (L) on the surface of the CAM; M, mesenchyme; A, allantoic endoderm; scale bar = 100 µm. (B) An Oil Red O cryostat section of a worm grown for 5 days on the CAM: note plug of chorionic ectoderm (C) associated with worm acetabulum; scale bar = 100 µm (from Fried and Nelson, 1978). Reproduced with the permission of the American Microscopical Society.

P. MICROPHALLIDAE

1. Microphallus pygmaeus

Free metacercariae of *Microphallus pygmaeus* were obtained from *Littorina saxatilis* snails in County Antrim, Northern Ireland and transferred to the subchorioallantoic space of 7- to 10-day old fertile chick eggs (Irwin and Saville, 1988b). Some eggs received daily injections of 0.2 ml chicken serum via the subchorioallantoic space. All the worms maintained in chick embryos for 6 days were ovigerous and there were considerably more eggs in digeneans that had been maintained in embryos which had serum added. Although there was no evidence that this trematode was capable of penetrating the CAM to obtain blood, it is likely that the parasite absorbs or ingests material from the undersurface of the CAM or from material in the subchorioallantoic space.

Q. MONOGENEA

1. Polystomoides *sp.*

Fried (1965) maintained *Polystomoides* sp. from the oral mucosa of painted turtles, *Chrysemys picta bellei*, for up to 7 days on the CAM at $30 \pm 1°C$. This procedure may be applicable for cultivation studies on other monogenetic trematodes. Specimens from the oral mucosa contained relatively unbranched straight intestinal ceca, whereas those grown on the CAM had highly diverticulate ceca. The shape of the ceca is influenced by nutritional and physicochemical factors of the habitat. The results of this study suggest that *Polystomoides oris* is a probable synonym of *P. coronatum* and that cecal shape is not a reliable characteristic for taxonomy in this group.

R. CESTODA

1. Spirometra mansonoides

Mueller (1966) noted that sparganum (plerocercoid) of *Spirometra mansonoides* can be introduced into the chorioallantoic space of 12-day-old developing chick embryos and recovered from the chick after hatching. Observations on worm development were not made.

2. Taenia hydatigera

Parmeter and Gemmell (1974) used the chick embryo as an experimental host to culture larvae of *Taenia hydatigera.* Onchospheres were inoculated onto the CAM surface, the allantoic fluid and the yolk sac of 8- to 9-day-old chick embryos. Metacestodes were collected from embryos and chicks that were allowed to hatch. Of 64 metacestodes recovered, 36 of the developing cysticerci were recovered from embryos prior to hatching and 28 were taken from hatched chicks. Viable metacestodes were seen in the CAM, the allantois, the yolk sac, the amniotic fluid and the body cavities of embryos.

Considerable growth and development was observed in many of the developing cysticerci removed from various chick embryo sites.

S. NEMATODA

1. Syngamus trachea

Winward (1976) used chick eggs to propagate the larval stages of *Syngamus trachea*. Larvae were inoculated into the albumen of 1-day-old eggs or the allantoic sac of 17-day-old embryos. Although extensive antibiotic treatment was used on the larvae, less than 10% of the inoculated chick embryos survived (most died from bacterial contaminants carried in or on the nematode larvae). However, some nematode larvae inoculated into either the albumen or allantoic sac survived the embryonic period and were recovered from chicks that hatched. In fact, some larvae migrated to the trachea of developing chicks (normal site of development for this nematode) and became sexually mature there.

IV. Structure and Function of Chick Extraembryonic Membranes

In this section, the structure and function of chick extraembryonic membranes are examined. Pertinent studies on these topics are related to experimental helminthology whenever possible.

A. EGG FORMATION AND EARLY DEVELOPMENT

The hen's egg is released from the ovary as a spheroid of yolk surrounded by a thin layer of cytoplasm that forms a bleb about 3 mm in diameter, the blastodisc, at the animal pole of the egg. The blastodisc contains the egg nucleus arrested in second meiotic division. The whole is enclosed by the tightly applied, non-cellular vitelline membrane. The egg is drawn into the opening of the oviduct where within 15 min sperm entry occurs, meiosis is completed and the zygotic nucleus formed. During the next 24–28 hours before the egg is laid, it is moved by peristaltic contractions through the oviduct where the auxiliary materials—albumen, shell membranes and shell—are laid down around it. The blastodisc is converted by cleavage into a plate of cells, the blastoderm (for details of these processes, see Conrad and Scott, 1938; Lillie, 1952; Romanoff, 1960).

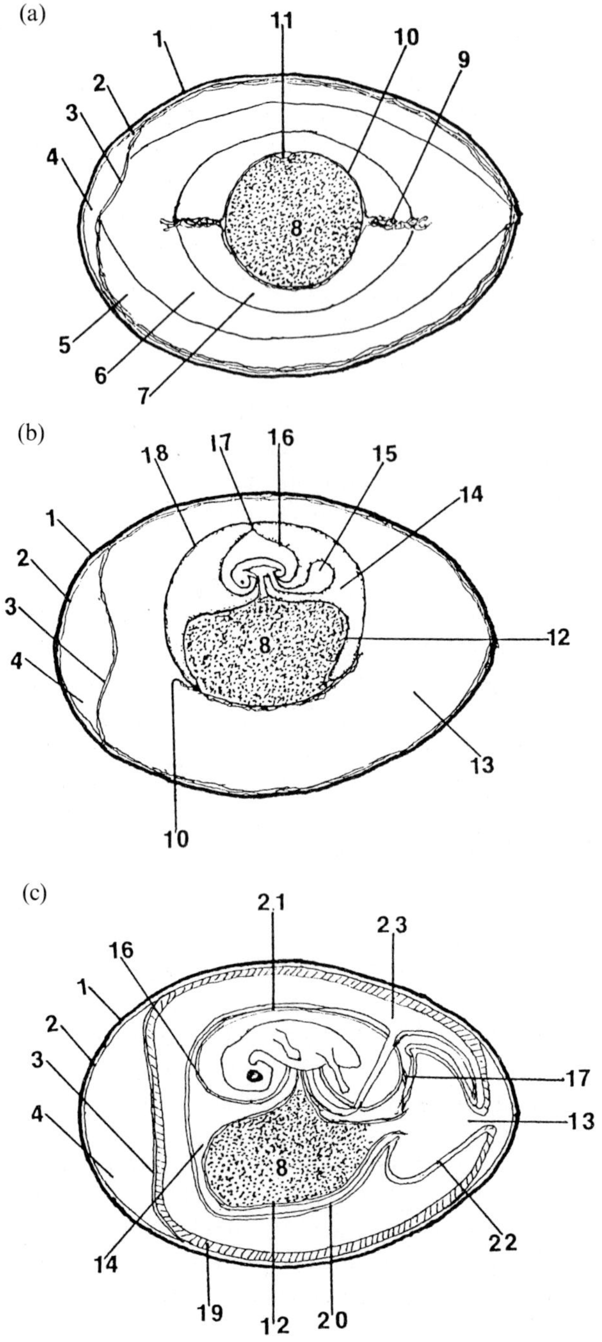
(a)
1
2
3
4
5
6
7
8
9
10
11
(b)
1
2
3
4
8
10
12
13
14
15
16
17
18
(c)
1
2
3
4
8
12
13
14
16
17
19
20
21
22
23

The newly laid egg is shown diagrammatically in Fig. 11a. Surrounding the egg are three albuminous layers: an inner thin albumen layer that allows the egg to rotate and orient with reference to gravity; a middle dense albumen layer, laid down as ropes of mucilaginous gel that were caused to spiral in passage through the oviduct and whose twisted ends are presumably the chalazae (Conrad and Scott, 1938); and an outer thin albumen layer. The two shell membranes are closely associated with each other and with the shell except at the blunt end of the shell where an air space lies between the membranes. The shell consists of densely packed $CaCO_3$ crystals with its outer surface covered by a thin peptide-carbohydrate cuticle (Freeman and Vince, 1974).

B. THE CLEIDOIC EGG AND EXTRAEMBRYONIC MEMBRANES

The chicken egg represents the highest state of perfection of the cleidoic (boxed) egg, one where the embryo is virtually shut off from the external environment and develops in a world of its own. Except for oxygen, the chick embryo has available all the material—organic and inorganic—it needs to develop into a self-feeding, almost independent hatchling. The yolk provides proteins, fats, sterols, vitamins and minerals. The albumen is the major reservoir of water for the embyro but also provides vitamins and minerals; furthermore, it yields as much protein as the yolk (Romanoff, 1967). The albumen also is bactericidal to some species of bacteria. This property may derive from the increase in pH in the albumen, from 7.6 to 8.5, during the first 2 days of incubation (Feeney and Allison, 1969) or from sequestration of vital bacterial food needs by various proteins along the albumen chains (Freeman and Vince, 1974). The shell is the major source of calcium for the embryo. Specialized extraembryonic structures are required to convert stored food into assimilable form, to provide a large, highly vascularized surface for exchange of respiratory gases, and to allow safe storage of excretory products. These structures are the extraembryonic membranes.

The chick embryo develops four such membranes: yolk sac, amnion,

---

FIG. 11. Development of the chick embryo and its extraembryonic membranes. (a) Newly laid egg; (b) after 90 hours incubation; (c) after 12 days incubation. 1, shell; 2, outer shell membrane; 3, inner shell membrane; 4, air space; 5, outer liquid albumen; 6, dense albumen; 7, inner liquid albumen; 8, yolk; 9, chalaza; 10, vitelline membrane; 11, blastoderm; 12, yolk sac; 13, albumen; 14, extraembryonic coelom; 15, allantois; 16, amnion; 17, sero-amniotic connection; 18, chorion; 19, chorioallantoic membrane (CAM); 20, inner allantoic wall; 21, fused inner allantoic wall and amnion; 22, albumen sac; 23, allantoic cavity (after Burton and Tullett, 1985). Not to scale.

chorion (serosa) and allantois. Later in development, the last two fuse to form the chorioallantoic membrane (CAM), of particular interest to helminthologists (see Section III).

C. THE YOLK SAC

The yolk sac is a higly vascularized membrane whose chief function is alimentation: the digestion of yolk and the absorption and distribution of its products. It is the major site of erythropoiesis during development and, prior to chorioallantois formation, acts as a respiratory surface for the embryo. It also carries on some excretory functions (Clark and Fischer, 1957).

Early in the first day of incubation, the disc-like blastoderm becomes bilaminar through delamination of cells from its lower surface and by polyinvagination of upper-level cells. A series of cell shifts and movements (gastrulation) takes place so that the upper layer becomes presumptive ectoderm, the lower layer endoderm and between the two migrating cells establish the extraembryonic mesoderm (for gastrular movements, see Fontaine and Le Douarin, 1977; Stern and Canning, 1990). The mesoderm forms two layers through a complex process: mesodermal cells gather in groups to form vesicles that fuse with each other, and then the whole splits to form an upper and lower layer enclosing the extraembryonic coelom (Kessel and Fabian, 1985). The upper or somatic mesoderm joins with the overlying ectoderm to form the somatopleure. The lower or splanchnic mesoderm that contains a developing capillary network in which are clumps of hemocytoblasts ("blood islands") fuses with the underlying endoderm to form the splanchnopleure, the incipient yolk sac.

The yolk sac spreads over the yolk and, after 5–7 days of incubation, completely envelops it except for a small area at its lower pole, the yolk sac umbilicus. Extension of vascularization moves more slowly, not reaching the edge of the yolk sac until day 14 of incubation. The vascular pattern undergoes a series of transformations that increase circulatory efficiency and continue until hatching (for details, see Romanoff, 1960).

The yolk sac endoderm is continuous with that forming the gut in the embryo but no significant amount of yolk moves from the yolk sac to the embryonic gut. The endoderm is a simple columnar epithelium that is underlain by mesenchymatous connective tissue containing the blood vessels, and a serosa. In 3-day-old embryos, the endoderm starts to develop longitudinal folds that project into the yolk. The folds deepen and continue to increase in number, achieving maximal surface in minimal space. Within the axes of the folds are arterioles and the erythropoietic tissue.

Digestion is intracellular during the first 2 days of incubation (Williams,

1967). From day 3 on, however, digestion is largely extracellular. The endoderm or, better, the digestive epithelium, secretes into the yolk enzymes that break down large molecules into their absorbable simple components. By day 9 of incubation, it becomes clear that some intracellular digestion continues. Epithelial cells, by pinocytosis and phagocytosis, take in large molecules (e.g. proteins) and pass some of them, unchanged, into the circulation; others, along with engulfed yolk particles, are broken down by lysosomes. The yolk sac epithelium also synthesizes some proteins that go to the embryo via the circulation (for more extensive treatment of the yolk sac and alimentation, see Lambson, 1970; Juurlink and Gibson, 1973; Freeman and Vince, 1974).

Erythropoiesis begins in the yolk sac 36 hours after the beginning of incubation. It reaches a peak on day 13 and then slowly declines until hatching. This decline is compensated for by the appearance of the bone marrow as a definitive erythropoietic tissue (Freeman and Vince, 1974). There are two distinct populations of erythrocytes: the primitive or embryonic and the definitive or adult. The primitive arises from the division of the mesodermal cells in the blood islands; these erythroid cells continue mitotic divisions until days 5 or 6 of incubation when division ceases (Campbell *et al.*, 1971). At this time, the definitive population appears from a yet to be determined source and grows rapidly. Members of the primitive population at any given time are at about the same stage of maturity, whereas the definitive population shows a variety of stages. The definitive population gradually supplants the primitive; by day 16, less than 1% of circulating erythrocytes are primitive (Bruns and Ingram, 1973).

Primitive and definitive erythrocytes differ not only in structure but also in the number and kinds of hemoglobin (Hb) they contain. Hb synthesis begins after about 35 hours of incubation (Weintraub *et al.*, 1971) and continues at a rapid pace during the series of mitotic divisions in the primitive erythrocytes. When mitosis stops after 5–6 days of incubation, the primitive erythrocytes are in an immature stage, and therefore Hb synthesis continues into day 7 to almost triple the Hb content of the cells (Fraser, 1966). The literature indicates fair agreement that definitive erythrocytes contain two Hbs that persist into adult life; however, Fraser *et al.* (1972) found 11 Hbs that rise and fall during embryogenesis. There is disagreement on primitive Hbs with numbers ranging from 2 (Fraser *et al.*, 1972; Bruns and Ingram, 1973) to 6 (Schalekamp *et al.*, 1972). Some of this variation may be attributed to the number and elegance of the analytical techniques used. In any case, during this 7- to 14-day incubation period, embryonic Hb is largely replaced by adult Hb, the number of erythrocytes steadily rises (Romanoff, 1967) and the oxygen-carrying capacity of the blood increases (Tazawa, 1980), the increase being co-ordinated with the replacement of embryonic by adult erythrocytes (Bauman *et al.*, 1987). This period of flux in the blood coincides with the period most generally used for culturing worms on the CAM (see Section II.D).

D. AMNION AND CHORION

The amnion and chorion arise in conjunction with each other from folds in the extraembryonic somatopleure that elevate, meet and fuse. After about 30 hours' incubation, the amniotic head fold appears just anterior to the embryo and starts to extend posteriad. Soon thereafter the amniotic tail fold appears posterior to the embryo and extends forward, accompanied by lateral amniotic folds moving toward the midline over the embryo. After 3 days of incubation, these folds fuse to establish the amnion, closely applied to the embryo, and the overlying chorion. The chorion has an outer ectodermal and an inner mesodermal layer, the amnion an outer mesodermal and an inner ectodermal layer. Between the chorion and amnion is the extraembryonic coelom that will soon be occupied by the expanded allantois.

Very shortly after amnion formation, amniotic fluid begins to be secreted, rapidly expanding the amnion and forming a "little pond" (Needham, 1942), the aquatic environment in which the embryo develops and is protected from desiccation and mechanical shocks (Fig. 11b,c). By the end of 4 days of incubation, non-inervated muscle fibers have differentiated and amniotic contractions begun; the contractions become rhythmical on day 6 and cease on about day 13 of incubation. At this time, the small strip of tissue maintaining a connection between amnion and chorion, the seroamniotic raphe, ruptures to allow some albumen to enter the amnion, raising the volume of amniotic fluid to its maximum. The embryo then begins to drink the fluid which drops precipitously in volume and disappears just before hatching. (For a more extensive discussion of the amnion, see Romanoff, 1960. Further discussion of the chorion is reserved for Section IV.F.)

E. ALLANTOIS

The last extraembryonic membrane to appear is the allantois, whose primary function is to collect excretory fluids and secondarily to absorb the remaining albumen late in development. Overshadowing these functions, however, is the contribution of vascularization to the CAM, the chief respiratory surface for the embryo.

Toward the end of day 2 of incubation, an out-pocketing of the endoderm appears in that portion of the archenteron destined to form the hindgut (Zwilling, 1946). This endodermal diverticulum has an outer layer of embryonic mesoderm and grows out into the extraembryonic coelom. During day 4 of incubation, it becomes clearly visible as a vascularized, stalked sac, the allantois (Fig. 11b). Growth from day 4 on requires the mechanical stimulus of excretory fluid collecting in the sac (Romanoff, 1960). The allantois expands rapidly and, on day 5, the mesodermal surface of its outer wall makes contact and starts to fuse with the mesoderm of the chorion; this is the beginning of CAM formation. The inner allantoic wall presses against the amnion and on day 7 fuses with it. The CAM grows out in a complex pattern to envelope the amnion containing the embryo and the shrinking yolk sac. By the end of day 9, the extending CAM starts forming the albumen sac around the remaining albumen (Fig. 11c). By the end of day 16, the albumen is completely absorbed (for more extensive discussion of allantois formation and CAM growth, see Lillie, 1952; Romanoff, 1960).

The allantois not only stores excretory products but also plays a role in water conservation. Although uric acid is the chief nitrogenous waste and is usually associated with a hypertonic urine, the allantoic fluid is actually hypotonic (Stewart and Terepka, 1969). From day 5 of incubation, the allantoic fluid increases in volume until it peaks on day 13 (Romanoff, 1967; Romanoff and Hayward, 1943). At this time the allantoic endoderm starts to resorb water by active transport, precipitating the nitrogenous wastes as urates, reducing the fluid to a minimum by day 18 (Stewart and Terepka, 1969).

Needham (1931) presented a succession of changes in nitrogen excretion in the chick embryo, from ammonotelic through ureotelic finally to a uricotelic mode. Although this view enjoyed popularity for some years, subsequent work (Clark and Fischer, 1957; Fisher and Eakin, 1957; Fisher, 1967) has shown it to be incorrect. Ammonia increases in the allantois up to day 11 of incubation, but this is simply due to the increase in volume of the allantoic fluid. After 11 days, ammonia accumulates in the fluid but there is no early peak and shift to urea. Similarly, urea continues to increase in volume but there is no shift to the uricotelic mode. Uric acid, like ammonia and urea, is found at all times in the allantoic fluid. It rises continuously in amount, particularly after day 11, and is the dominant nitrogenous waste at all stages (for a more extensive discussion of excretion, see Fisher, 1967).

F. CHORIOALLANTOIC MEMBRANE (CAM)

The CAM consists of three tissue layers: the inner allantoic epithelium derived from the allantoic endoderm; the middle layer of areolar connective tissue containing the blood vessels and derived from the fusion of allantoic and chorionic mesoderm (see Section IV.E); and the outer chorionic epithelium derived from the ectoderm of the chorion. The allantoic epithelium starts as a cuboidal epithelium 1–2 cells thick. After 11 days of incubation, the cells vary from squamous to cuboidal, but by 14 days they are predominantly cuboidal and bear microvilli on their apical surfaces (Borysko and Bang, 1953; Rangan and Sirsat, 1962; Leeson and Leeson, 1963; Ganote *et al.*, 1964). The allantoic epithelium is of primary interest to virologists cultivating viruses (see Borysko and Bang, 1953; Buddingh, 1952) and those who study active transport (e.g. Stewart and Terepka, 1969; Saleudden *et al.*, 1976).

As the CAM forms and matures, changes occur in the areolar connective tissue and its vascularization. From days 4–10 of incubation, rapidly dividing endothelial cells in the developing capillaries and blood vessels (see below) synthesize glycoconjugates—glycoproteins, glycosaminoglycans (GAGs) and glycolipids—which are retained within the cells (Ausprunk, 1982). From 4–14 days of incubation, there is a shift from a high concentration of hyaluronic acid (HA) to a high concentration of sulfated glycoconjugants, secreted by the endothelial cells, in the matrix; the HA and sulfated GAGs probably play a role in the differentiation of arterioles and venules (Ausprunk, 1986). Collagen fibers, associated with stellate fibrocytes, increase in number and thickness, concentrating around arterioles and venules and under the chorionic epithelium and the allantoic epithelium (Ganote *et al.*, 1964).

The allantoic capillary network in the just developing CAM is superseded by capillaries and small vessels formed within the CAM 4–15 days after the beginning of incubation. Allantoic endothelial cells lack a basement membrane and microfilaments, have abundant free ribosomes and rough endoplasmic reticulum, but pinocytotic vesicles are rare; chorioallantoic endothelial cells have a basement membrane, abundant microfilaments and pinocytotic vesicles, but free ribosomes and rough endoplasmic reticulum are reduced (Sethi and Brookes, 1971).

As the vascular network matures, it invades the chorionic epithelium and was once thought to come to lie on top of the epithelium in direct contact with the inner shell membrane (Danchakoff, 1917; Borysko and Bang, 1953). It is now clear that thin, flat cytoplasmic projections from some of the epithelial cells are interpolated between the capillary endothelium and the shell membranes (Ganote *et al.*, 1964; Sethi and Brookes, 1971; Dunn and Fitzharris, 1979; Burton and Palmer, 1989). The exact nature of this respiratory vascularization has been interpreted differently. Fülleborn (1895) as quoted in Lillie (1952, p. 278), described it as "...a great blood sinus...". Most subsequent authors including Lillie (1952) have referred to it as a capillary network (Danchakoff, 1917; Romanoff, 1960; DeFouw *et al.*, 1989) or a capillary plexus (Ganote *et al.*, 1964; Tazawa, 1980). Sethi and Brooks (1971) used the terms capillary and sinusoid interchangeably. Leeson and

Leeson (1963) called the CAM vascularization a sinus, but the strong proponent of this characterization is Narbaitz (1977). Equally vehement in opposing use of the term sinus and insisting on capillary plexus are Burton and Palmer (1989). This contretemps is largely semantic, stemming from the definition of sinus and sinusoid used. The important thing is that the vascular network or sinus provides an excellent respiratory surface that increases in efficiency as it matures.

Capillary formation is virtually complete by day 11 of incubation (Ausprunk *et al.*, 1974). Oxygen enters the blood by diffusion through pores in the shell, the shell membranes, the cytoplasmic projection of chorionic cells and the capillary endothelium. Between days 10 and 14, the resistance of this barrier is reduced three times, thus facilitating the passage of O_2 (Bissonnette and Metcalfe, 1978). These developments, along with changes in erythrocyte count, hemoglobin, etc. (see Section IV.C), bring the respiratory activity to its peak.

Maturation of the chorionic epithelium is correlated with its particular function of removing Ca^{2+} from the shell for transmission to the embryo. Transport of Ca^{2+} begins on day 14 of incubation (Coleman and Terepka, 1972a; Tuan, 1987) and reaches its maximum on days 17–18 (Garrison and Terepka, 1972) or day 19 (Tuan, 1987).

The chorionic epithelium at 7 days consists of two layers of low cuboidal cells on a basal lamina (Leeson and Leeson, 1963; Coleman and Terepka, 1972a). As the CAM spreads to cover the embryo it comes into increasingly close contact with the inner shell membrane; by day 11, the epithelium is so closely associated with the inner shell membrane they are practically inseparable (Coleman and Terepka, 1972a). As day 14 of incubation approaches, two specialized cells associated with Ca^{2+} transmission differentiate. The first to be described in the literature (Skalinsky, 1965) is a cell with many mitochondria, apical vesicles and long microvilli. It has been variously characterized as "intercalated" or "bulbous" (Skalinsky, 1965), "calcium-absorbing" (CAC) (Owczarzak, 1971) or "villus cavity" (VC) (Coleman and Terpeka, 1972a), the last being the most commonly used. Coleman and Terepka (1972a) described the other specialized cell, the "capillary covering" cell (CC), as characterized by thin cytoplasmic projections between the capillary wall and the inner shell membrane (see above). Narbaitz (1977) and his followers refer to this cell as "sinus covering" (SC). In addition to the specialized cells, Coleman and Terepka (1972a) described two other chorionic cells: (1) "...a basal or undifferentiated type of cell..." (p. 121) and (2) a "dark" cell, noting "...whether this is a viable cell in the living membrane is not known" (p. 123).

The function of the CC and VC cells in Ca^{2+} transmission is moot. There is agreement that the VC cells produce H^+ to solubilize the Ca^{2+} (Leeson and Leeson, 1963; Owczarzak, 1971; Coleman and Terepka, 1972a; Narbaitz *et al.*, 1981). The vesicles in the apical region of both cells and studies of active transport in the CAM indicate that pinocytosis and intracellular compartmentalization are involved in Ca^{2+} transmission (Terepka *et al.*, 1969; Coleman and Terepka 1972b; Dunn and Fitzharris, 1987). As to the cells taking up the Ca^{2+}, Coleman and Terepka (1972a) opted for the CC cells, but most workers (e.g. Skalinsky, 1965; Owczarzak, 1971; Narbaitz *et al.*, 1981) have opted for VC cells. The evidence to support either of these views is indirect and insufficient (Narbaitz, 1987). More recent work tends to ignore the specialized cells, emphasising the molecular approach and treating the chorionic epithelium as simply an epithelium (see Tuan, 1987).

The changes in the chorionic epithelium associated with Ca^{2+} transmission could be of significance in experiments where tissues or organisms are grown on the CAM beyond day 13 of incubation. Further, when shell-less cultures of embryos are used as hosts as in the Auerbach *et al.* (1974) procedure (see Section II.D), the embryo becomes calcium-deficient by day 9 of incubation and retarded in growth after 13 days; although pieces of shell added to the CAM may raise total Ca^{2+}, the embryo will still be retarded (Dunn *et al.*, 1987).

The chorionic epithelium, although normally a simple cuboidal epithelium, has the capacity to become a keratinized, stratified squamous epithelium. Moscona (1959), using the "dropped membrane" technique of Woodruff and Goodpasture (1931; see Section II.D), found that if the opening in the shell was left uncovered, or if a covering window was improperly sealed, the chorionic epithelium became stratified squamous and keratinized. Further study indicated that increased O_2 in atmospheric air as compared with the normal milieu of the CAM was responsible for the transformation (Moscona and Carneckas, 1959). Similar modifications of the epithelium have been reported in studies of tissue pieces grown on the CAM. Danchakoff (1918) noted "cornification" of the epithelium near splenic tissue growing on the CAM. Huxley and Murray (1924) and Willier (1927) observed stratification and keratinization of the epithelium when embryonic pieces were grown on the CAM. Ebert (1959) obtained similar results when he injected a mixture of Rous sarcoma virus and cardiac microsomes into the CAM. Fishbein *et al.* (1985), culturing a trematode on the CAM, found the CAM responded by forming tissue plugs associated with the acetabulum and oral sucker (Fig. 5); the epithelium of the plugs and that surrounding them was described as "hypertrophied ectoderm" (p. 270). Clearly, the chorionic epithelium is able to respond to a variety of stimuli.

The CAM has been used in studies of excretion, respiration, active transport, the movement of ions, Ca^{2+} transmission and toxicity in cosmetics and household products. It has been used as the cultural substrate for some viruses, bacteria, protozoans and helminths; for developing embryonic primordia and growing tumors. In many of these studies, the CAM has been considered a model for some adult membrane or as consisting of "relatively undifferentiated cells of all three germ layers" (Rich *et al.*, 1965, p. 96).

Coleman and Terepka (1972a), struck by the morphological similarity between the allantoic portion of the CAM and the toad's urinary bladder and their common origin from hindgut endoderm, ascribed similar functions and ionic movements to each. Although acknowledging some morphological and functional similarities between the two, Pácha *et al.* (1985) found significant electrophysiological differences. Ebert (1959) injected a virus–microsome mixture into the CAM of 9- to 11-day-old embryos (see above) and tentatively concluded that muscle fibers had been induced in what he called "the mesenchyme". According to Rich *et al.* (1965), the spindle-shaped cells that Ebert (1959) found were not induced muscle fibers but were a common reaction of mesenchymatous tissue to a variety of stimuli. Representative of a recent movement to use the CAM for cosmetic testing (see above) is the publication of Leighton *et al.* (1985), who found that the CAM showed the same degrees of inflammation to graded doses of a given substance as does the conjunctiva of the rabbit's eye. Parish (1985), however, found necrosis not inflammation in tests on the CAM. Kobayashi *et al.* (1989) used the CAM as a model of the oral and vaginal mucosa to study the invasion of these membranes by *Candida albicans*. In these diverse studies, virtually no attention was paid to the changing nature of the CAM over time, its relation to the developing embryo, nor to the possible influence of the CAM itself.

The particular strain of chicken eggs used may affect the results of an experiment. Using embryos from two strains of chicken, White Leghorn and a hybrid White Leghorn × California Grey, Ebert (1959) found twice as many successful inoculations of the virus–microsome mixture on the CAM in the White Leghorn embryos as compared with the hybrids. Stephenson and Tompkins (1965) transplanted embryonic limb cartilage and bone from various tetrapod classes onto the CAM and found the results to be correlated with the degree of phyletic relationship to the chick host. Avian bone and cartilage from distantly related species continued to grow on the chick CAM; in transplants from amphibian, reptilian and mammalian donors, only the cartilage continued to grow. An as yet unexplained quality of the CAM has been raised by McLachlan and Phoplonker (1988) who found that pieces of CAM or allantois grafted to the anterior margin of the embryonic chick limb induced formation of an extra digit.

The changes over time affecting the extraembryonic membranes, particularly the CAM, are summarized in Table 2. The changes up to day 13 of incubation represent differentiation and maturation; from that point on, the changes are senescent (Ganote *et al.*, 1964). In any case, when organisms are cultured on the CAM the changes occurring during the period of cultivation should be reckoned with in interpreting the results.

TABLE 2 *Events in the development of the chick embryo and its extraembryonic membranes*[a]

Incubation day	*Age (hours)*	*Event*
2	25–28	Paired heart primordia appear
	26–29	Blood islands forming
	29–33	Heart primordia begin fusion
	32–36	Bending and contraction of heart
	36–38	Primitive hemoglobin synthesis begins; erythropoiesis of primitive cells begins
	40–45	Amniotic head fold appears
	45–49	Circulatory system developing; amniotic tail fold appears
3	49–50	Allantoic bud forms
	65–69	Allantois clearly a small sac
	70–72	Embryo surrounded by amnion
4	84	Amniotic contractions begin
	96	Chorion and allantois begin fusion to form CAM
5	108–120	Synthesis of adult hemoglobins begins; erythropoiesis of definitive cells begins; erythropoiesis of primitive cell reduces; four-chambered heart formed
6	120	Amniotic contractions become rhythmical
	144	Active transport of amino acids by yolk sac
8	180–192	Mineralization of bone begins
9	192–216	CAM completed, fuses with shell membrane; hematopoiesis begins in bone marrow
12	264–288	Absorption of albumen begins; calcium absorption from shell begins; imbibition of amniotic fluid begins
13	288–312	Absorption of allantoic fluid begins; yolk sac transport of lipids accelerates; allantoic epithelium matures; amniotic contractions cease; sero-amniotic connection ruptures
19	432–456	Absorption of allantoic fluid completed; yolk sac withdrawal begins
20	456–480	Inner shell membrane pierced; shell pipped; chorionic epithelium degenerates; CAM circulation reduces
21	480–504	Chick hatches

[a] Derived from Freeman and Vince (1974, appendix I).

V. THE CHICK EMBRYO AS A HABITAT FOR HELMINTHS

A. NORMAL HABITATS OF HELMINTHS GROWN *IN OVO*

Most worms which can be cultivated *in ovo* from the metacercaria to the ovigerous adult normally develop in various enteric and non-enteric sites in avian hosts. Representative examples and their sites of development are as follows: *Amblosoma suwaense*, cloaca and rectum; *Clinostomum marginatum*, mouth cavity; *Diplostomum spathaceum*, upper part of the small intestine; *Leucochloridiomorpha contantiae*, bursa of Fabricius; *Leucochloridium variae*, cloaca and rectum; *Microphallus pygmaeus*, upper part of small intestine; *Philophthalmus hegeneri* nictitating membrane, orbit and Harderian gland of the eye; *Posthodiplostomum minimum*, upper part of small intestine; and *Sphaeridiotrema globulus*, lower part of the small intestine.

The CAM and some other extraembryonic sites of the domestic chick embryo can substitute for various *in vivo* habitats. This section examines the chick embryo as a habitat for helminths and wherever possible comparisons are made between worms grown in the normal *in vivo* site and the chick embryo. Unless otherwise stated, the discussion refers to trematodes that have become ovigerous on the CAM.

B. DEVELOPMENTAL STIMULI

Most parasites need a "trigger" or developmental stimuli, usually provided by the host, to pass from the larval to adult stage. In most *in ovo* cultivation studies on helminths, excysted or free metacercariae have been placed on the CAM. Thus, potential stimuli for excystation or larval transformation from the CAM have been precluded. However, for one species, *Himasthla quissetensis*, grown from cercaria to adult on the CAM, trigger mechanisms from the CAM may have allowed for cercarial tail loss, release of cystogenous material and subsequent development to the adult without the formation of an encysted metacercaria (Fried and Groman, 1985). It is not known what factors from the CAM, if any, were responsible for such transformation phenomena.

Development on the CAM from the excysted metacercaria to the ovigerous adult is an orderly process for several species studied to date. Body and tegumentary changes of CAM-worms are similar to those seen in worms from normal sites (Fried and Mishkind, 1985; Fried and Fujino, 1984). Host-induced developmental stimuli, whether from a normal or ectopic site such as the CAM, are poorly understood and need further study.

One trigger of the CAM environment is change in temperature (Δt). Larvae are usually obtained from a cold-blooded site (typically a freshwater snail or fish) at 4–15°C and subsequent development on the CAM occurs at higher temperatures (usually 37–41°C). However, in most cases where trematodes have been successfully cultivated in chick embryos, Δt alone is not sufficient to induce significant post-metacercarial development. Most trematode larvae that have been cultivated on the CAM do not show post-metacercarial development *in vitro* in either nutrient or non-nutrient saline solutions maintained at 37–41°C. (Fried and Huffman, 1982; Fried and Ramundo, 1987). Thus, temperature alone is not a sufficient trigger for development to the ovigerous stage on the CAM, even for the progenetic trematodes studied to date.

C. NUTRITIONAL REQUIREMENTS

Helminths grown on the CAM usually derive nutrients from that site by feeding on tissue, cell exudate, secretions or blood from the CAM. The relative nutritional contributions of tissue, secretions, exudate and blood for any helminth cultivated *in ovo* are not known. Moreover, worms in the subchorioallantoic space or allantois (see Irwin and Saville, 1988a,b) or in the albumen (Larson and Uglem, 1990) probably obtain nourishment from these habitats. Development of the cestode, *Taenia hydatigera*, from egg to metacestode in chick embryos (Parmeter and Gemmell, 1974) must involve transtegumentary feeding. The relative role of transtegumentary *vs* gut feeding is not well-known even for trematodes maintained *in vivo* (see review by Pappas, 1988); but trematodes cultivated on the CAM probably obtain some nutrient by transtegumentary feeding. Studies on several trematodes cultivated in chick embryos show that these worms feed on blood and tissue from the CAM (Fried and Huffman, 1982; Wisnewski *et al.*, 1986). The gut of these parasites may contain either red or black material, usually reflecting different degrees of hemoglobin metabolism of erythrocytes from the blood meal. An examination of cellular material in the gut of helminths is difficult, because ingesta are often rapidly lysed in the intestinal ceca of worms. Erythrocytes in various phases of digestion have been seen in the intestinal ceca of several CAM-grown trematodes (Fried and Huffman, 1982; Wisnewski *et al.*, 1986). It is probable that intrinsic digestive enzymes from the gastrodermis of CAM-grown worms initiate extracellular digestion. Detailed information on the feeding and digestion of trematodes in normal sites is sparse (Smyth and Halton, 1983) and non-existent for digeneans cultured *in vitro* or *in ovo*.

D. OXYGEN TENSION

Variation must exist in the oxygen tension in the numerous enteric and non-enteric sites occupied by the avian parasites mentioned in Section V.A.

Although the oxygen tensions are not known for most of the sites that harbor avian parasites, it is probable that values for the mouth cavity and nictitating membrane are high (about 100 mmHg) compared to relatively low values in the gut environment (about 0.5–30 mmHg). Moreover, because the avian enteric parasites that have been cultured *in ovo* normally live in association with the intestinal mucosa, the O_2 tension in their gut habitat is probably near 30 mmHg (Barrett, 1981). The CAM is undoubtedly an aerobic site with O_2 tensions of about 60 mmHg (Freeman and Vince, 1974). Parasites grown on the CAM probably adapt well to an aerobic environment with a mean O_2 tension of 60 mmHg, despite normal *in vivo* values that vary from below 30 to above 100 mmHg. Because there is no information on the metabolism of helminths grown in chick embryos, it is not possible to say what metabolic pathways are used by CAM-grown parasites. Parasites grown *in vivo* use a combination of both aerobic and anaerobic metabolic pathways (Barrett, 1981). A similar situation probably exists for CAM-grown worms.

E. CARBON DIOXIDE TENSION

Carbon dioxide plays an important role in the biology of parasites: it initiates activation and trigger phenomena, it is an attractant in some nematodes and catabolism of carbohydrates may involve CO_2 fixation in some parasites (Barrett, 1981). The adult stages of parasites cultivated in chick embryos normally live in diverse habitats (see Section V.A) where the CO_2 tension may vary considerably. Thus, parasites from the mouth or nictitating membrane or intestine of avian hosts may face CO_2 tensions ranging from 20 to 600 mmHg (Barrett, 1981). The venous pCO_2 on the CAM ranges from 10 to 50 mmHg from days 10 to 20 post-embryonation and the arterial pCO_2 ranges from 20 to 60 mmHg for the same time (Romanoff, 1967). The effects of CO_2 tension on parasites cultured *in ovo* need to be determined.

F. OXIDATION–REDUCTION POTENTIAL

Information on redox potential in animals should be used carefully (Barrett, 1981). Redox values play a role in certain activation phenomena, i.e. hatching of eggs, excystation of larvae. Because most parasites implanted into chick embryos have been activated prior to implantation, redox potential may be less significant *in ovo* than *in vivo*. However, the redox rate may play a role in helminth cytochrome chains (Barrett, 1981) and should be examined for a possible function in helminth metabolism. The redox potential of aerobic sites such as the eye and mouth is between -150 and $+200$ mV, whereas intestinal contents tend to have a reduced redox potential of about -100 mV (Barrett, 1981). Redox values of various sites in chick embryos are quite variable (Romanoff, 1967), but tend to be about $+200$ mV. The significance of redox potential to helminths grown *in ovo* remains to be determined.

G. HYDROGEN ION CONCENTRATION

The pH of parasitic habitats can vary considerably. Thus, avian intestinal parasites normally experience gut fluid pH values that are acidic (ca. 4–6), although neutrality may be reached in the lower small intestine and rectum (Duke, 1986). The pH of body fluids and tissue is usually neutral (ca. 6.8–7.4; Barrett, 1981). Therefore, avian eyeflukes and mouth flukes normally experience pH values around neutrality. Although pH values of the CAM are not available, values for the embryo and amniotic fluid 7–17 days post-embryonation are ca. pH 6–8 (Romanoff, 1967). It is probable that the pH of the CAM approaches neutrality. Information on the pH of the CAM infected with parasites is not available. Because parasites often release acidic secretions, it is probable that the pH of infected membranes is different than that of uninfected controls. However, the pH of the CAM could well be buffered by bicarbonate ions from the chick embryo.

H. OSMOTIC PRESSURE

Changes in osmotic pressure may play a role in activation phenomena in the biology of helminth larvae. Larvae placed in embryos are usually activated prior to their implantation, and therefore the role of osmotic pressure in relationship to parasites cultivated in chick embryos may not be important. Osmotic pressure values of embryos from 5 to 15 days post-embryonation are about 270–300 mosmolal and tend to rise to about 350 mosmolal at hatching (Romanoff, 1967). The osmoconcentration of chick embryo tissue and amniotic fluid is similar to avian and mammalian fluids and tissue (about 300 mosmolal). Metacercariae cultured in chick embryos are usually obtained from freshwater snails where the osmotic pressure of the water is less than 10 mosmolal and that of the snail tissue about 40–120 mosmolal (Barrett, 1981). As in most *in vivo* environments, the CAM also presents a hyperosmotic environment to the parasite relative to either the freshwater habitat or the hemolymph of the molluscan intermediate host.

J. TEMPERATURE

The optimal temperature for the cultivation of chicken embryos at a relative humidity of 40–60% is 37.8°C (Freeman and Vince, 1974). Good development can still be achieved within the temperature range 37–40°C and a wide range of relative humidity. Although an oxygen tension of about 160 mmHg (that of air) is usually used during incubation, considerable latitude exists as to the gaseous environment used. Chick embryos can be cultured under variable pO_2, pCO_2, pN or other gaseous environments. The effects of altered gas environments at the optimal development temperature of the embryo can be used in studies on the *in ovo* culture of parasites.

Considerable variation in temperature can be achieved by altering the conditions of incubation (both temperature and humidity controls on the incubator). Helminths have been maintained on CAMs as high as 43°C and as low as 28–30°C (Fried and Foley, 1969; Fried, 1965). As expected, helminths of many poikilothermic hosts will be killed by high temperatures on the CAM (usually those in excess of 32°C). The effects of Δt on worm development on the CAM have not been well-explored. The CAM provides an excellent site to study temperature tolerance of parasites, because older embryos (usually older than 12 days) are quite tolerant to changes in temperature.

K. IMMUNOLOGICAL CONSIDERATIONS

Relatively little information is available on the immunology of non-economically important trematodes grown *in vivo*, and there have been no studies to determine the effects of chick embryo humoral immune substances on trematodes.

Host cellular immune reactions have been observed in the presence of some trematodes grown on the CAM. Spirorchids and schistosomes are walled-off on the CAM within 3 days, but details of the cellular reaction leading to the death of these blood flukes on the CAM are not available (Fried and Tornwall, 1969; Fried *et al.*, 1982). Less marked cellular reactions that do not result in the death of the parasite on the CAM have been reported. Thus, Fried and Nelson (1978) showed the presence of a lesion containing cells and exudate in the CAM infected with *Zygocotyle lunata*. Hypertrophy of the chorionic ectoderm and acetabular worm plugs involving some cell reactions were seen in *Leucochloridiomorpha constantiae*-infected membranes (Fishbein *et al.*, 1985). The most detailed observation to date on cellular phenomena in a parasite–CAM relationship was shown by Huffman *et al.* (1984), who reported cellular inflammation, an acellular eosinophilic response and the presence of heterophils at the interface between *Sphaeridiotrema globulus* and the CAM surface. Eggs from the parasite were also encapsulated on the surface of the CAM.

McGhee *et al.* (1977) examined *Plasmodium gallinaceum in ovo* and stated that neither avian complement nor its components exist in chicken embryos until the time of hatching and that the embryo has little or no ability to form antibodies. Other studies have not corroborated the statements of McGhee *et al.* (1977) in regard to complement or immune substances. According to Gabrielsen *et al.* (1973), complement can be detected in the chick embryo as early as 10 days post-incubation. In quail embryos, *Coturnix coturnix*, the C_3 component of complement was detected as early as 7 days post-embryonation and functional maturation of C_3 (conversion of C_3 to C_{3b}) was present by day 15 (Kai *et al.*, 1985). With regard to the ability of the chick embryo to form antibodies, Kincade and Cooper (1971) showed the presence of IgM in the bursa of Fabricius, cecal tonsil and the spleen of chick embryos by 13, 17 and 19 days post-embryonation, respectively. The development of humoral immunity in the chick embryo has been reviewed by Seto (1981) and Glick (1986).

L. RELATIONSHIP OF EMBRYO DEVELOPMENT TO HELMINTH CULTURE

Numerous events unfold during the development of the chick embryo. These events have been detailed in Section IV and in Table 2. Table 3 provides some of the highlights of development on a temporal basis and speculations on how these events may relate to the *in ovo* cultivation of helminths.

VI. SUMMARY AND CONCLUSIONS

A total of 23 species from 14 families of the Digenea have been studied in chick embryos, mainly on the chick chorioallantoic membrane (CAM). Most species for which cultivation has produced ovigerous adults in chick embryos have been avian digeneans with progenetic metacercariae. Less success has been obtained with the hermaphroditic, non-progenetic, economically important trematodes such as fasciolids and echinostomatids, although post-metacercarial development has been achieved for *Fasciola hepatica* and *Echinostoma trivolvis* (synonym of *E. revolutum*) (see Fried and Butler, 1979; Fried and Pentz, 1983). Success with human or animal blood flukes has been minimal, although adults of *Schistosoma mansoni* and *Spirorchis* spp. (turtle blood flukes) have at least been maintained on the CAM (see Fried *et al.*, 1982; Fried and Tornwall, 1969). Schistosome cercariae and *in vitro* transformed cercariae (schistosomules) should be tested in chick embryos. Marine avian schistosomes in the genera *Austrobilharzia* and *Ornithobilharzia*, along with the freshwater avian schistosome *Trichobilharzia*, would provide useful material for avian embryo studies on non-human schistosomes.

Studies on trematodes in chick embryos have been done mainly to gain basic biological information on these parasites. That is, to identify species for which definitive hosts are not available; for studies on worm-intraspecific variation, growth and development; for studies on worm feeding and digestion; and for studies on worm-mediated chemoattraction and worm site location on the CAM (Fried, 1989).

TABLE 3 *Relationship of embryo development to possible effect on worm cultivation* in ovo

Age of embryo (days)	*Event in the life of the embryo*[a]	*Possible effect on helminths in the embryo*
2	Beginning of hemoglobin synthesis	Worms in embryo often contain metabolized hemoglobin in their intestinal ceca
3	Allantois develops	Will form CAM when fused with chorion; CAM is the most used site for helminth development
4	Embryo surrounded by amnion	Amniotic cavity may serve as potential site for worm development
5	Erythrocytes produced along with synthesis of adult hemoglobins	Erythrocytes and hemoglobin often ingested by helminths
6	Rhythmic contractions of amnion	May help to ventilate CAM and other extraembryonic membranes; useful for gaseous exchange by helminths
9	CAM is fixed in position relative to shell	Maximum expansion of CAM; embryo now well-suited to receive worms
12	Beginning of absorption of albumen	Increased availability of nutrients from chick blood
12	Calcium absorption from shell	Availability of calcium and other nutrients from shell and shell membrane for worms
13	Acceleration of lipid transport from the yolk sac	Increased availability of lipids for worms
16	Bursa of Fabricius secretions become apparent	Possible immunogenic effects on developing worms
20	CAM ectoderm degenerates and CAM circulation declines[b]	CAM no longer available as a site for helminths

[a] Modified from Freeman and Vince (1974).
[b] In studies with trematodes on the CAM, degeneration of the membrane may begin by day 17 (see Fried, 1962a).

Sites other than the upper surface of the CAM have not been well-explored for digeneans, although Irwin and Saville (1988a,b) have examined the subchorioallantois and allantois as habitats for stigeids and microphallids. They have also studied the effects of serum supplements to the embryo on the enhancement of worm growth and development. Irwin and Saville's work should be extended to other helminths. The albumen in the hen's egg is a good site for the development of *Clinostomum marginatum* and *Amblosoma sawaense* (see Larson and Uglem, 1990; Fried *et al.*, 1981), but the reasons for the better growth of these parasites in the albumen than on the CAM are not known. The inoculation of trematode larvae into CAM blood vessels, the yolk sac, the amnion and the embryo proper have not been explored and may provide useful avenues of research.

Only a single study has been done with a monogenean trematode, *Polystomoides* sp. (see Fried, 1965), in which worms were grown on the CAM at 30°C. The study provided evidence of blood-feeding and intra-specific variation of a monogenean and should provide an impetus for other work on the monogenetic trematodes in chick embryos.

The easiest technique to use to prepare eggs for helminth transplant studies is the Zwilling (1959) procedure as modified by Fried (1973), whereby eggs are opened on day 3 and worms implanted on the CAM on days 6–8. The simplicity of this technique should favor its use over the more difficult Woodruff and Goodpasture (1931) procedure as described by Fried (1962a) and used recently by Leno and Holloway (1986) and Larson and Uglem (1990). Once helminthologists master the Zwilling procedure for opening fertile eggs, they will not return to the more difficult Woodruff and Goodpasture method. The technique of Auerbach *et al.* (1974) for the *in vitro* cultivation of chick embryos is simple to use, but requires large numbers of eggs because of heavy post-operative losses. The technique has the obvious advantage of allowing the helminthologist to examine worms on a flat CAM in the absence of a shell. It is a particularly useful procedure for studies on worm feeding and behavior.

Little information is available on the use of chick embryos to culture cestodes, other than Parmeter and Gemmell's (1974) study on the development of onchospheres of *Taenia hydatigera* to metacestodes and the brief observations on pseudophyllidean larval development by Smyth (1958, 1959) and Mueller (1966). Because hymenolepid and other cylophyllidean cysticercoids are relatively easy to excyst *in vitro* (Rothman, 1959), larval tapeworms may provide interesting material for study. Likewise, nematode work in chick embryos is sparse except for the post-larval development study of *Trichinella spiralis* (see McCoy, 1936) and that of *Syngamus trachea* (see Winward, 1976). Further studies using larval and adult nematodes are needed.

The chick embryo has not been well-explored in medical or veterinary helminthology as a site for anthelmintic studies, although the effects of X-irradiation and antibiotic tolerance on the brachylaimid trematode, *Leucochloridiomorpha constantiae*, have been examined (Fried, 1989). The potential to test anthelmintics on worms grown in chick embryos is an untapped resource.

This chapter has examined the significant literature on the development, structure and function of chick extraembryonic membranes along with the physicochemical characteristics of the developing chick embryo; wherever possible, we have attempted to relate developmental events in the embryo to situations that might exist in the normal definitive sites of helminths that have been cultured in chick embryos.

It is apparent that information on the histopathology, pathobiochemistry and immunobiology of chick embryos infected with helminths is minimal. Hopefully, this chapter will stimulate further studies on the use of the chick embryo for basic and applied studies in helminthology.

References

Auerbach, R., Kuban, L., Knighton, D. and Folkman, J. (1974). A simple procedure for the long-term cultivation of chicken embryos. *Developmental Biology* **41,** 391–394.

Ausprunk, D. H. (1982). Synthesis of glycoproteins by endothelial cells in embryonic blood vessels. *Developmental Biology* **90,** 79–90.

Ausprunk, D. H. (1986). Distribution of hyaluronic acid and sulfated glycosaminoglycans during blood-vessel development in the chick chorioallantoic membrane. *American Journal of Anatomy* **177,** 313–331.

Ausprunk, D. H., Knighton, D. R. and Folkman, J. (1974). Differentiation of vascular endothelium in the chick chorioallantois: A structural and autoradiographic study. *Developmental Biology* **38,** 237–248.

Austin, R. M. and Fried, B. (1972a). Effects of penicillin and streptomycin on *in vitro* survival of the metacercaria of *Leucochloridiomorpha constantiae* (Mueller, 1935) (Trematoda) and on its development on the chick chorioallantois. *Proceedings of the Helminthological Society of Washington* **39,** 262–263.

Austin, R. M. and Fried, B. (1972b). Survival, wound healing, and development in laboratory-injured trematode metacercariae of *Leucochloridiomorpha constantiae. Proceedings of the Helminthological Society of Washington* **39,** 258–261.

Barrett, J. (1981). "Biochemistry of Parasitic Helminths". University Park Press, Baltimore, Maryland.

Bauman, R., Fischer, J. and Engelke, M. (1987). Functional properties of primitive and definitive red cells from chick embryo: Oxygen-binding characteristics, pH and membrane potential, and responses to hypoxia. *Journal of Experimental Zoology* **1,** supplement 1, 227–238.

Berntzen, A. K. and Macy, R. W. (1969). *In vitro* cultivation of the digenetic trematode *Sphaeridiotrema globulus* (Rudolphi) from the metacercarial stage to egg production. *Journal of Parasitology* **55,** 136–139.

Bissonnette, J. M. and Metcalfe, J. (1978). Gas exchange of the fertile hen's egg: Components of resistance. *Respiration Physiology* **34,** 209–218.

Blaškovič, D. and Styk, B. (1967). Laboratory methods of virus transmission in multicellular organisms. *In* "Methods in Virology" (K. Maramorosch and H. Koprowski, eds), pp. 163–233. Academic Press, London and San Diego.

Borysko, E. and Bang, F. B. (1953). The fine structure of the chorio-allantoic membrane of the normal chick embryo. *Bulletin of the Johns Hopkins Hospital* **92,** 257–289.

Brackett, S. and Beckman, A. J. (1942). The fate of some species of schistosome cercariae in chick embryos. *American Journal of Hygiene* **36,** 216–223.
Bruns, G. A. P. and Ingram, V. M. (1973). Erythropoiesis in the developing chick embryo. *Developmental Biology* **30,** 455–459.
Buddingh, G. J. (1952). Chick-embryo techniques. *In* "Viral and Rickettsial Infections of Man" (T. M. Rivers, ed.), 2nd edn, pp. 109–125. J. B. Lippincott, Philadelphia.
Burton, F. G. and Tullet, S. G. (1985). Respiration of avian embryos. *Comparative Biochemistry and Physiology* **82A,** 735–744.
Burton, G. J. and Palmer, N. E. (1989). The chorioallantoic capillary plexus of the chicken egg: A microvascular corrosion casting study. *Scanning Microscopy* **3,** 549–558.
Campbell, G. LeM., Weintraub, H., Mayall, B. H. and Holtzer, H. (1971). Primitive erythropoiesis in early chick embryogenesis. II. Correlation between hemoglobin synthesis and the mitotic history. *Journal of Cell Biology* **50,** 669–681.
Clark, H. and Fischer, D. (1957). A reconsideration of nitrogen excretion by the chick embryo. *Journal of Experimental Zoology* **136,** 1–15.
Clegg, J. A. and Smith, M. A. (1987). Trematoda. In "*In Vitro* Methods for Parasite Cultivation" (A. E. R. Taylor and J. R. Baker, eds), pp. 254–281. Academic Press, London and San Diego.
Coleman, J. R. and Terepka, A. R. (1972a). Fine structural changes associated with the onset of calcium, sodium and water transport of the chick chorio-allantoic membrane. *Journal of Membrane Biology* **7,** 111–127.
Coleman, J. R. and Terepka, A. R. (1972b). Electron probe analysis of the calcium distribution in cells of the embryonic chick chorioallantoic membrane. II. Demonstration of intracellular location during active transcellular transport. *Journal of Histochemistry and Cytochemistry* **20,** 414–424.
Conrad, R. M. and Scott, H. M. (1938). The formation of the egg of the domestic fowl. *Physiological Reviews* **18,** 481–494.
Danchakoff, V. (1917). The position of the respiratory vascular net in the allantois of the chick. *Americal Journal of Anatomy* **21,** 407–413.
Danchakoff, V. (1918). Equivalence of different hematopoietic anlages [*sic*]. (By method of stimulation of their stem-cells). II. Grafts of adult spleen on the allantois and response of the allantoic tissues. *American Journal of Anatomy* **24,** 127–190.
Davies, C. and Smyth, J. D. (1978). *In vitro* cultivation of *Fasciola hepatica* metacercariae and of partially developed flukes recovered from mice. *International Journal for Parasitology* **8,** 125–131.
DeFouw, D. O., Rizzo, V. J., Steinfeld, R. and Feinberg, R. N. (1989). Mapping of the microcirculation in the chick chorioallantoic membrane during normal angiogenesis. *Microvascular Research* **38,** 136–147.
Duke, G. E. (1986). Alimentary canal: Secretion and digestion, special digestive functions, and absorption. *In* "Avian Physiology" (P. D. Sturkie, ed.), 4th edn, pp. 289–302. Springer-Verlag, New York.
Dunn, B. E. and Fitzharris, T. P. (1979). Differentiation of the chorionic epithelium of chick embryos maintained in shell-less culture. *Developmental Biology* **71,** 216–227.
Dunn, B. E. and Fitzharris, T. P. (1987). Endocytosis in the embryonic chick chorionic epithelium. *Journal of Experimental Zoology* **1,** supplement 1, 75–79.

Dunn, B. E., Clark, N. B. and Scharf, K. E. (1987). Effect of calcium supplementation on growth of shell-less cultured chick embryos. *Journal of Experimental Zoology* **1,** supplement 1, 33–37.

Ebert, J. D. (1959). The formation of muscle and muscle-like elements in the chorioallantoic membrane following inoculation of a mixture of cardiac microsomes and Rous sarcoma virus. *Journal of Experimental Zoology* **142,** 587–613.

Feeney, R. E. and Allison, R. G. (1969). "Evolutionary Biochemistry of Proteins". Interscience, New York.

Ferguson, M. S. (1940). Excystment and sterilization of metacercariae of the avian strigeid trematode *Posthodiplostomum minimum*, and their development into adult worms in sterile culture. *Journal of Parasitology* **26,** 359–372.

Fishbein, J. M., Fried, B. and Stableford, L. T. (1985). Histopathology of chick extraembryonic membranes experimentally infected by *Leucochloridiomorpha constantiae* (Trematoda). *Transactions of the American Microscopical Society* **104,** 267–271.

Fisher, J. R. (1967). Patterns of nitrogen metabolism and excretion in some vertebrate embryos. *In* "The Biochemistry of Animal Development" (R. Weber, ed.), pp. 413–436. Academic Press, London and San Diego.

Fisher, J. R. and Eakin, R. E. (1957). Nitrogen excretion in developing chick embryos. *Journal of Embryology and Experimental Morphology* **5,** 215–224.

Font, W. F. (1980). *Amblosoma suwaense* (Trematoda : Brachylaimidae : Leucochloridiomorphinae) from *Campeloma decisum* in Wisconsin. *Journal of Parasitology* **66,** 861–862.

Font, W. F. and Wittrock, D. D. (1980). Scanning electron microscopy of *Leucochloridiomorpha constantiae* during development from metacercariae to adult. *Journal of Parasitology* **66,** 955–964.

Fontaine, J. and Le Douarin, N. M. (1977). Analysis of endoderm formation in the avian blastoderm by the use of quail-chick chimaeras. *Journal of Embryology and Experimental Morphology* **41,** 209–222.

Fraser, R. C. (1966). The rate of hemoglobin chain formation in developing chick embryos. *Experimental Cell Research* **44,** 195–200.

Fraser, R., Horton, B., Dupourque, D. and Chernoff, A. (1972). The multiple hemoglobins of the chick embryo. *Journal of Cellular Physiology* **80,** 79–88.

Freeman, B. M. and Vince, M. A. (1974). "Development of the Avian Embryo". Chapman and Hall, London.

Fried, B. (1962a). Growth of *Philophthalmus* sp. (Trematoda) on the chorioallantois of the chick. *Journal of Parasitology* **48,** 545–550.

Fried, B. (1962b). Growth of *Philophthalmus* sp. (Trematoda) in the eyes of chicks. *Journal of Parasitology* **48,** 395.

Fried, B. (1965). Transplantation of a monogenetic trematode, *Polystomoides* sp., to the chick chorioallantois. *Journal of Parasitology* **51,** 983–986.

Fried, B. (1969). Transplantation of trematodes to the chick chorioallantois. *Proceedings of the Pennsylvania Academy of Sciences* **43,** 232–234.

Fried, B. (1970). Excystation of metacercariae of *Posthodiplostomum minimum minimum* Hoffman, 1958, and their development in the chick and on the chorioallantois. *Journal of Parasitology* **56,** 944–946.

Fried, B. (1973). The use of 3-day-old chick embryos for the cultivation of *Leucochloridium variae* McIntosh, 1932 (Trematoda). *Journal of Parasitology* **59,** 591–592.

Fried, B. (1978). Trematoda. *In* "Methods of Cultivating Parasites *in Vitro*" (A. Taylor and J. R. Baker, eds), pp. 151–192. Academic Press, London and San Diego.
Fried, B. (1981). Thermal activation and inactivation of the metacercaria of *Philophthalmus hegeneri* Penner and Fried, 1963. *Zeitschrift für Parasitenkunde* **65,** 359–360.
Fried, B. (1986). Chemical communication in hermaphroditic digenetic trematodes. *Journal of Chemical Ecology* **12,** 1659–1677.
Fried, B. (1989). Cultivation of trematodes in chick embryos. *Parasitology Today* **5,** 3–5.
Fried, B. and Bradford, J. D. (1984). Histochemical and thin-layer chromatographic analyses of neutral lipids in various host sites infected with *Leucochloridiomorpha constantiae* (Trematoda). *Comparative Biochemistry and Physiology* **78B,** 175–177.
Fried, B. and Butler, M. S. (1978). Infectivity, excystation and development on the chick chorio-allantois of the metacercaria of *Echinostoma revolutum* (Trematoda). *Journal of Parasitology* **64,** 175–177.
Fried, B. and Butler, C. S. (1979). Excystation, development on the chick chorio-allantois and neutral lipids in the metacercaria of *Fasciola hepatica* (Trematoda). *Revista Ibérica de Parasitologia* **79,** 395–400.
Fried, B. and Caruso, J. H. (1970). Histologic observations on the intestine of *Echinostoma revolutum* and its ingesta. *Transactions of the American Microscopical Society* **89,** 110–112.
Fried, B. and Contos, N. (1973). *In vitro* cultivation of *Leucochloridiomorpha constantiae* (Trematoda) from the metacercaria to the ovigerous adult. *Journal of Parasitology* **59,** 936–937.
Fried, B. and Davis, J. R. (1972). Infectivity and development of X-irradiated metacercariae of *Leucochloridiomorpha constantiae* (Trematoda) in the chick and the chorioallantois. *Transactions of the American Microscopical Society* **91,** 208–212.
Fried, B. and Diaz, V. (1987). Site-finding and pairing of *Echinostoma revolutum* (Trematoda) on the chick chorioallantois. *Journal of Parasitology* **73,** 546–548.
Fried, B. and Foley, D. A. (1969). Temperature tolerance of *Echinostoma revolutum* on the chick chorioallantois. *Journal of Parasitology* **55,** 982–984.
Fried, B. and Foley, D. A. (1970). Development of *Clinostomum marginatum* (Trematoda) from frogs in the chick and on the chorioallantois. *Journal of Parasitology* **56,** 332–335.
Fried, B. and Fujino, T. (1984). Scanning electron microscopy of *Echinostoma revolutum* (Trematoda) during development in the chick embryo and the domestic chick. *International Journal for Parasitology* **14,** 75–81.
Fried, B. and Groman, G. (1985). Cultivation of the cercaria of *Himasthla quissetensis* (Trematoda) on the chick chorioallantois. *International Journal for Parasitology* **15,** 219–223.
Fried, B. and Holmes, M. (1979). Further studies on the development of *Leucochloridiomorpha constantiae* (Trematoda) metacercariae on the chick chorioallantois. *Proceedings of the Helminthological Society of Washington* **46,** 70–73.
Fried, B. and Huffman, J. F. (1982). Excystation and development in the chick and on the chick chorioallantois of *Sphaeridiotrema globulus* (Trematoda). *International Journal for Parasitology* **12,** 427–431.
Fried, B. and Kramer, M. D. (1968). Histochemical glycogen studies on *Echinostoma revolutum*. *Journal of Parasitology* **54,** 942–944.

Fried, B. and Mishkind, S. H. (1985). Scanning electron microscopy of *Leucochloridiomorpha constantiae* (Trematoda) from the chick chorioallantois and bursa of Fabricius. *Transactions of the American Microscopical Society* **104,** 100–103.

Fried, B. and Nelson, P. D. (1978). Development of *Zygocotyle lunata* (Trematoda) metacercariae on the chick chorioallantois. *Transactions of the American Microscopical Society* **97,** 402–405.

Fried, B. and Pentz, L. (1983). Cultivation of excysted metacercariae of *Echinostoma revolutum* (Trematoda) in chick embryos. *International Journal for Parasitology* **13,** 219–223.

Fried, B. and Ramundo, G. P. (1987). Excystation and cultivation *in vitro* and *in ovo* of *Cyathocotyle bushiensis* (Trematoda) metacercariae. *Journal of Parasitology* **73,** 541–545.

Fried, B. and Roberts, T. (1972). Pairing of *Leucochloridiomorpha constantiae* (Mueller, 1935) (Trematoda) *in vitro*, in the chick and on the chorioallantois. *Journal of Parasitology* **58,** 88–91.

Fried, B. and Tornwall, R. E. (1969). Survival and egg laying of turtle blood flukes (Trematoda : Spirorchiidae) on the chick chorioallantois. *Proceedings of the Helminthological Society of Washington* **36,** 86–88.

Fried, B. and Weaver, L. J. (1969). Cultivation of a frog lung fluke, *Haematoloechus* sp., on the chick chorioallantois. *Turtox News* **47,** 39–40.

Fried, B., Weaver, L. J. and Kramer, M. D. (1968). Cultivation of *Echinostoma revolutum* (Trematoda) on the chick chorioallantois. *Journal of Parasitology* **54,** 939–941.

Fried, B., Foley, D. A. and Kern, K. D. (1970). *In vitro* and *in vivo* excystation of *Clinostomum marginatum* (Trematoda) metacercariae. *Proceedings of the Helminthological Society of Washington* **37,** 222.

Fried, B., Barber, L. W. and Butler, M. S. (1978). Growth and development of the tetracotyle of *Cotylurus strigeoides* (Trematoda) in the chick, on the chorioallantois and *in vitro*. *Proceedings of the Helminthological Society of Washington* **45,** 162–166.

Fried, B., Fine, R. H. and Felter, B. L. (1980). Growth, development and pairing of *Leucochloridiomorpha constantiae* (Trematoda) metacercariae on the chorioallantois of chick embryos cultivated *in vitro*. *Parasitology* **81,** 41–45.

Fried, B., Heyer, B. L. and Pinski, A. K. (1981). Cultivation of *Amblosoma suwaense* (Trematoda : Bachylaimidae) in chick embryos. *Journal of Parasitology* **67,** 50–52.

Fried, B., Eveland, L. K. and Cohen, L. M. (1982). Preliminary observations on survival and pairing of *Schistosoma mansoni* on the chick chorioallantois. *Proceedings of the Helminthological Society of Washington* **49,** 330–331.

Fried, B., LeFlore, W. B. and Bass, H. S. (1985). Histochemical localization of alkaline phosphatase activity in *Leucochloridiomorpha constantiae* (Trematoda) cultivated on the chick chorioallantois. *Journal of Parasitology* **71,** 510–512.

Fülleborn, F. (1895). Beiträge zur Entwicklung der Allantois der Vogel. Inaugural dissertation, Berlin.

Gabrielsen, A. E., Pickering, R. J., Linna, T. J. and Good, R. A. (1973). Haemolysis in chicken serum. II. Ontogenetic development. *Immunology* **25,** 179–184.

Ganote, C. E., Beaver, D. L. and Moses, H. L. (1964). Ultrastructure of the chick chorioallantoic membrane and its reaction to inoculation trauma. *Laboratory Investigation* **13,** 1575–1589.

Garrison, J. C. and Terepka, A. R. (1972). Calcium-stimulated respiration and active calcium transport in the isolated chorioallantoic membrane. *Journal of Membrane Biology* **7,** 128–145.

Glick, B. (1986). Immunophysiology. *In* "Avian Physiology" (P. D. Sturkie, ed.), 4th edn, pp. 87–101. Springer-Verlag, New York.

Harris, K. R., Fried, B. and Mayer, D. H. (1972). Infectivity, growth and development of *Leucochloridiomorpha constantiae* (Trematoda) in the chick and on the chorioallantois. *Journal of Parasitology* **58,** 213–216.

Huffman, J. E. and Fried, B. (1990). *Echinostoma* and echinostomiasis. *Advances in Parasitology* **29,** 215–269.

Huffman, J. E., Fried, B., Roscoe, D. E. and Cali, A. (1984). Comparative pathologic features and development of *Sphaeridiotrema globulus* (Trematoda) infections in the mute swan and domestic chicken chorioallantois. *American Journal of Veterinary Research* **45,** 187–391.

Huxley, J. S. and Murray, P. D. F. (1924). A note on the reactions of chick chorioallantois to grafting. *Anatomical Record* **28,** 385–389.

Irwin, S. W. B. and Saville, D. H. (1988a). An alternative method for the culture of *Diplostomum spathaceum* (Trematoda) in chick embryos. *Journal of Parasitology* **74,** 504–505.

Irwin, S. W. B. and Saville, D. H. (1988b). Cultivation and development of *Microphallus pygmaeus* (Trematoda : Microphallidae) in fertile chick eggs. *Parasitology Research* **74,** 396–398.

Irwin, S. W. B., Saville, D. H. and Chubb, J. C. (1989). Cultivation and development of *Diplostomum* metacercariae on chick embryonic membranes. *Program and Abstracts of the British Society for Parasitology*, p. 31, University of Southampton.

Juurlink, B. H. J. and Gibson, M. A. (1973). Histogenesis of the yolk sac in the chicken. *Canadian Journal of Zoology* **51,** 509–519.

Kai, C., Yoshikawa, K., Yamanouchi, K., Okada, H. and Morikawa, S. (1985). Ontogeny of the third component of complement of Japanese quails. *Immunology* **54,** 463–470.

Kannangara, D. W. W. and Smyth, J. D. (1974). *In vitro* cultivation of *Diplostomum spathaceum* and *Diplostomum phoxini* metacercariae. *International Journal for Parasitology* **4,** 667–673.

Kessel, J. and Fabian, B. C. (1985). Graded morphogenetic patterns during the development of the extraembryonic blood system and coelom of the chick blastoderm: A scanning electron microscope and light microscope study. *American Journal of Anatomy* **173,** 99–112.

Kincade, P. W. and Cooper, M. D. (1971). Development and distribution of immunoglobin-containing cells in the chicken. *Journal of Immunology* **106,** 371–376.

Kobayashi, I., Kondon, Y., Shimizu, K. and Tanaka, K. (1989). A role of secreted proteinase of *Candida albicans* for the invasion of chick chorioallantoic membrane. *Microbiology and Immunology* **33,** 709–719.

Lambson, R. O. (1970). An electron microscopic study of the entodermal cells of the yolk sac of the chick during incubation and after hatching. *American Journal of Anatomy* **129,** 1–20.

Larson, O. R. and Uglem, G. L. (1990). Cultivation of *Clinostomum marginatum* (Digenea : Clinostomatidae) metacercariae *in vitro*, in chick embryo and in mouse coelom. *Journal of Parasitology* **76,** 505–508.

La Rue, G. R. (1957). The classification of digenetic Trematoda: A review and a new system. *Experimental Parasitology* **6,** 306–349.

Leeson, T. A. and Leeson, C. R. (1963). The chorio-allantois of the chick: Light and electron microscopic observations at various times of incubation. *Journal of Anatomy* **97,** 585–595.

Leighton, J., Nassauer, J. and Tchao, R. (1985). The chick embryo in toxicology: An alternative to the rabbit eye. *Food and Chemical Toxicology* **23,** 293–298.

Leno, G. H. and Holloway, H. L. Jr (1986). The culture of *Diplostomum spathaceum* metacercariae on the chick chorioallantois. *Journal of Parasitology* **72,** 555–558.

Lewis, P. D. (1974). Helminths of terrestrial molluscs in Nebraska. II. Life cycle of *Leucochloridium variae* McIntosh, 1932 (Digenea: Leucochloridiidae). *Journal of Parasitology* **60,** 251–255.

Lillie, F. R. (1952). "Development of the Chick: An Introduction to Embryology", 3rd edn (revised by H. L. Hamilton). Henry Holt, New York.

Long, P. L. (1978). Chicken embryos. *In* "Methods of Cultivating Parasites *in Vitro*" (A. E. R. Taylor and J. R. Baker, eds), pp. 129–147. Academic Press, London and San Deigeo.

Madsen, L. R. and Johnson, A. D. (1974). Development of *Alaria* mesocercariae on the chick chorioallantoic membrane and in mice and chicks. *Transactions of the American Microscopical Society* **93,** 106–112.

McCoy, O. R. (1936). The development of trichinae in abnormal environments. *Journal of Parasitology* **22,** 54–59.

McGhee, R. B., Dingh, S. D. and Lushbaugh, W. B. (1977). *Plasmodium gallinaceum*: Changing virulence patterns of malaria parasites during adaptation from neonate chick to chicken embryo. *Experimental Parasitology* **43,** 220–230.

McLachlan, J. C. and Phoplonker, M. H. (1988). Limb reduplicating effects of chorio-allantoic membrane and its components. *Journal of Anatomy* **158,** 147–156.

Moscona, A. (1959). Squamous metaplasia and keratinization of chorionic epithelium of the chick embryo in egg and in culture. *Developmental Biology* **1,** 1–23.

Moscona, A. and Carneckas, Z. I. (1959). Etiology of keratogenic metaplasia in the chorioallantoic membrane. *Science* **129,** 1743–1744.

Mueller, J. F. (1966). Host–parasite relationships as illustrated by the cestode *Spirometra mansonoides*. *In* "Host–Parasite Relationships" (J. E. Macauley, ed.), pp. 15–58. Oregon State Unversity Press, Corvallis.

Narbaitz, R. (1977). Structure of the intra-chorionic blood sinus in the chick embryo. *Journal of Anatomy* **124,** 347–354.

Narbaitz, R. (1987). Role of vitamin D in the development of the chick embryo. *Journal of Experimental Zoology* **1,** supplement 1, 15–23.

Narbaitz, R., Kacew, S. and Sitwell, L. (1981). Carbonic anhydrase activity in the chick embryo chorioallantois: Regional distribution and vitamin D distribution. *Journal of Embryology and Experimental Morphology* **65,** 127–137.

Needham, J. (1931). "Chemical Embryology". Cambridge University Press, Cambridge.

Needham, J. (1942). "Biochemistry and Morphogenesis". Cambridge University Press, Cambridge.

Owczarzak, A. (1971). Calcium-absorbing cell of the chick chorioallantoic membrane. I. Morphology, distribution and cellular interactions. *Experimental Cell Research* **68,** 113–129.

Pácha, J., Ujec, E., Popp, M. and Capek, K. (1985). Sodium transport and electrical properties of the chick chorioallantoic membrane. *General Physiology and Biophysics* **4,** 367–374.

Pappas, P. W. (1988). The relative roles of the intestines and external surfaces in the nutrition of monogeneans, digeneans and nematodes. *Parasitology* **96,** 5105–5121.

Parish, W. E. (1985). Ability of *in vitro* (corneal injury—eye organ—and chorio-allantoic membrane) tests to represent histopathological features of acute eye inflammation. *Food and Chemical Toxicology* **23,** 215–227.

Parmeter, S. N. and Gemmell, M. A. (1974). The chick embryo as an experimental host for developing larvae of *Taenia hydatigera. Journal of Parasitology* **60,** 1048–1049.

Paul, J. (1960). "Cell and Tissue Culture", 2nd edn. Williams and Wilkins, Baltimore, Maryland.

Pipkin, A. C. and Jensen, D. V. (1958). Avian embryos and tissue culture in the study of parasitic protozoa. I. Malarial parasites. *Experimental Parasitology* **7,** 491–530.

Rangan, S. R. S. and Sirsat, S. M. (1962). The fine structure of the normal chorio-allantoic membrane of the chick embryo. *Quarterly Journal of Microscopical Science* **103,** 17–23.

Rich, R. R., Rogers, D. K. and Leaders, F. E. (1965). Mesenchymal metaplasia of the chick chorioallantoic membrane: A non-specific response to selected stimuli. *Experimental Cell Research* **40,** 96–103.

Romanoff, A. L. (1960). "The Avian Embryo". Macmillan, New York.

Romanoff, A. L. (1967). "Biochemistry of the Avian Embryo". John Wiley, New York.

Romanoff, A. L. and Hayward, F. W. (1943). Changes in volume and physical properties of allantoic and amniotic fluids under normal and extreme temperatures. *Biological Bulletin* **84,** 141–147.

Romanoff, A. L. and Romanoff, A. J. (1949). "The Avian Egg". John Wiley, New York.

Rothman, A. H. (1959). Studies on the excystment of tapeworms. *Experimental Parasitology* **8,** 336–364.

Rugh, R. (1962). "Experimental Embryology", 3rd edn. Burgess, Minneapolis.

Saleuddin, A. S. M., Kyriakides, C. P. M., Peacock, A. and Simkiss, K. (1976). Physiological and ultrastructural aspects of ion movements across the chorio-allantois. *Comparative Biochemistry and Physiology* **54A,** 7–12.

Schalekamp, M., Schalekamp, M., Van Goor, D. and Slingerland, R. (1972). Re-evaluation of the presence of multiple hemoglobins during the ontogenesis of the chicken. *Journal of Embryology and Experimental Morphology* **28,** 681–713.

Sethi, N. and Brookes, M. (1971). Ultrastructure of the blood vessels in the chick allantois and chorioallantois. *Journal of Anatomy*. **109,** 1–15.

Seto, F. (1981). Early development of the avian immune system. *Poultry Science* **60,** 1981–1985.

Shimazu, T. (1974). A new digenetic trematode, *Amblosoma suwaense* sp. nov., the morphology of its adult and metacercariae (Trematoda, Brachylaimidae). *Japanese Journal of Parasitology* **23,** 100–105.

Skalinsky, E. I. (1965). Functional morphology of the intercalated cells of the chick embryo chorio-allantois. *Doklady Akademe Nauk, USSR* **161,** 730–732.

Smyth, J. D. (1958). Cultivation and development of larval cestode fragments *in vitro. Nature* **181,** 1119–1122.

Smyth, J. D. (1959). Maturation of larval pseudophyllidean cestodes and strigeid trematodes under axenic conditions. The significance of nutritional levels in platyhelminth development. *Annals of the New York Academy of Sciences* **77,** 189–238.

Smyth, J. D. and Halton, D. W. (1983). "The Physiology of Trematodes", 2nd edn. Cambridge University Press, Cambridge.

Stephenson, N. G. and Tompkins, J. K. N. (1965). Transplantation of embryonic cartilage and bone onto the chorioallantois of the chick. *Journal of Embryology and Experimental Morphology* **12,** 825–839.

Stern, C. D. and Canning, D. R (1990). Origin of cells giving rise to mesoderm and endoderm. *Nature (London)* **343,** 273–275.

Stewart, M. E. and Terepka, A. R. (1969). Transport functions of the chick chorioallantoic membrane. I. Normal histology and evidence for active transport from the allantoic fluid *in vivo*. *Experimental Cell Research* **58,** 93–106.

Tazawa, H. (1980). Oxygen and CO_2 exchange and acid–base regulation in the avian embryo. *American Zoologist* **20,** 395–404.

Terepka, A. R., Stewart, M. E. and Merkel, N. (1969). Transport functions of the chick chorioallantoic membrane. II. Active calcium transport *in vitro*. *Experimental Cell Research* **58,** 107–117.

Tuan, R. S. (1987). Mechanism and regulation of calcium transport by the chick embryonic chorioallantoic membrane. *Journal of Experimental Zoology* **1,** supplement 1, 1–13.

Weintraub, H., Campbell, G. LeM. and Holtzer, H. (1971). Primitive erythropoiesis in early chick embryogenesis. I. Cell cycle kinetics and the control of cell division. *Journal of Cell Biology* **50,** 652–668.

Williams, J. (1967). Yolk utilization. *In* "The Biochemistry of Animal Development" (R. Weber, ed.), pp. 341–382. Academic Press, London and San Diego.

Willier, B. H. (1927). The specificity of sex, of organization, and of differentiation of embryonic chick gonads as shown by grafting experiments. *Journal of Experimental Zoology* **46,** 409–465.

Winward, L. D. (1976). *Syngamus trachea*: Infections produced by prenatal inoculations of chicken embryos. *Experimental Parasitology* **40,** 74–76.

Wisnewski, N., Fried, B. and Halton, D. W. (1986). Growth and feeding of *Echinostoma revolutum* on the chick chorioallantois and in the domestic chick. *Journal of Parasitology* **72,** 684–689.

Woodruff, A. M. and Goodpasture, E. W. (1931). The susceptibility of the chorioallantoic membrane of chick embryos to infection with the fowl-pox virus. *American Journal of Pathology* **7**, 209–222.

Zwilling, E. (1946). Regulation in the chick allantois. *Journal of Experimental Zoology* **101**, 445–454.

Zwilling, E. (1959). A modified chorioallantoic grafting procedure. *Transplantation Bulletin* **6,** 115–116.

Infection Characteristics of *Schistosoma japonicum* in Mice and Relevance to the Assessment of Schistosome Vaccines

G. F. MITCHELL

The Walter and Eliza Hall Institute of Medical Research, Melbourne, Victoria 3050, Australia

W. U. TIU AND E. G. GARCIA

College of Public Health, University of the Philippines, Manila, Ermita 1000, Philippines

I. Introduction

Over the past few years and generally through application of gene cloning techniques, a large number of schistosome proteins, or portions of proteins, have become available. These have found use in a whole range of immunobiological studies including serology and the assessment of vaccinating efficacy. Animal species used for vaccine assessment include inbred mice and rats, monkeys and, to a lesser extent, rabbits and guinea pigs, with the bulk of experiments being performed with *Schistosoma mansoni*. The mouse and rat models of *S. mansoni* and/or *S. japonicum* find wide application because of the ready availability at low cost of genetically defined mouse and rat

ADVANCES IN PARASITOLOGY VOL. 30
ISBN 0-12-031730-3

strains, the potential for detailed immunological dissection of vaccine-based resistance, and the possibility of identifying clearly the site(s) of worm attrition in resistant individuals. Infection characteristics in these small laboratory hosts are generally regarded to be reasonably well defined; from examination of the extensive literature, the term "reasonably" is appropriate.

Several groups have pursued the objective of a defined-antigen vaccine against *S. japonicum* with emphasis on mice for assessment of vaccinating effects of antigens expressed in *Escherichia coli*. In our own laboratories, attention has been directed towards *E. coli*-derived proteins of *S. japonicum* (Philippines) (Saint *et al.*, 1986; Smith *et al.*, 1986; Mitchell *et al.*, 1988; Henkle *et al.*, 1990) that may induce resistance against initial or continuing infection or resistance against severe immunopathologic disease. The latter approach involves sensitization to induce anti-embryonation immunity in which immune responses inhibit maturation of the egg and accelerate destruction of the source of immunopathological antigens (i.e. the mature, miracidium-containing egg) (Garcia and Mitchell, 1982, 1987; Mitchell, 1990a). The importance of the mouse model for all these studies has increased the need to clearly identify characteristics of infection with Philippines isolates of *S. japonicum* and to identify any "peculiarities" relative to other geographical isolates ("strains") of this human schistosome, and relative to the better studied parasite, *S. mansoni*.

II. Mouse Infections with *S. japonicum*

Parasites designated as "*Schistosoma japonicum*" are very different from *S. mansoni* and several geographical strains are recognized—Chinese, Philippines, Japanese, Formosan and Indonesian. These strains differ in the subspecies of *Oncomelania hupensis* snail intermediate host required and their ability to "cross-infect" across these oncomelanian subspecies (Chiu, 1968; Chi *et al.*, 1971; DeWitt, 1954; Li Hsu and Hsu, 1960, 1968). Differences have been reported in the pathogenicity (Swanson and Williams, 1963; Warren and Berry, 1972), drug susceptibility (Li Hsu *et al.*, 1963) and respiration of the miracidia (Bruce *et al.*, 1971) of these "national" *S. japonicum* strains. Even within *S. japonicum* (Philippines), isoenzyme differences have been well documented (Woodruff *et al.*, 1987; Merenlender *et al.*, 1987). Preliminary data indicate that sequence differences may exist in the genes of the Mr 28 000 glutathione *S*-transferase enzymes of the Mindoro and Sorsogon geographic strains of *S. japonicum* (Philippines) (K. M. Davern and K. J. Henkle, unpublished observations). Geographic strains of *S. japonicum* (Chinese) have been shown to differ in their ability to infect one particular geographic strain of *O. h. hupensis* snails (Xu and Ni, 1987).

Comparative studies by Hsu and Li Hsu (1960a) established the high infectivity for mice of Philippine and Japanese isolates of *S. japonicum* relative to Chinese and Formosan isolates. Infection with a single male plus a single female cercaria was more successful (50–60% *vs* 30%) and the proportion of adult worms establishing after exposure to 50 male plus 50 female cercariae was approximately twice as great. In our own laboratories using *S. japonicum* (Philippines), exposure of mice at the College of Public Health (CPH), to 20–25 cercariae of Mindoro isolates results routinely in at least 50% establishing as adult worms, but at the Walter and Eliza Hall Institute (WEHI), a comparable number of cercariae from a Sorsogon isolate maintained in the laboratory results in around 30% establishing as adults in the portal system. In a recent comparative study (Table 1), mice were infected at CPH with cercariae from Mindoro field snails or from a laboratory (WEHI)-maintained Sorsogon isolate cycled for several years through mice and *O. h. quadrasi* snails collected originally in Sorsogon. The data indicate that a significant difference in infectivity exists. Ratios of male to female worms were comparable, the figure of 3 for this ratio being somewhat higher than what is usually recorded in our laboratories (see below). Far more information is required before *genetically* based differences in infectivity of *S. japonicum* isolates and geographic strains can be emphasized (see Jones and Kusel, 1989, in regard to *S. mansoni*).

TABLE 1 *Infectivity and susceptibility to PZQ of two isolates of* Schistosoma japonicum *(Philippines) in mice given 20 cercariae per mouse percutaneously*

			Worm burdens	
Parasite	*Treatment*	*No. of mice*	*No.*	*♂:♀ ratio*
Field, Mindoro	—	26	15.3 ± 0.8	3.0 ± 0.7
Laboratory, Sorsogon	—	27	9.1 ± 0.6	3.0 ± 0.5
Field, Mindoro	PZQ[a]	7	4.1 ± 1.5[b]	—
Laboratory, Sorsogon	PZQ[a]	8	0.6 ± 0.2[b]	—

[a] 250 mg praziquantel (PZQ) kg^{-1} body weight administered orally in corn oil daily on days 35, 36 and 37 of infection, perfusion day 42. Control mice (untreated) perfused on day 40 of infection; these mice consisted of BALB/c and (C57BL/6 × 129/J)F_1 hybrids, the PZQ treated mice being F_1 hybrids only. Arithmetic mean ± S.E.M. indicated.
[b] Majority of worms stunted; all female (♀) worms that were present (4 + 1 total in the two groups) were stunted as were all male (♂) worms in the group given Sorsogon cercariae.

Exposure to more than 20–25 cercariae per mouse with *S. japonicum* (Philippines) results in high mortality at about 2 months of infection; in contrast, 100 cercariae are used routinely for the infection of mice with *S. mansoni* (Puerto Rico) with infectivity varying markedly but probably averaging around 25% and few deaths occurring before 2 months. It remains to be determined whether there are any important consequences (e.g.

accelerated induction of primary or memory responses to released antigens) of large numbers of dying *S. mansoni* schistosomula relative to *S. japonicum* schistosomula in experimental challenge infections with the two species.

Important differences in infection characteristics between *S. japonicum* (Japanese and Chinese) and *S. mansoni* (Puerto Rico) were documented in a classic study by Olivier (1952). After exposure to high numbers of cercariae, *S. mansoni* schistosomula were *readily detected* in lungs over a 2-week period, the peak recoveries being recorded on days 5–8. In contrast, schistosomula of *S. japonicum* were *rarely detected* in the lungs. Lung haemorrhages (petechiae) were prominent in *S. japonicum*-infected mice during the first week but were rare in *S. mansoni*-infected mice. The schistosomula of *S. japonicum* first appeared in liver perfusates in large numbers late in the first week of infection, but with *S. mansoni* the parasites first appeared in the liver during the second week of infection. Haemorrhagic spots have also been detected in the lungs of rabbits and monkeys infected with *S. japonicum* (Chou and Tsung, 1966). Clearly, *S. japonicum* does not linger in the lungs and the transit time for this parasite, whether recirculating in the bloodstream or not, is very short relative to *S. mansoni*.

A detailed study has been performed by Giron-Garcia (1980) using *S. japonicum* (Philippines) (Leyte and Mindoro) in mice and the essential

---

FIG. 1. Numbers of *S. japonicum* parasites in perfusates, tissues and body cavities as well as lung petechiae in young adult outbred white mice exposed percutaneously on day 0 to 100 cercariae per mouse (data of Giron-Garcia, 1980). Five heparinized mice were killed per day over a 30-day period, and the pleural and peritoneal cavities were flushed with citrate saline and washings aspirated. The vena cava was doubly ligated and severed above and below the diaphragm and the portal vein ligated and severed close to the liver. The liver was removed and perfused through the hepatic vein. For perfusion of mesenteric vessels, the iliac aorta was ligated and the needle for perfusion inserted into the aorta. For heart–lung perfusion, the vena cavae and aorta were ligated, heart and lungs removed and perfused through the right ventricle. After perfusion, the lungs were teased. All the samples were examined under a stereomicroscope, and the mean number of parasites over a 2-day period determined (i.e. point 2 days = mean no. of worms per mouse on days 1 + 2; point 4 days = mean no. of worms per mouse on days 3 + 4, etc.). Some details of the stage of development of schistosomula and schistosomes in the perfusate of the different organs and cavities are as follows: early schistosomula only in pleural and peritoneal cavities with no pigment in caeca and no suckers; in lung–heart perfusate and teased lung to day 16, only early schistosomula with no pigment in caeca, but signs of developing suckers, thereafter advanced schistosomula and pre-adults; in liver perfusate, all development stages recorded from early schistosomule (starting day 4 and still present until day 13) to sexually differentiated adults after day 15 though some pre-adults still recorded at day 30; in mesenteric perfusate, only advanced schistosomula (with pigment in caeca and developed suckers) from days 7 to 15, thereafter pre-adults and adults. The details of the techniques used can be found in Yolles *et al.* (1947) and Olivier (1952).

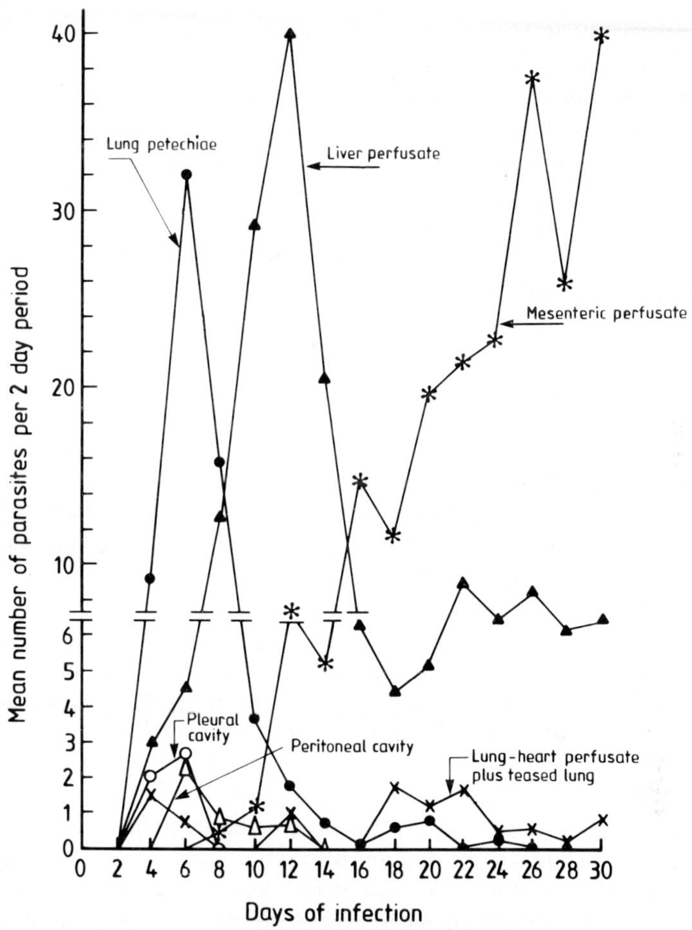

results are present in Fig. 1. As in the studies reported by Olivier (1952), schistosomula were rare in lung–heart perfusates and teased lungs, lung petechiae were obvious, and some parasites were detected in liver perfusates early in the course of infection. Interestingly, some maturing schistosomes were found in the lungs at late time points when the majority appear to be moving from the liver to spend more time in the extrahepatic portal system (see below). The number of lung petechiae at days 4–6 approximates the number of adult worms subsequently perfused (Olivier, 1952; Giron-Garcia, 1980; Garcia *et al.*, 1984).

The study by Giron-Garcia (1980) was intended to throw light on the controversial question of whether *S. japonicum* migrates from the lungs to liver predominantly through the bloodstream or predominantly by a tissue route (Sueyasu, 1920; Miyagawa and Takemoto, 1921; Faust and Meleney, 1924; Sadun *et al.*, 1958; Wilks, 1967; Georgi *et al.*, 1987). It failed to do that, but detection of some parasites in the pleural cavity combined with lesions in the diaphragm and an absence of haemorrhages in abdominal locations where schistosomula should be exiting from terminal arterial vessels into the portal system (i.e. capillary beds of the gastrointestinal tract), were taken as some evidence for tissue migration. The question remains entirely open. If a substantial number of *S. japonicum* schistosomula migrate through tissues from the lung to liver, then this could represent a difference between *S. japonicum* and *S. mansoni* (Wheater and Wilson, 1979; Bloch, 1980; Miller and Wilson, 1980). *S. mansoni* is said to have an " . . . entirely intravascular migration, with potentially several passages around the pulmonary-systemic circulation, before chance entry into arteries leading to the hepatic portal system".

The only study with *S. japonicum* in which a significant recovery of schistosomula has been reported from lungs was that of Usawattanakul *et al.* (1982), using in excess of 200 cercariae of *S. japonicum* (Japanese); the maximum recovery was 12% of administered cercariae. Tracking experiments using radiolabelled parasites and organ autoradiography have not been applied as systematically to *S. japonicum* as they have for studies on the migration of *S. mansoni* (e.g. Georgi *et al.*, 1987). However, in a personal communication from D. A. Dean and B. L. Mangold, radiolabelled *S. japonicum* schistosomula are said to have a short transit time in the lungs of mice.

The apparent short residency time of *S. japonicum* schistosomula in the lungs may severely limit the opportunities for the immune-mediated arrest of migrating larvae in this site despite substantial tissue damage manifesting as petechial haemorrhages. As discussed below, the lung is receiving increased attention as an organ where worm attrition may occur in sensitized *S. mansoni*-infected mice, principally through a cellular inflammatory response with the interesting possibility that some parasites may even be expelled into the airways. Up to 10 references on the presence of schistosomula in alveoli and other air spaces are cited by Dean *et al.* (1987). It is stated that with *S. mansoni*: "There is a period of development in this organ (the lungs), during which the schistosomulum may be a 'sitting target' for immune effector mechanisms" (Wilson and Coulson, 1989). Another quotation from Dean *et al.* (1987) is relevant to the comparative aspects of *Schistosoma* spp.: ". . . a direct relationship is seen between rate of passage through the lungs and chance of surviving to adulthood".

The perfusion data of Giron-Garcia (1980) clearly indicate that pre-adult schistosomes of *S. japonicum* (Philippines) are concentrated in the liver up to about 15 days of infection. As indicated above, the lungs may not be an appropriate site to mount a protective response against *S. japonicum*, unless schistosomes can be diverted back to the lung as a result of peculiarities in the hepatoportal system of the type described below. In *S. japonicum* infection, the liver may be a more appropriate site for the expression of vaccine-based resistance (see McLaren, 1989, in regard to liver-phase immunity to *S. mansoni* in vaccinated guinea pigs; see also Pearce and James, 1986). But, how immunologically competent is the liver, or at least, what inflammatory immune effector mechanisms can be mounted in this site to assist any direct efforts that antibodies might have on the mobile worms? What cytokines are produced by endogenous Kupffer cells, endothelial cells and liver parenchymal cells or by inflammatory cells brought into the organ? The liver has traditionally been viewed as a suboptimal site for immune *induction* with antigens delivered by the portal route being relatively "tolerogenic". This area of research is in need of re-examination, not only because of its innate interest and the paucity of "hard data", but because it is likely to be an essential prerequisite to any real understanding of the immunology of schistosomiasis. The presence of a population of inflammatory cells including lymphocytes in the liver (as granulomas, etc.) in response to eggs may also significantly alter the immune responses induced to antigens derived from adult worms and delivered to the liver from the portal system.

Differences in infection characteristics between *S. japonicum* (Chinese) and *S. japonicum* (Philippines) may exist, even though a relevant comparative study in a single laboratory has not been performed to our knowledge. Because relatively large differences can be demonstrated *between* Philippine isolates (Table 1), it is predictable that differences between a Philippine isolate and other geographical isolates will be demonstrable. Faecal egg excretion in mice infected with *S. japonicum* (Chinese) commences at 7 weeks (Moloney and Webbe, 1983). In *S. japonicum* (Philippines), egg-laying females in copula can be present at 24 days. Faecal egg output can therefore be expected at 5 weeks of infection allowing for a 12 to 14-day maturation time for immature eggs deposited by the female into mature miracidium-containing eggs (Cheever and Deb, 1989; Garcia *et al.*, 1983b). As judged by the number of *S. japonicum* (Chinese) cercariae used by various workers in the People's Republic of China (PRC) to infect mice, the infectivity of this parasite is less than *S. japonicum* (Philippines), as highlighted by Hsu and Li-Hsu (1960a). Clearly, *S. japonicum* (Philippines) can be highly infective, highly pathogenic (see below) and a rapidly maturing parasite in mice.

Differences between *S. mansoni* and *S. japonicum* and the infections they cause in humans and experimental animals have been summarized by Cheever (1987). The principal features of *S. japonicum* are as follows: eggs are deposited in clusters and worm pairs themselves tend to cluster and remain in the one location. Egg production per worm pair per day is at least four times greater for *S. japonicum* (it may approach 10 times; Moore and Warren, 1967) and eggs may persist longer in tissues (though this statement does *not* refer to the persistence of *viable* eggs). Because *S. japonicum* eggs do not contain an obvious hepatotoxic antigen (see below), infected hypothymic nude mice can survive longer than *S. mansoni*-infected nude mice. *S. japonicum*-infected nude mice pass normal numbers of eggs in faeces compared with *S. mansoni*-infected nude mice and the reasons for this difference remain unexplained. Heavily-infected mice have smaller circumoval granulomas than do mice lightly infected with *S. japonicum*, whereas no such difference may be apparent in *S. mansoni* infections. As discussed below, granuloma modulation occurs early in infection and, although T cells are clearly involved in the induction of granuloma formation in both *S. mansoni*- and *S. japonicum*-infected mice, serum antibodies may be the principal modulating entities in murine schistosomiasis japonica. In contrast, regulatory T-cell effects appear to dominate in granuloma modulation in chronic murine schistosomiasis mansoni.

In general, *S. japonicum* has a wider range of definitive hosts than *S. mansoni*, indicating that it may be more resilient in the face of untoward immunological and physiological factors. It is said that *S. japonicum*, unlike *S. mansoni*, uses relatively non-specific cues for attachment, orientation and penetration (Haas *et al.*, 1987).

In regard to the use of therapeutic drugs in schistosomiasis mansoni and japonica in humans *vs* mice, it is becoming clear that praziquantel (PZQ) is much more potent in humans than in mice. Doses of PZQ per kg body weight required in mice are more than 10 times that required in humans. While this could obviously reflect differences in drug absorption and metabolism (bioavailability) in the two host species, the good evidence for immune-mediated facilitation of PZQ action (Doenhoff *et al.*, 1988; Brindley *et al.*, 1989) raises a more interesting possibility. Laboratory infections almost invariably involve a one-off exposure to cercariae, whereas infections in humans (other than tourists) must be multiple. Thus immune responses in humans induced by trickle infections and repeated exposure to cercariae must be very different from immune responses in laboratory mice exposed once to cercariae. These considerations highlight the likely limitations of much of the experimental work on schistosomiasis (e.g. challenge of vaccinated animals, examination of antibody specificities during infection, time course of granuloma modulation) quite apart from the fact that one single worm pair in the mouse is likely to be a huge worm burden relative to the average worm burden in humans (Cheever, 1969). This issue is also important in the interpretation of resistance in animals exposed to male-only infections. Vogel and Minning (1953) demonstrated impressive resistance to mixed cercarial challenge in rhesus monkeys exposed to trickle infections of male-only cercariae (see also Lin and Sadun, 1959). Exposure of mice to a single male-only infection does not lead to significant resistance in either the *S. japonicum* (Garcia *et al.*, 1984; Moloney *et al.*, 1984) or *S. mansoni* systems (Dean *et al.*, 1978a,b; Bickle *et al.*, 1979; Colley and Freeman, 1981).

III. Resistance to Reinfection with *S. japonicum*

Chronically infected mice are highly resistant to homologous challenge with schistosomes. The results of a large number of studies with both *S. mansoni* and *S. japonicum* are compatible with the notion that egg-induced liver pathology contributes in large part to resistance to reinfection in this small laboratory host. It is now widely accepted that much of the resistance to infection—but not all (Smithers *et al.*, 1987)—in chronically infected mice containing a bisexual infection, and thus many eggs in tissues (Dean *et al.*, 1978a; Harrison *et al.*, 1982; Moloney *et al.*, 1984), is a consequence of a "leaky" portal system resulting from egg-induced pathology in the liver [Wilson *et al.*, 1983; Pons *et al.*, 1989; Dean *et al.*, 1981b; see also Moloney *et al.*, 1984, 1987c and Garcia *et al.*, 1984 in the *S. japonicum* (Chinese and Philippine) systems]. Interest in an immunological basis for high-level resistance to reinfection in mice has therefore waned and has been replaced by the examination of the immunological basis of high-level resistance to infection following exposure of mice to irradiated cercariae in which egg-induced liver pathology can play no part (Taylor and Bickle, 1986). The death-knell for the chronically infected mouse for detailed analysis of the immunology of resistance to reinfection in this model really came as early as 1981 with a highly informative experiment in parabiotic mice conducted by Dean *et al.* (1981b).

It is stated in the older literature that "Mice can be protected much more readily against *S. japonicum* than against *S. mansoni*" (Sadun, 1963). Sadun emphasized that this comment did not refer to "dead vaccines" that were generally ineffective with both parasites. If pathology is more severe during acute murine schistosomiasis japonica because of increased egg production, then it might be expected that homologous resistance to reinfection will be more obvious in *S. japonicum* infections than in *S. mansoni* infections of mice. It is of interest to note that, because of increased egg output and the severity of immunopathology, schistosomiasis japonica is believed to be a far more serious disease than schistosomiasis mansoni. Yet field studies in the Philippines, that can be compared with others in Kenya on *S. mansoni*, do not support this contention (Domingo *et al.*, 1980; Pesigan *et al.*, 1958; Lewert *et al.*, 1979). The multitude of references on *in vivo* and *in vitro* cellular reactions in schistosomiasis mansoni contrasts with the limited number in schistosomiasis japonica, e.g. cellular responses to challenge schistosomula of *S. japonicum* (Chinese) in the skin of infected mice (Wu *et al.*, 1989) and ADCC reactions to schistosomula *in vitro* (Zhang and Zhu, 1989).

The lung has become a recent focus of attention for the expression of resistance to schistosomes. A debate on the relative importance of this organ in schistosome attrition has appeared recently (Wilson and Coulson 1989; McLaren, 1989). Early studies were inconclusive in regard to the lung as a site of attrition of *S. mansoni* schistosomula. Von Lichtenberg *et al.* (1973, 1977) and von Lichtenberg and Byram (1980) demonstrated an accelerated inflammatory response and killing of schistosomula in sensitized compared with naive mice, but only to intravenously injected transformed schistosomula rather than to those arriving as more mature schistosomula through the normal transcutaneous route. Others have reported inflammatory responses (Wheater and Wilson, 1979) or the killing of schistosomula (Magalhaes-Filho, 1959) in the lungs of *S. mansoni*-infected mice. An early study by Smith *et al.* (1975) strongly suggested that cells infiltrating the lungs in response to schistosomule challenge in sensitized animals need be activated (? non-specifically) for the expression of anti-schistosomule effects. Dean *et al.* (1978b) suggested that an inflammatory response to eggs in the lungs was helpful in militating against the successful negotiation of the lungs by challenge schistosomula, though this was not supported by earlier studies by von Lichtenberg *et al.* (1963). One of the first indications that challenge worm attrition in mice vaccinated with irradiated cercariae was occurring in the lung or even post-lung stages came from this same group (Dean *et al.*, 1984; see also Mangold and Dean, 1986; Georgi *et al.*, 1987). With *S. japonicum* (Japanese), Lin and Sadun (1959) observed some effect on challenge schistosomula associated with an accelerated lung inflammatory response in challenged infected mice (see also Sadun, 1963). In the study of Usawattanakul *et al.* (1982), some—but not all—mouse strains showed a reduced recovery of lung schistosomula upon challenge of mice infected with *S. japonicum* (Japanese). A contribution of delayed migration of schistosomula to the lung has not been discounted in this phenomenon or, for that matter, accelerated migration. We have found no effect on challenge worm burdens in egg-sensitized mice subsequently injected intravenously with *S. japonicum* (Philippines) eggs prior to cercarial challenge (Garcia *et al.*, 1984). However, Moloney *et al.* (1987a) demonstrated a slight protective effect in mice sensitized with soluble egg antigens (SEA) and challenged intravenously with eggs before cercarial challenge with *S. japonicum* (Chinese).

More recently, the case for the lung as an important site of worm attrition has strengthened considerably, at least in *S. mansoni* infection of mice (reviewed in Wilson and Coulson, 1989; see also Mountford *et al.*, 1989). A study by Vignali *et al.* (1989) with *S. mansoni* in rats (see Ford *et al.*, 1984) strongly suggested that the *speed* of immune recognition of migrating schistosomula is critical in lung trapping and the expression of resistance in the lungs of sensitized hosts. Similarly, Ward and McLaren (1989) demonstrated an immune serum-dependent trapping and damage of *S. mansoni* schistosomula in the lungs of rats. In the mouse, in which cell-mediated immunity rather than antibody has been emphasized in recent years in the expression of resistance (James *et al.*, 1987; Sher *et al.*, 1989; Capron *et al.*, 1987; Wilson and Coulson, 1989), it would appear that T-cell-dependent cellular inflammatory responses can damage lung schistosomula.

Antibody-mediated effects in lung-stage killing remain confusing. For example, two recent studies highlight differences between the mouse/*S. mansoni* and mouse/*S. japonicum* (Chinese) systems in the analysis of passive transfer of resistance with sera from mice repeatedly exposed to irradiated cercariae. Using CBA/Ca mice, McLaren and Smithers (1988) demonstrated that immune mouse serum given at day 5 was less effective than when given at day 0, the time of challenge of naive recipients with *S. mansoni* cercariae. In contrast, Moloney *et al.* (1987b), using CBA/H mice exposed to *S. japonicum* (Chinese), showed that immune mouse serum at day 5 was more effective than when given at the time of cercarial challenge (day 0). This suggests that the partial resistance mediated by mouse serum (antibodies ?) is expressed pre-lung with *S. mansoni*, but attrition is post-lung with *S. japonicum*.

The protective effects of irradiated cercariae are not only species-specific, as demonstrated by many studies, but are also strain-specific in the case of *S. japonicum* (Chinese) and *S. japonicum* (Philippines) (Moloney *et al.*, 1985). Thus, mice exposed to the irradiated cercariae of *S. japonicum* (Chinese) were 50% resistant to homologous challenge but were not resistant to challenge with *S. japonicum* (Philippines). The important reciprocal experiment has not been performed and it remains a possibility that *S. japonicum* (Philippines) parasites are less susceptible to immune responses induced by irradiated homologous or heterologous cercariae. Certainly, in limited studies performed to date, any protective effects of irradiated *S. japonicum* (Philippines) cercariae have been highly variable (E. G. Garcia and W. U. Tiu, unpublished observations) and, in one recent study, ^{60}Co-irradiated *S. japonicum* (Philippines) did not vaccinate against homologous challenge (Mitchell *et al.*, in press b).

Chronically infected mice treated with PZQ and rechallenged with *S. japonicum* several weeks after drug administration are *not* resistant. This was demonstrated clearly by Moloney *et al.* (1987c), and recent data from these laboratories (Mitchell *et al.*, 1990) reinforce this. As discussed at the end of Section II, it will be important to repeat such experiments using trickle infections prior to PZQ administration *and* also for challenge in order to simulate more closely what may happen in natural infections. It will also be important to conduct vaccination studies with cloned antigens in previously infected mice that have been drug-cured. Initial phase-III human vaccination trials (testing for efficacy with natural exposure) are likely to involve the immunization of youngsters after drug treatment of existing infection with resistance to reinfection being monitored.

Immune responses elicited by several weeks or months of infection in mice are obviously not sufficient for the expression of any resistance to reinfection of a type exploitable by way of vaccination. Resistance to reinfection (of the immunopathology-based type) wanes rapidly after curative drug treatment of mice infected for 7–10 weeks with either *S. japonicum* (Chinese) (Moloney *et al.*, 1987c) or *S. mansoni* (Doenhoff *et al.*, 1980; Andrade and Azevado de Brito, 1982). In this case, the immunogenic stimuli of *both* chronic infection (resulting in much IgG_1 production: Sher *et al.*, 1977; Chapman *et al.*, 1979) and drug-killed worms must be substantial. Whether "blocking antibodies" contribute in any way to a lack of resistance to reinfection in drug-cured, previously infected mice has not been established.

IV. Mouse Strain Variation in Susceptibility to *S. japonicum*

After the exposure of different inbred mouse strains to the cercariae of *S. japonicum* (Philippines), the number of adult worms establishing can vary (Mitchell *et al.*, 1981). A consistent difference from the majority of mouse strains has been thoroughly documented in two strains—namely, WEHI 129/J and, to a lesser extent, C57BL/6 mice. The difference is that a high proportion (50–100%) of WEHI 129/J, and a lower proportion of C57BL/6 (20%), contain no worms in liver and portal perfusates several weeks after percutaneous exposure to cercariae (Mitchell *et al.*, 1981; Garcia *et al.*, 1983a; Davern *et al.*, 1987; Mitchell 1989a). Schistosomula establish in WEHI 129/J as evidenced by lung petechiae at day 5 of infection (Garcia *et al.*, 1983a). Relatively mature worms can be found in lungs at later time points in the course of infection at a time when none is present in the portal system, and focal deposits of large amounts of pigment indicative of worm death can be found in the liver and lungs. This latter phenomenon with respect to the lungs has been documented by Elsaghier *et al.* (1989) using *S. mansoni* in 129/Ola mice. 129/J mice obtained directly from the Jackson Laboratory are entirely permissive to infection (Fanning and Kazura, 1984; Mitchell *et al.*, in press a; D. Bout, E. Skamene and G. F. Mitchell, unpublished observations), as are all F1 hybrids of WEHI 129/J crossed with any other mouse strain including C57BL/6 maintained at WEHI (Wright *et al.*, 1988a; Mitchell *et al.*, in press a).

It is now clear that liver and portal system peculiarities contribute to the apparent resistance of WEHI 129/J and C57BL/6 mice. Thus, using the technique devised by Wilson *et al.* (1983) of injecting microbeads into a mesenteric vein, a high proportion of uninfected WEHI 129/J and a low proportion of uninfected C57BL/6 mice trap the beads in the *lungs* rather than the liver. In other mouse strains, including WEHI 129/J F1 hybrids, the 20 μm beads are trapped quantitatively in the liver (Mitchell, 1989b; Mitchell *et al.*, in press a). Similar observations have been made with 129/Ola mice in the UK (Elsaghier *et al.*, 1989; Coulson and Wilson, 1989; Elsaghier and McLaren, 1989). In these mice, terminal branches of the intrahepatic portal venous system appear to be occluded. Intrahepatic anastosomes between the portal and hepatic venous systems (and distinct from the extrahepatic porto-systemic shunting via a collateral circulation in chronic schistosomiasis) presumably account for diversion of injected microbeads into the lungs. Schistosomes and schistosomula arriving in the portal system will therefore experience problems in maintaining residency in the liver and be returned to the lungs (Wilson *et al.*, 1983; Pons *et al.*, 1989). No other laboratory has reported peculiarities in C57BL/6 mice. In (BALB/c × WEHI 129/J)F1 × WEHI 129/J backcross mice, the numbers of resistant mice and numbers of mice showing a shunting of beads from the portal system to the lungs are comparable (20–25%) and indicate the possible participation of two genes in the phenomena (Mitchell *et al.*, in press a; Wright *et al.*, 1988a).

The reasons for these observations in WEHI 129/J and 129/Ola mice, and some C57BL/6 mice maintained at WEHI, remain unknown. We have previously emphasized an infectious cause (Mitchell, 1989a), as discussed by others (Van Snick and Masson, 1980; Coutelier *et al.*, 1986), to account for certain immunological disturbances in some 129/J mice. However, it is equally likely that nutritional factors may play some role and we are currently looking into the possibility that hypervitaminosis A contributes to liver peculiarities in some but not all individuals in some but not all mouse strains maintained on a particular diet (Mitchell, 1990a; Mitchell *et al.*, in press a).

WEHI 129/J mice are high and/or early responders to the Mr 26 000 glutathione *S*-transferases of *S. japonicum* and *S. mansoni* (Mitchell, 1989a; Wright *et al.*, 1988b), but the contribution of these responses to resistance (killing of worms in the liver and lungs) remains in doubt (Davern *et al.*, 1987) with the discovery of hepatoportal abnormalities in these mice. WEHI 129/J and C57BL/6 mice sensitized with eggs produce large granulomas in the lungs following the intravenous injection of eggs. Of all mouse strains (Mitchell *et al.*, 1983), they are probably the highest responders in this acute granulomatous hypersensitivity response to *S. japonicum* (Philippines) eggs

(W. U. Tiu and G. F. Mitchell, unpublished observations). It is possible that this results, in part, from insult of the lungs with gut-derived toxins diverted from the liver by the intrahepatic portosystemic shunts in these mice. A confusing situation exists in regard to mouse strain variation in the size of *S. mansoni* egg granulomas and other disease manifestations of schistosomiasis mansoni (compare Cheever *et al.*, 1987; Dean *et al.*, 1981a; Claas and Deelder, 1979; Fanning *et al.*, 1981; Fanning and Kazura, 1985; Colley and Freeman, 1981, 1983; Metzger and Peterson, 1988; Jones *et al.*, 1983). Contributing to the confusion is undoubtedly variable husbandry conditions (including intercurrent infections) for mice in different laboratories. What is clear is that there is no association between granuloma size and hepatic fibrosis in these comparative mouse studies. In terms of granuloma formation to *S. japonicum* eggs, C57BL/6 are known to be high responders (Mitchell *et al.*, 1981, 1983; Cheever *et al.*, 1984; Stavitsky, 1987; see also Allen *et al.*, 1977), although a fibrotic response may not be prominent in this strain (Cheever *et al.*, 1984). Mouse strain variations (i.e. BALB/c *vs* C57BL/6) in the rate of killing of eggs of *S. mansoni* may exist (Mitchell *et al.*, 1990).

V. Granuloma Formation and Modulation: Immunoregulation in Egg-induced Pathology

Much of our current knowledge on the immunopathology of chronic hepatosplenic disease in schistosomiasis derives from studies in mouse models of *S. mansoni* and *S. japonicum* infection that are considered excellent representations of the human diseases (Warren, 1982; von Lichtenberg, 1987; Smithers and Doenhoff, 1982). As discussed below, liver granuloma formation cannot be considered entirely pathologic, a small compact granuloma—in which the impacted egg usually dies within a month or so of deposition (Reis and Andrade, 1987)—being a desirable host response that spares hepatocytes from hepatotoxic egg products. An additional twist is the beneficial effects in terms of resistance to reinfection, at least in small laboratory hosts, of a "leaky" portal system resulting from severe egg-initiated immunopathology in the liver (see above).

Basic immunological events in granuloma formation and modulation have recently been outlined (Mitchell, 1990a): elaboration of soluble products (i.e. aqueously soluble egg antigens known as SEA), uptake by antigen-presenting cells, association of fragments with class II MHC, recognition by $CD4^+$ T cells of both Th1 and Th2 subpopulations (though details of what lymphokines operate are sketchy), production of inflammatory lymphokines and attraction ($\pm$activation) of inflammatory cells (facilitated by factors from eggs themselves and any tissue damage) such as eosinophils, mononuclear phagocytes, fibroblasts, lymphocytes, neutrophils, plasma cells and mast cells in this order of decreasing prevalence (von Lichtenberg, 1987). Collagen, fibronectin and glycosaminoglycans, the principal extracellular matrix components, accumulate in the granulomas as well as along portal tracts (Olds *et al.*, 1986; Grimaud *et al.*, 1987), there being apparent differences between humans and mice in the contributions of phlebitis with attendant perivascular fibrosis and egg-initiated periovular fibrosis to total portal fibrosis (Grimaud, 1983; Grimaud *et al.*, 1987). Death of the miracidium in the egg terminates SEA production and the process of lesion resolution commences, the timing of which must depend on calcification of the egg and intermolecular cross-linking of collagen isotypes (Andrade and Grimaud, 1988) with collagenases as the key enzymes.

With the above as an outline for egg-initiated immunopathology in schistosomiasis, many differences in detail have nevertheless been identified according to where the egg is located, the host species and the geographic variant of the one parasite species (e.g. Warren and Berry, 1972; Cheever, 1987; Li Hsu *et al.*, 1972; Edungbola *et al.*, 1982; Meleney *et al.*, 1953). A very interesting recent finding indicates that a degree of autoimmune T-cell reactivity to denatured collagen may be involved in increasing the size of *S. mansoni* egg granulomas in a "high responder" mouse strain, C57BL/6 (Wyler *et al.*, 1987). An unexplained observation is the reduction in *S. mansoni* egg granuloma size in baboons vaccinated with irradiated schistosomula (Damian *et al.*, 1984; Damian, 1984).

One of the disappointing aspects of granuloma formation (and modulation) is the extremely slow progress being made on the identification of immunopathological egg antigens. Using mouse models, a large number of studies that commenced in the 1970s, were aimed at identifying such antigens in SEA (reviewed in Mitchell and Cruise, 1986). SEA from either *S. mansoni* or *S. japonicum* eggs is a very complex mixture of molecules (Norden and Strand, 1984), and one antigen of particular interest in *S. japonicum* eggs is a glycoprotein of Mr 140 000 that is a major immunogen. Monoclonal antibodies to this component of SEA modulate granuloma formation in mice (Sidner *et al.*, 1987). Other glycoproteins of *S. mansoni* eggs have been described (e.g. Pelley *et al.*, 1976; Lustigman *et al.*, 1985). The strong possibility exists that carbohydrates/glycolipids of *S. mansoni* and *S. japonicum* eggs may be immunopathological in mice (Weiss *et al.*, 1987; Tiu *et al.*, 1989).

The most intriguing aspect of the immunopathology of chronic schistosomiasis, for an immunologist at least, is the decrease in granuloma size and disease abatement that occurs as infection proceeds in both humans and mice (Andrade and Warren, 1964; Domingo and Warren, 1968). This immune-mediated event is said to be induced earlier in murine *S. japonicum* infection than in *S. mansoni* infection (see below), but mouse-strain differences are likely to be as obvious as any parasite species differences. A decrease in granuloma size has been documented following repeated injections of small doses of the SEA of *S. japonicum* (Chinese) (Zhang *et al.*, 1983).

The mechanisms underlying granuloma modulation (i.e. down-regulation of granulomatous disease or endogenous desensitization) are believed to be (a) the reduced production of *fertile* eggs by the worm pairs ("anti-fecundity immunity"), (b) inhibition of egg maturation ("anti-embryonation immunity") coupled with accelerated destruction of the miracidium, and (c) immune deviation or suppression of the $CD4^+$ T cell responsible for promoting granuloma formation. Mechanism (a), at least in terms of a reduction in the total number of eggs produced by schistosome worm pairs (Bushara *et al.*, 1983; Vogel and Minning, 1953), is unlikely to be a major contributing factor to granuloma modulation in chronic disease (Damian *et al.*, 1986). Mechanism (b) is discussed in some detail below. Mechanism (c), with its emphasis on suppressor T-cell involvement, has been the basis of much work in the *S. mansoni*/mouse system (Stavitsky, 1987; Olds, 1989; Boros, 1986; Phillips and Lammie, 1986; Abe and Colley, 1984; Mahmoud, 1983) and with little recent progress coinciding with the raising of serious questions about mechanisms that hang over the entire "suppressor T cell" story. Decreased lymphocyte responses to schistosome antigens using crude bulk culture approaches have certainly been demonstrated in chronically infected humans and mice (Colley *et al.*, 1977; Ottesen *et al.*, 1978; Ellner *et al.*, 1981; Garb and Stavitsky, 1984). T cells with reactivity to anti-SEA antibodies have been described in some clinically defined patient groups in Brazilian schistosomiasis mansoni (Parra *et al.*, 1988), but the actual *in vivo* immunoregulatory role of such T cells remains unknown in humans and mouse models.

Despite early evidence to the contrary, granuloma formation in schistosomiasis japonica in the mouse, like schistosomiasis mansoni, is a classical T-cell-dependent inflammatory response of the broadly defined DTH type (Garcia and Mitchell, 1987; Stavitsky, 1987; Mitchell, 1990a). Differential vulnerability of major T-cell-stimulating antigens to degradation during egg extraction from tissues is a satisfactory explanation for the early emphasis on differences between *S. mansoni* and *S. japonicum* infections in regard to egg-initiated granulomatous hypersensitivity.

Many studies commencing as early as the 1950s have suggested that antibody responses to egg antigens are more prominent in *S. japonicum* infections than in *S. mansoni* infections even though antibodies *per se* are most unlikely to contribute to granuloma formation (Cheever *et al.*, 1985a).

Other likely non-contributors are mast cell products, histamine sensitivity of tissue, C5 levels, the beige mutation and X-linked B-cell defects in CBA/N mice (Cheever *et al.*, 1987). Disease caused by *S. japonicum* can be more severe than that caused by *S. mansoni* (Meleney *et al.*, 1953; Li Hsu *et al.*, 1972; but see Section II), presumably reflecting high egg production per worm pair, a clustering of eggs and host responses that—at least early in infection—may be focally more destructive (von Lichtenberg *et al.*, 1973; Warren *et al.*, 1975). Other evidence exists that hepatotoxic effects of egg antigens are *less* obvious in *S. japonicum*-infected nude mice (Cheever *et al.*, 1985b) compared with *S. mansoni*-infected nude or other T-cell-deprived mice (Dunne and Doenhoff, 1983; Damian, 1987; Buchanan *et al.*, 1973; Byram *et al.*, 1979; Doenhoff *et al.*, 1985). This suggests that *S. japonicum* eggs are relatively deficient in hepatotoxins or that the *S. japonicum* granuloma, even when of small size as in T-cell-deficient mice, can very efficiently sequester this type of activity within the granuloma. It is difficult to determine from the literature whether single eggs of *S. mansoni* are innately more able to induce large granulomas in various sensitized and unsensitized hosts than are single eggs of the various *S. japonicum* strains. Certainly, von Lichtenberg *et al.* (1973) and Erickson *et al.* (1974) have emphasized the intrinsically low granuloma-forming potential of single *S. japonicum* eggs in hamsters compared with single *S. mansoni* eggs or *S. japonicum* egg clusters in this host species. An answer to the question of whether in the mouse, and egg-for-egg, *S. mansoni* is a richer source of T-cell-stimulating antigens than eggs of the various *S. japonicum* strains, must await the identification of the full range of such antigens and, more importantly, their $CD4^+$ T-cell (Th1 + Th2-cell)-stimulating epitopes.

A study by Warren and Berry (1972) on the pathogenicity of various strains of *S. japonicum* in mice, with liver and spleen weight, portal pressure, haematocrits and liver pathology as the indices, results in the following ranking from high to low: Philippine → Formosan → Japanese → Chinese. As in the studies of Hsu and Li Hsu (1960b), the highest egg output occurred with *S. japonicum* (Philippines).

Experiments by Olds *et al.* (1982) demonstrated clearly an effect of immune serum on granuloma modulation in *S. japonicum*-infected C57BL/6 mice that was later attributed to IgG_1 antibodies (Olds and Stavitsky, 1986) with a contribution from anti-idiotype antibodies (Olds and Kressina, 1985). Mouse serum will not modulate granuloma formation using lyophilized eggs, suggesting an effect of antibodies on egg viability (Mitchell *et al.*, 1983). Human sera can reduce lung granuloma size in mice (Garcia *et al.*, 1985) and Sidner *et al.* (1987) have demonstrated a granuloma-modulating effect of chronically infected mouse serum and monoclonal antibodies. Antibodies or at least immunoglobulins can be demonstrated within *S.*

japonicum eggs (Hirata *et al.*, 1986; von Lichtenberg *et al.*, 1973; Kawanaka *et al.*, 1983). It is abundantly clear that antibody-mediated effects dominate in granuloma modulation in chronic *S. japonicum* infections in mice, whereas antibodies are likely to contribute in only a minor way to granuloma modulation in mice chronically infected with *S. mansoni.*

ADCC reactions with egg destruction and involving eosinophils and other granulocytes together with anti-egg antibodies (of unknown specificity) have been demonstrated using *S. mansoni* eggs *in vitro* (James and Colley, 1976, 1978; Olds and Mahmoud, 1980; de Brito *et al.*, 1984). Such effects on *S. japonicum* eggs have not been described and we have postulated that the principal effect of anti-egg antibodies in granuloma modulation in schistosomiasis japonica is the inhibition of maturation of eggs (i.e. anti-embryonation immunity) leading to destruction of the embryo before it matures into a miracidium and a rich source of T-cell-stimulating immunopathologic antigens (Garcia and Mitchell, 1982, 1987). Immunization of mice with living immature eggs increases the proportion of killed eggs in the tissues of mice after infection and disease is less severe (Garcia *et al.*, 1987, 1989). The prolonged embryonation time of *S. japonicum* eggs (9–10 days: Vogel, 1942) relative to the 5–7 days for *S. mansoni* (Gönnert, 1955) may increase opportunities for immune-mediated inhibitory effects on embryonation. It is reported that, within the first 2 months of infection in mice, no visible effects on miracidia can be seen in mice that are or are not able to produce normal levels of anti-egg antibodies (Cheever *et al.*, 1985a). This demonstration, of course, does not bear on the validity of the anti-embryonation hypothesis in *chronic* infections or in egg-immunized infected mice.

The good evidence that anti-egg antibodies contribute to granuloma modulation in schistosomiasis japonica is particularly encouraging for those aiming to develop prototype vaccines for disease prevention, i.e. amelioration of hepatosplenic disease through the induction of granuloma modulation. Sensitization for antibody production to antigens of immature eggs, for example, should be far simpler than induction of immunoregulatory T cells as may be required for schistosomiasis mansoni (Mathew and Boros, 1986; Phillips and Lammie, 1986). Antigen manipulation to increase antibody production in genetic low responders by conjugation of epitopes to T-cell-stimulating carriers should be achievable, whereas induction of a particular immunoregulatory T-cell response in genetic low responders may be difficult (Mitchell, 1990b). Such considerations have been the basis of our efforts to identify antigens of immature and maturing eggs and to clone the genes for these antigens for production of a disease-inhibitory vaccine against schistosomiasis japonica. To date, no molecular data exists to account for the phenomenon of anti-embryonation immunity. We have been unable to induce anti-embryonation with homogenates of immature eggs, suggesting that substantial quantities of antigens of the maturing egg are critical (E. G. Garcia, unpublished observations).

Another potential role for antibodies comes from the work of Olds, Stavitsky and colleagues. Anti-Id antibodies may inhibit the activities of T

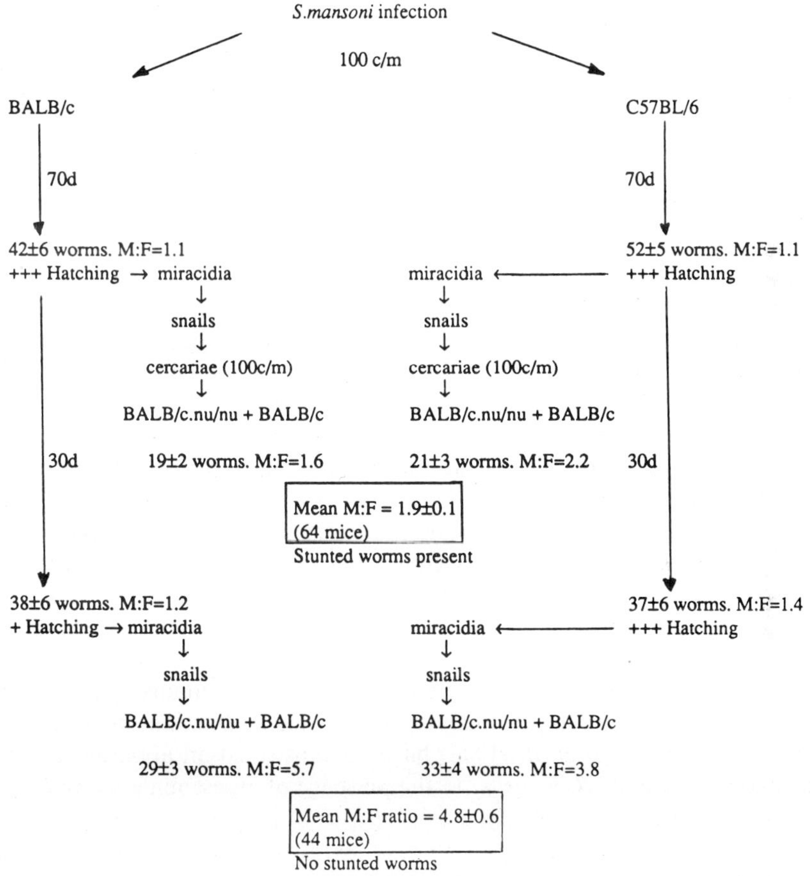

FIG. 2. Experimental outline and essential results of a study to determine the influence of time of infection (day 70 *vs* day 100) with *S. mansoni* on the sex ratio of worms derived eventually from eggs. Two donor mouse strains were used, an apparent decrease in hatchability of liver eggs being recorded in BALB/c at the late time point. Two recipient mouse types were also used—BALB/c and BALB/c.nu/nu—but pooled data on worm counts at about day 50 of infection (±S.E.M.) and the ♂:♀ sex ratio (with a mean ±S.E.M. also indicated) are presented. The reasons why stunted worms were only present in recipients of cercariae derived from miracidia from eggs of 70-day infected mice remain unknown. Data from Mitchell *et al.* (1990).

cells involved in induction of granuloma formation but this, like a comparable role for anti-Id T cells, remains speculative for the moment (Stavitsky, 1987; Olds, 1989; Mitchell, 1989c).

In recent experiments, we have described a phenomenon in schistosomiasis that may proceed in parallel with granuloma modulation if not related more causally. The sex ratios of worms in schistosome infections are invariably biased towards males (the comprehensive literature on this is reviewed by Liberatos, 1987). By taking eggs from mouse livers *late* in the course of *S. mansoni* infection, the bias towards males is further accentuated (Fig. 2). That is, miracidia from chronically infected mice, used to infect snails for the generation of cercariae for the infection of recipient mice, result in higher male:female worm ratios than when miracidia are harvested from eggs of short-term infected mice (Mitchell *et al.*, 1990). Males predominate over females in infections in trapped rats in a schistosomiasis endemic area of the Philippines (Sorsogon) and snails collected in such endemic areas contain cercariae that result in many more male than female worms in laboratory mice. It is conceivable, therefore, that anti-egg immune responses in chronically infected hosts may act preferentially against the heterogametic (WZ) female *miracidium* (? W-chromosome encoded antigens). If this is a richer source of immunopathological antigens than the male miracidium (and there is *no* evidence for this as yet), then the *average* size of granulomas late in infection will be reduced. Moreover, if male miracidia predominate over female miracidia with correspondingly higher numbers of male cercariae in endemic areas, then many individuals may be exposed to trickle infections with male worms only. In the *S. japonicum*/rhesus monkey system at least, this leads to substantial resistance to subsequent mixed cercarial challenge (Vogel and Minning, 1953; Lin and Sadun, 1959). Whether this is the case in humans remains unknown but must be worthy of detailed analysis. Interestingly, Damian and Chapman (1983) and Harrison *et al.* (1982) have demonstrated an apparent reduced fecundity of female *S. mansoni* in the presence of excess male worms.

VI. CONCLUSIONS

Determination of the efficacy of any cloned schistosome gene product as a vaccine is often made in the first instance in inbred strains of mice. The importance of the mouse model therefore dictates that aspects and peculiarities of infection with the different schistosomes be identified and characterized. Moreover, studies on induced immune responses that reduce the severity of egg-initiated disease, including anti-embryonation immunity, rely heavily on the use of mice. Finally, with the availability of appropriate antisera and other reagents, analyses of the effects of immunological manipulation of the host on schistosome infections can be performed most readily in inbred mouse strains and their mutants (in this regard, systematic studies on the effects of anti-cytokine antibodies are eagerly awaited). An important feature of mouse schistosomiasis models in regard to human infections is that parasites taken directly from endemic areas generally establish well in these hosts and do not require adaptation through passage and selection for something quite different from what may be infecting humans.

Despite obvious similarities between *S. japonicum* and the better studied *S. mansoni* parasite (in, for example, broad life-cycle characteristics coupled with the dominance of immunopathology in disease processes), there are many equally obvious differences. Extrapolations of experimental findings from one to the other parasite are rendered even more hazardous by the existence of different geographic strains of *S. japonicum*. The clear demonstration of antibody-mediated effects in granuloma modulation in *S. japonicum* (Philippines), with the maturing egg as an attractive target, offers good prospects for the development of an anti-disease vaccine. Whether antibody is the principal driving force in granuloma modulation in infection with other *S. japonicum* strains is presently unknown.

In this chapter, we have emphasized differences between *S. japonicum* (Philippines) and *S. mansoni* in particular, in regard to lung stages and the potential contribution of inflammatory responses in this organ to the expression of resistance. Compared with *S. mansoni*, the lung schistosomule may be more of a "moving target" in *S. japonicum*. Similarities and differences between the two principal *S. japonicum* pathogens for humans these days—*S. japonicum* (Philippines) and *S. japonicum* (Chinese)—are not entirely clear. From the work of the Hsus, and the Moloney group in London, on *S. japonicum* (Chinese), irradiated cercariae can be potent vaccinating entities. The situation may or may not be different for *S. japonicum* (Philippines). Certainly, mice chronically infected with both parasites show impressive resistance to reinfection, presumably reflecting an effect of egg-initiated liver pathology with a resultant shunting of challenge schistosomula back to the lungs. If this is the site of their demise, then clearly the mouse *is* able to prejudice survival of schistosomula and pre-adults in the lungs provided they are continually returned to that site. A combination of shunting mediated by an altered hepatoportal system and inflammatory responses in the lungs (and perhaps liver) may be the explanation for resistance demonstrated in many naive 129/J mice and some naive C57BL/6 mice.

It may well evolve that analyses of immune responses in the mouse model of *S. japonicum* will be more useful in regard to anti-disease vaccines (e.g. induction of anti-embryonation immunity) than in the *primary* identification of resistance to infection. In regard to the latter, more rapid progress may come from the analysis of serological correlates of resistance to reinfection following drug cure in humans, and then the testing in the mouse model of the cloned antigenic targets of the antibody specificities better expressed in the resistants compared with susceptibles. Of course, the mouse should also be entirely satisfactory for the examination of vaccinating effects of novel antigens/epitopes, i.e. molecules serving critical functions in the parasite that are not of high innate immunogenicity, but which nevertheless can be rendered immunogenic, and that are available to induced neutralizing antibodies in particular. As discussed here, the liver may be an appropriate site for the expression of resistance to *S. japonicum*, a highly pathogenic, highly infectious and rapidly maturing parasite that does not linger in the lungs.

Acknowledgements

Studies on schistosomiasis in the Melbourne and Manila laboratories are supported currently by the Australian National Health and Medical Research Council (NHMRC), the Australian International Development Assistance Bureau (AIDAB), the Edna McConnell Clark Foundation, the UNDP/World Bank/WHO Special Programme for Research and Training–Rockefeller Foundation Health Sciences for the Tropics/Partnerships in Research programme, the Schistosomiasis component of the above-mentioned Special Programme, the John D. and Catherine T. MacArthur Foundation Program on Research and Research Training on the Biology of Parasitic Diseases, and the US Agency for International Development (USAID).

References

Abe, T. and Colley, D. G. (1984). Modulation of *Schistosoma mansoni* egg-induced granuloma formation. III. Evidence for an anti-idiotype, I-J positive, I-J restricted soluble T suppressor factor. *Journal of Immunology* **132,** 2084–2088.

Allen, E. M., Moore, V. L. and Stevens, J. O. (1977). Strain variation in BCG-induced chronic pulmonary inflammation in mice. I. Basic model and possible genetic control by non-H-2 genes. *Journal of Immunology* **119,** 343–347.

Andrade, Z. A. and Azevedo de Brito, P. (1982). Curative chemotherapy and resistance to reinfection in murine schistosomiasis. *American Journal of Tropical Medicine and Hygiene* **31,** 116–121.

Andrade, Z. A. and Grimaud, J. A. (1988). Morphology of chronic collagen resorption: A study on the late stages of schistosomal granuloma involution. *American Journal of Pathology* **132,** 389–399.

Andrade, Z. A. and Warren, K. S. (1964). Mild prolonged schistosomiasis in mice: Alterations in host response with time and development of portal fibrosis. *Transactions of the Royal Society of Tropical Medicine and Hygiene,* **58,** 53–57.

Bickle, Q., Bain, J., McGregory, A. and Doenhoff, M. (1979). Factors affecting the acquisition of resistance against *Schistosoma mansoni* in the mouse. III. The failure of primary infection with cercariae of one sex to induce resistance to reinfection. *Transactions of the Royal Society of Tropical Medicine and Hygiene* **73,** 37–41.

Bloch, E. H. (1980). *In vivo* microscopy of schistosomiasis. II. Migration of *Schistosoma mansoni* in the lungs, liver and intestines. *American Journal of Tropical Medicine and Hygiene* **29,** 62–70.

Boros, D. L. (1986). Immunoregulation of granuloma formation in murine schistosomiasis mansoni. *Annals of the New York Academy of Sciences* **465,** 313–323.

Brindley, P. J., Strand, M., Norden, A. P. and Sher, A. (1989). Role of host antibody in the chemotherapeutic action of praziquantel against *Schistosoma mansoni*: Identification of target antigens. *Molecular and Biochemical Parasitology* **34,** 99–108.

Bruce, J. I., Ruff, M. D., Chiu, J. K. and Howard, L. (1971). *Schistosoma mansoni* and *Schistosoma japonicum* oxygen uptake by miracidia. *Experimental Parasitology* **30,** 124–131.

Buchanan, R. D., Fine, D. P. and Colley, D. G. (1973). *Schistosoma mansoni* infection in mice depleted of thymus-dependent lymphocytes. II. Pathology and altered pathogenesis. *American Journal of Pathology* **71,** 107–218.

Bushara, H. O., Hussein, M. F., Majid, M. A., Musa, B. E. H. and Taylor, M. G. (1983). Observations on cattle schistosomiasis in the Sudan, a study in comparative medicine, IV. Preliminary observations on the mechanisms of naturally acquired resistance. *American Journal of Tropical Medicine and Hygiene* **32,** 1065–1070.

Byram, J. E., Doenhoff, M. J., Mussallam, R., Brink, L. H. and von Lichtenberg, F. (1979). *Schistosoma mansoni* infections in T-cell deprived mice and the ameliorating effect of administering homologous chronic infection serum. II. Pathology. *American Journal of Medicine and Hygiene* **28,** 274–285.

Capron, A., Dessaint, J. P., Capron, M., Ouma, J. H. and Butterworth, A. E. (1987). Immunity to schistosomes. Progress towards vaccine. *Science* **238,** 1065–1072.

Chapman, C. B., Knopf, P. H., Hicks, J. D. and Mitchell, G. F. (1979). IgG_1 hypergammaglobulinaemia in chronic parasite infections in mice: Magnitude of the response in mice infected with various parasites. *Australian Journal of Experimental Biology and Medical Science* **57,** 369–387.

Cheever, A. W. (1969). Quantitative comparisons of the intensity of *Schistosoma mansoni* infections in man and experimental animals. *Transactions of the Royal Society of Tropical Medicine and Hygiene* **63,** 781–795.

Cheever, A. W. (1987) Comparison of pathologic changes in mammalian hosts infected with *Schistosoma mansoni, S. japonicum* and *S. haematobium. Memorias do Instituto Oswaldo Cruz, Rio de Janeiro* **82,** supplement IV, 39–46.

Cheever, A. W. and Deb, S. (1989). Persistence of hepatic fibrosis and tissue eggs following treatment of *Schistosoma japonicum* infected mice. *American Journal of Tropical Medicine and Hygiene* **40,** 620–628.

Cheever, A. W., Duvall, R. H. and Hallack, T. A., Jr (1984). Differences in hepatic fibrosis and granuloma size in several strains of mice infected with *Schistosoma japonicum. American Journal of Tropical Medicine and Hygiene* **33,** 602–607.

Cheever, A. W., Byram, J. E., Hieny, S., von Lichtenberg, F., Lunde, M. N. and Sher, A. (1985a). Immunopathology of *Schistosoma japonicum* and *S. mansoni* infection in B cell depleted mice. *Parasite Immunology* **7,** 399–413.

Cheever, A. W., Byram, J. E. and von Lichtenberg, F. (1985b). Immunopathology of *Schistosoma japonicum* infection in athymic mice. *Parasite Immunology* **7,** 387–398.

Cheever, A. W., Duvall, R. H., Hallack, T. A. Jr, Minker, R. G., Malley, J. D. and Malley, K. G. (1987). Variation of hepatic fibrosis and granuloma size among mouse strains infected with *Schistosoma mansoni. American Journal of Tropical Medicine and Hygiene* **37,** 85–97.

Chi, L. W., Wagner, E. D. and Wold, N. (1971). Susceptibility of *Oncomelanian* hybrid snails to various geographic strains of *Schistosoma japonicum. American Journal of Tropical Medicine and Hygiene* **20,** 89–94.

Chiu, J. K. (1968). Cercaria production of geographic strains of *Schistosoma japonicum* in *Oncomelania hupensis chiui. Journal of the Formosan Medical Association* **67,** 259–265.

Chou, K.-C. and Tsung, Y.S. (1966). Roentologic and pathologic studies of experimental lesions of schistosomiasis japonica. *Chinese Medical Journal (Peking)* **85,** 87–95.

Claas, F. H. J. and Deelder, A. M. (1979). H-2 linked immune response to murine experimental *Schistosoma mansoni* infection. *Journal of Immunogenetics* **6,** 167–175.

Colley, D. G. and Freeman, G. L., Jr (1981). Differences in adult *Schistosoma mansoni* worm burden requirements for the establishment of resistance to reinfection in inbred mice. I. CBA/J and C57BL/6 mice. *American Journal of Tropical Medicine and Hygiene* **29,** 1279–1285.

Colley, D. G. and Freeman, G. L., Jr (1983). Differences in adult *Schistosoma mansoni* worm burden requirements for the establishment of resistance to reinfection in inbred mice. II. C57BL/KsJ, SWR/J, SJL/J, BALB/c.AnN, DBA/2N, B10.A(3R) and B10.A(5R) mice. *American Journal of Tropical Medicine and Hygiene* **32,** 543–549.

Colley, D. G., Hieny, S. E., Bartholomew, R. K. and Cook, J. A. (1977). Immune responses during human schistosomiasis mansoni. III. Regulatory effects of patient sera on human lymphocyte blastogenic responses to schistosomal antigen preparations. *American Journal of Tropical Medicine and Hygiene* **26,** 917–920.

Coulson, P. S. and Wilson, R. A. (1989). Portal shunting and resistance to *Schistosoma mansoni* in 129 strain mice. *Parasitology* **99,** 383–389.

Coutelier, J.-P., van der Logt, J. M., Heessen, F. W. A., Warner, G. and van Snick, J. (1986). Rheumatoid factor production in 129/Sv mice: Involvement of an intestinal infectious agent. *Journal of Immunology* **137,** 337–340.

Damian, R. T. (1984). Immunity in schistosomiasis: A holistic view. *Contemporary Topics in Immunobiology* **12,** 359–420.

Damian, R. T. (1987). The exploitation of host immune responses by parasites. *Journal of Parasitology* **73,** 1–13.

Damian, R. T. and Chapman, R. W. (1983). The fecundity of *Schistosoma mansoni* in baboons, with evidence for a sex ratio effect. *Journal of Parasitology* **69,** 987–989.

Damian, R. T., Roberts, M. L., Powell, M. R., Clark, J. K., Lewis, F. A. and Stirewalt, M. A. (1984). *Schistosoma mansoni* egg granuloma size reduction in challenged baboons after vaccination with irradiated cryopreserved schistosomula. *Proceedings of the National Academy of Sciences (USA)* **81,** 3552–3556.

Damian, R. T., Rawlings, C. A. and Bosshardt, S. C. (1986). The fecundity of *Schistosoma mansoni* in chronic nonhuman primate infections and after transplantation into naive hosts. *Journal of Parasitology* **72,** 741–747.

Davern, K. M., Tiu, W. U., Morahan, G., Wright, M. D., Garcia, E. G. and Mitchell, G. F. (1987). Responses in mice to Sj26, a glutathione *S*-transferase of *Schistosoma japonicum* worms. *Immunology and Cell Biology* **65,** 473–482.

Dean, D. A., Minard, P., Stirewalt, M. A., Vannier, W. E. and Murrell, K. D. (1978a). Resistance of mice to secondary infection with *Schistosoma mansoni.* I. Comparison of bisexual and unisexual initial infections. *American Journal of Tropical Medicine and Hygiene* **27,** 951–956.

Dean, D. A., Minard, P., Murrell, K. D. and Vannier, W. E. (1978b). Resistance of mice to secondary infection with *Schistosoma mansoni.* II. Evidence of a correlation between egg deposition and worm elimination. *American Journal of Tropical Medicine and Hygiene* **27,** 957–965.

Dean, D. A., Bukowski, M. A. and Cheever, A. W. (1981a). Relationship between acquired resistance, portal hypertension and lung granulomas in ten strains of mice infected with *Schistosoma mansoni. American Journal of Tropical Medicine and Hygiene* **30,** 806–814.

Dean, D. A., Bukowski, M. A. and Clark, S. S. (1981b). Attempts to transfer the resistance of *Schistosoma mansoni*-infected and irradiated cercariae immunized mice by means of parabiosis. *American Journal of Tropical Medicine and Hygiene* **30,** 113–120.

Dean, D. A., Mangold, B. L., Georgi, J. R. and Jacobson, R. H. (1984). Comparison of *Schistosoma mansoni* migration patterns in normal and irradiated cercaria-immunized mice by means of autoradiographic analysis. *American Journal of Tropical Medicine and Hygiene* **33,** 89–96.

Dean, D. A., Mangold, B. L., Kassim, O. O. and von Lichtenberg, F. (1987). Sites and mechanisms of schistosome elimination. *Memorias do Instituto Oswaldo Cruz, Rio de Janeiro* **82,** supplement IV, 31–37.

de Brito, P. A., Kazura, J. W. and Mahmoud, A. A. F. (1984). Host granulomatous response in schistosomiasis mansoni: Antibody and cell-mediated damage of parasite eggs *in vitro. Journal of Clinical Investigation* **74,** 1715–1723.

DeWitt, W. B. (1954). Susceptibility of snail vectors to geographic strains of *Schistosoma japonicum. Journal of Parasitology* **40,** 453–456.

Doenhoff, M., Bickle, Q., Bain, J., Webbe, G. and Nelson, G. (1980). Factors affecting the acquisition of resistance against *Schistosoma mansoni* in the mouse. V. Reduction in the degree of resistance to reinfection after chemotherapeutic elimination of recently patent primary infections. *Journal of Helminthology* **54,** 7–16.

Doenhoff, M. J., Hassounah, O. A. and Lucas, S. B. (1985). Does the immunopathology induced by schistosome eggs potentiate parasite survival? *Immunology Today* **6,** 203–206.

Doenhoff, M. J., Modha, J. and Lambertucci, J. R. (1988). Anti-schistosome chemotherapy enhanced by antibodies specific for a parasite esterase. *Immunology* **65,** 507–510.

Domingo, E. O. and Warren, K. S. (1968). Endogenous desensitization: Changing host granulomatous response to schistosome eggs of different stages of infection with *Schistosoma mansoni. American Journal of Pathology* **52,** 369–377.

Domingo, E. O., Tiu, E., Peters, P. A., Warren, K. S., Mahmoud, A. A. F. and Houser, H. B. (1980). Morbidity in schistosomiasis japonica in relation to intensity of infection: Study of a community in Leyte, Philippines. *American Journal of Tropical Medicine and Hygiene* **29,** 858–867.

Dunne, D. W. and Doenhoff, M. J. (1983). *Schistosoma mansoni* egg antigens and hepatocyte damage in infected T cell-deprived mice. *Contributions in Microbiology and Immunology* **7,** 22–29.

Edungbola, L. D., Cha, Y. N., Bueding, E. and Schiller, E. L. (1982). Granuloma formation around exogenous eggs of *Schistosoma mansoni* and *Schistosoma japonicum* in mice. *African Journal of Medicine and Medical Science* **11,** 75–79.

Ellner, J. J., Olds, G. R., Osman, G. S., Elkholy, A. and Mahmoud, A. A. F. (1981). Dichotomies in reactivity to worm antigen in human schistosomiasis mansoni. *Journal of Immunology* **126,** 309–312.

Elsaghier, A. A. F. and McLaren, D. J. (1989). *Schistosoma mansoni*: Evidence that vascular abnormalities correlate with the "nonresponsiveness" trait in 129/Ola mice. *Parasitology* **99,** 377–381.

Elsaghier, A. A. F., Knopf, P. M., Mitchell, G. F. and McLaren, D. J. (1989). *Schistosoma mansoni*: Evidence that nonpermissiveness in 129/Ola mice involves worm relocation and attrition in the lungs. *Parasitology* **99,** 365–375.

Erickson, D. G., Jones, C. E. and Tang, D. B. (1974). Schistosomiasis mansoni, haematobia and japonica in hamsters: Liver granuloma measurements. *Experimental Parasitology* **35,** 425–433.

Fanning, M. M. and Kazura, J. W. (1984). Genetic-linked variation in susceptibility in mice to schistosomiasis mansoni. *Parasite Immunology* **6,** 95–103.

Fanning, M. M. and Kazura, J. W. (1985). Further studies on genetic variation of hepatosplenic disease and modulation in murine schistosomiasis mansoni. *Parasite Immunology* **7,** 213–222.

Fanning, M. M., Peters, P. A., Davis, R. S., Kazura, J. W. and Mahmoud, A. A. F. (1981). Immunopathology of murine infection with *Schistosoma mansoni*: Relationship of genetic background to hepatosplenic disease and modulation. *Journal of Infectious Diseases* **144,** 148–153.

Faust, E. C. and Meleney, H. E. (1924). Studies on schistosomiasis japonica. *American Journal of Hygiene, Monograph Series* 1–339.

Ford, M. J., Bickle, Q. D., Taylor, M. G. and Andrews, B. J. (1984). Passive transfer of resistance and the site of immune-dependent elimination of the challenge infection in rats vaccinated with highly irradiated cercariae of *Schistosoma mansoni*. *Parasitology* **89,** 461–482.

Garb, K. S. and Stavitsky, A. B. (1984). Depressed *in vitro* and *in vivo* antibody response and adoptive transfer of delayed hypersensitivity to myoglobin with spleen cells of mice chronically infected with *Schistosoma japonicum* injected with myoglobin. *Infection and Immunity* **43,** 1097–1099.

Garcia, E. G. and Mitchell, G. F. (1982). Vaccination against severe hepatosplenic disease in schistosomiasis japonica: An hypothesis. *Acta Medicina Philippina* **18,** 107–112.

Garcia, E. G. and Mitchell, G. F. (1987). Immunology of resistance to infection and disease in schistosomiasis japonica. *In* "Immunology, Immunopathology and Immunoprophylaxis of Parasitic Infections" (E. J. L. Soulsby, ed.), Vol. II, pp. 35–48. CRC Press, Boca Raton, Florida.

Garcia, E. G., Tiu, W. U. and Mitchell, G. F. (1983a). Innate resistance to *Schistosoma japonicum* in a proportion of 129/J mice. *Journal of Parasitology* **69,** 613–615.

Garcia, E. G., Mitchell, G. F., Tapales, F. P. and Tiu, W. U. (1983b). Reduced embryonation of *Schistosoma japonicum* eggs as a contributory mechanism in modulation of granuloma in chronically sensitized mice. *Southeast Asian Journal of Tropical Medicine and Public Health* **14,** 272–273.

Garcia, E. G., Mitchell, G. F., Espinas, F. J. M., Tapales, F. P., Quicho, L. P. and Tiu, W. U. (1984). Further studies on resistance to reinfection with *Schistosoma japonicum* in mice. *Asian Pacific Journal of Allergy and Immunology* **2,** 27–31.

Garcia, E. G., Mitchell, G. F., Beall, J. A. and Tiu, W. U. (1985). *Schistosoma japonicum*: The modulation of lung granuloma and inhibition of maturation of eggs in mice using human sera. *Asian Pacific Journal of Allergy and Immunology* **3,** 156–160.

Garcia, E. G., Mitchell, G. F., Rivera, P. T., Evardome, R. R., Almonte, R. E. and Tiu, W. U. (1987). Evidence of anti-embryonation immunity and egg destruction in mice sensitized with immature eggs of *Schistosoma japonicum. Asian Pacific Journal of Allergy and Immunology* **5,** 137–141.

Garcia, E. G., Rivera, P. T., Mitchell, G. F., Evardome, R. R., Almonte, R. E. and Tiu, W. U. (1989). Effects of induction of anti-embryonation immunity on liver granulomas, spleen weight and portal pressure in mice infected with *Schistosoma japonicum. Acta Tropica* **46,** 93–99.

Georgi, J. R., Wade, S. E. and Dean, D. A. (1987). *Schistosoma mansoni*: Mechanism of attrition and routes of migration from lungs to hepatic portal system in the laboratory mouse. *Journal of Parasitology* **73,** 706–711.

Giron-Garcia, R. (1980). A study on the migratory pathways of *Schistosoma japonicum* from the pulmonary circulation to the portal vein of white mice. M.Sc. thesis, Institute of Public Health, University of the Philippines.

Gönnert, R. (1955). Schistosomiasis—Studies II Uber die Eibildung bei *Schistosoma mansoni* und das Schichsal der Eier in Wirtsorganisms. *Zeitschrift für Tropenmedizen und Parasitologie* **6,** 33–52.

Grimaud, J. A. (1983). Immunopathology of hepatic fibrosis. *Contributions in Microbiology and Immunology* **7,** 190–197.

Grimaud, J. A., Boros, D. L., Takiya, C., Mathew, R. C. and Emonard, H. (1987). Collagen isotypes, laminin, and fibronectin in granulomas of the liver and intestines of *Schistosoma japonicum*-infected mice. *American Journal of Tropical Medicine and Hygiene* **37,** 335–344.

Haas, W., Granzer, M. and Garcia, E. G. (1987). Host identification by *Schistosoma japonicum* cercariae. *Journal of Parasitology* **73,** 568–577.

Harrison, R. A., Bickle, Q. and Doenhoff, M. (1982). Factors affecting the acquisition of resistance against *Schistosoma mansoni* in the mouse: Evidence that the mechanisms which mediate resistance during early patent infections may lack immunological specificity. *Parasitology* **84,** 93–110.

Henkle, K. J., Davern, K. M., Wright, M. D., Ramos, A. J. and Mitchell, G. F. (1990). Comparison of the cloned genes of the 26- and 28-kilodalton glutathione *S*-transferases of *Schistosoma japonicum* and *Schistosoma mansoni. Molecular and Biochemical Parasitology* **40,** 23–34.

Hirata, M., Hieda, T. and Tsutsumi, H. (1986). Immunoglobulins inside *Schistosoma japonicum* eggs from the livers of mice. *Journal of Parasitology* **72,** 578–582.

Hsu, H. F. and Li Hsu, S. Y. (1960a). Susceptibility of albino mice to different geographic strains of *Schistosoma japonicum. Transactions of the Royal Society of Tropical Medicine and Hygiene* **54,** 466–468.

Hsu, H. F. and Li Hsu, S. Y. (1960b). Distribution of eggs of different geographic strains of *Schistosoma japonicum* in the viscera of infected hamsters and mice. *American Journal of Tropical Medicine and Hygiene* **9,** 240–247.

James, S. L. and Colley, D. G. (1976). Eosinophil-mediated destruction of *Schistosoma mansoni* eggs. *Journal of the Reticuloendothelial Society* **20,** 359–374.

James, S. L. and Colley, D. G. (1978). Eosinophil-mediated destruction of *Schistosoma mansoni* eggs *in vitro*. II. The role of cytophilic antibody. *Cellular Immunology* **38,** 35–47.

James, S. L., Pearce, E. J., Lanar, D. and Sher, A. (1987). Induction of cell-mediated immunity as a strategy for vaccination against *Schistosoma mansoni. Acta Tropica* **44,** 50–54.

Jones, J. T. and Kusel, J. R. (1989). Intra-specific variation in *Schistosoma mansoni. Parasitology Today* **5,** 37–39.

Jones, J. T., McCaffery, D. M. and Kusel, J. R. (1983). The influence of the H-2 complex on responses to infection by *Schistosoma mansoni* in mice. *Parasitology* **86,** 19–30.

Kawanaka, M., Hayashi, S. and Ohtomo, H. (1983). Localization of antigen within eggs of *Schistosoma japonicum* that participate in circumoval precipitin (COP) reaction. *Journal of Parasitology* **69,** 444–446.

Lewert, R. M., Yogore, M. G., Jr and Blas, B. L. (1979). Schistosomiasis japonica in Barrio San Antonio, Basey, Samar, the Philippines. I. Epidemiology and morbidity. *American Journal of Tropical Medicine and Hygiene* **28,** 1010–1025.

Liberatos, J. D. (1987). *Schistosoma mansoni*: Male-biased sex ratios in snails and mice. *Experimental Parasitology* **64,** 165–177.

Li Hsu, S. Y. and Hsu, H. F. (1960). Infectivity of the Philippine strain of *Schistosoma japonicum* in *Oncomelania hupensis, O. formosana* and *O. nosophora. Journal of Parasitology* **46,** 793–796.

Li Hsu, S. Y. and Hsu, H. F. (1968). The strain complex of *Schistosoma japonicum* in Taiwan, China. *Zeitschrift für Tropenmedizin und Parasitologie* **19,** 43–59.

Li Hsu, S. Y., Chu, K. Y. and Hsu, H. F. (1963). Drug susceptibility of geographic strains of *Schistosoma japonicum. Zeitschrift für Tropenmedizin und Parasitologie* **14,** 37–40.

Li Hsu, S. Y., Hsu, H. F., Davis, J. R. and Lust, G. C. (1972). Comparative studies on the lesions caused by eggs of *Schistosoma japonicum* and *Schistosoma mansoni* in livers of albino mice and rhesus monkeys. *Annals in Tropical Medicine and Parasitology* **66,** 89–97.

Lin, S. S. and Sadun, E. H. (1959). Studies on the host–parasite relationships to *Schistosoma japonicum*. V. Reactions in the skin, lungs and liver of normal and immune animals following infection with *Schistosoma japonicum. Journal of Parasitology* **45,** 549–559.

Lustigman, S., Mahmoud, A. A. F. and Hamburger, J. (1985). Glycoproteins in soluble egg antigen of *Schistosoma mansoni*: Isolation, characterization and elucidation of their immunochemical and immunopathological relation to the major egg glycoprotein (MEG). *Journal of Immunology* **134,** 1961–1967.

McLaren, D. J. (1989). Will the real target of immunity to schistosomiasis please stand up? *Parasitology Today* **5,** 279–282.

McLaren, D. S. and Smithers, S. R. (1988). Serum from CBA/Ca mice vaccinated with irradiated cercariae of *Schistosoma mansoni* protects naive recipients through the recruitment of cutaneous effector cells. *Parasitology* **97,** 287–302.

Magalhaes-Filho, A. (1959). Pulmonary lesions in mice experimentally infected with *Schistosoma mansoni*. *American Journal of Tropical Medicine and Hygiene* **8,** 527–535.

Mahmoud, A. A. F. (1983). Regulation of immunopathology in parasitic infections. *In* "Regulation of the Immune Response" (P. O. Ogra and D. M. Jacobs, eds), pp. 267–277. Karger, Basel.

Mangold, B. L. and Dean, D. A. (1986). Passive transfer with serum and IgG antibodies of irradiated cercaria-induced resistance against *Schistosoma mansoni* in mice. *Journal of Immunology* **136,** 2644–2648.

Mathew, R. C. and Boros, D. I. (1986). Anti-L3T4 antibody treatment suppresses hepatic granuloma formation and abrogates antigen-induced interleukin-2 production in *Schistosoma mansoni* infection. *Infection and Immunity* **54,** 820–826.

Meleney, H. E., Sandground, J. H., Moore, D. V., Most, H. and Carney, B. H. (1953). The histopathology of experimental schistosomiasis. II. Bisexual infections with *S. mansoni, S. japonicum* and *S. haematobium*. *American Journal of Tropical Medicine and Hygiene* **2,** 883–915.

Merenlender, A. M., Woodruff, D. S., Upatham, E. S., Viyanant, V. and Yuan, H. C. (1987). Large genetic distance between Chinese and Philippine *Schistosoma japonicum*. *Journal of Parasitology* **73,** 861–863.

Metzger, J. M. and Peterson, L. B. (1988). Cyclosporin A enhances the pulmonary granuloma response induced by *Schistosoma mansoni* eggs. *Immunopharmacology* **15,** 103–116.

Miller, P. and Wilson, R. A. (1980). Migration of the schistosomula of *Schistosoma mansoni* from the lungs to the hepatic portal system. *Parasitology* **80,** 267–288.

Mitchell, G. F. (1989a). Glutathine *S*-transferases—Potential components of anti-schistosome vaccines? *Parasitology Today* **5,** 34–37.

Mitchell, G. F. (1989b). Portal system peculiarities may contribute to resistance in 129/J mice against schistosomiasis. *Parasite Immunology* **11,** 713–717.

Mitchell, G. F. (1989c). Vaccines and vaccination strategies: Helminths. *In* "Parasites: Molecular Biology, Drug and Vaccine Design" (N. Agabian and A. Cerami, eds), pp. 349–363. UCLA Symposium on Molecular and Cellular Biology, New Series, Vol. 131. Alan R. Liss, New York.

Mitchell, G. F. (1990a). Immunopathology of schistosomiasis. *In* "Reviews in Medical Microbiology" (R. J. Williams, ed.), Vol. 1, pp. 101–107. Churchill Livingstone, Edinburgh.

Mitchell, G. F. (1990b). The Wellcome Trust Lecture: Problems specific to parasite vaccines. *Parasitology* **98,** S19-S28.

Mitchell, G. F. and Cruise, K. M. (1986). Schistosomiasis: Antigens and host–parasite interactions. *In* "Parasite Antigens: Towards New Strategies for Vaccines" (T. W. Pearson, ed.), pp. 275–316. Marcel Dekker, New York.

Mitchell, G. F., Garcia, E. G., Anders, R. F., Valdez, C. A., Tapales, F. P. and Cruise, K. M. (1981). *Schistosoma japonicum*: Infection characteristics in mice of various strains and a difference in the response to eggs. *International Journal of Parasitology* **11,** 267–276.

Mitchell, G. F., Cruise, K. M., Garcia, E. G., Vadas, M. A. and Munoz, J. J. (1983). Attempts to modify lung granulomatous responses to *Schistosoma japonicum* eggs in low and high responder mouse strains. *Australian Journal of Experimental Biology and Medical Science* **61,** 411–424.

Mitchell, G. F., Garcia, E. G., Davern, K. M., Tiu, W. U. and Smith, D. B. (1988). Sensitization against the parasite antigen Sj26 is not sufficient for consistent expression of resistance to *Schistosoma japonicum. Transactions of the Royal Society of Tropical Medicine and Hygiene* **82,** 885–889.

Mitchell, G. F., Garcia, E. G., Wood, S. M., Diasanta, R., Almonte, R., Calica, E., Davern, K. M. and Tiu, W. U. (1990). Studies on the sex ratio of worms in schistosome infections. *Parasitology* **101,** 27–34.

Mitchell, G. F., Wright, M. D., Wood, S. M. and Tiu, W. U. (in press, a). Further studies on variable resistance of 129/J and C57BL/6 mice to infection with *Schistosoma japonicum* and *Schistosoma mansoni. Parasite Immunology.*

Mitchell, G. F., Davern, K. M., Wood, S. M., Wright, M. D., Argyropoulos, V. P., McLeod, K. S., Tiu, W. U. and Garcia, E. G. (in press, b). Attempts to induce resistance in mice to *Schistosoma japonicum* and *Schistosoma mansoni* by exposure to crude schistosome antigens plus cloned glutathione *S*-transferases. *Immunology and Cell Biology.*

Miyagawa, Y. and Takemoto, S. (1921). The mode of infection of *S. japonicum* and the principal route of its journey from the skin to the portal vein in the host. *Journal of Pathology and Bacteriology* **24,** 168–174.

Moloney, N. A. and Webbe, G. (1983). The host–parasite relationship of *Schistosoma japonicum* in CBA mice. *Parasitology* **87,** 327–342.

Moloney, N. A., Webbe, G. and Luty, A. J. (1984). Factors affecting the acquisition of resistance to *Schistosoma japonicum* in the mouse. I. The correlation between egg deposition and worm elimination. *Parasitology* **89,** 345–360.

Moloney, N. A., Garcia, E. G. and Webbe, G. (1985). The strain specificity of vaccination with ultra violet attenuated cercariae of the Chinese strain of *Schistosoma japonicum. Transactions of the Royal Society of Tropical Medicine and Hygiene* **79,** 245–247.

Moloney, N. A., Luty, A. J. F., Hinchcliffe, P. and Webbe, G. (1987a). The involvement of eggs and soluble egg antigens in resistance to *Schistosoma japonicum. Journal of Helminthology* **61,** 9–17.

Moloney, N. A., Hinchcliffe, P. and Webbe, G. (1987b). Passive transfer of resistance to mice with sera from rabbits, rats or mice vaccinated with ultra violet-attenuated cercariae of *Schistosoma japonicum. Parasitology* **94,** 497–508.

Moloney, N. A., Hinchcliffe, P. and Webbe, G. (1987c). Loss of resistance to reinfection with *Schistosoma japonicum* in mice after treatment with praziquantel. *Transactions of the Royal Society of Tropical Medicine and Hygiene* **81,** 247–254.

Moore, D. E. and Warren, K. S. (1967). Hepatosplenic schistosomiasis mansoni and japonica compared in mice each infected with one pair of worms. *Transactions of the Royal Society of Tropical Medicine and Hygiene* **61,** 104–109.

Mountford, A. P., Coulson, P. S., Saunders, N. and Wilson, R. A. (1989). Characteristics of protective immunity induced by drug-attenuated larvae of *Schistosoma mansoni*: Antigen localization and antibody responses. *Journal of Immunology* **143,** 989–995.

Norden, A. P. and Strand, M. (1984). *Schistosoma mansoni, S. haematobium* and *S. japonicum*: Identification of genus- and species-specific antigens in egg glycoproteins. *Experimental Parasitology* **58,** 333–344.

Olds, G. R. (1989). Immunopathology and resistance in helminth infections. *In* "Parasitic Infections" (J. H. Leech, M. A. Sande and R. K. Root, eds), pp. 61–80. Churchill Livingstone, New York.

Olds, G. R. and Kressina, T. F. (1985). Network interactions in *Schistosoma japonicum* infection: Identification and characterization of serologically-distinct immunoregulatory auto-anti-idiotypic antibody population. *Journal of Clinical*

Investigation **76,** 2338–2347.

Olds, G. R. and Mahmoud, A. A. F. (1980). Role of host granulomatous response in murine schistosomiasis mansoni. *Journal of Clincal Investigation* **66,** 1191–1199.

Olds, G. R. and Stavitsky, A. B. (1986). Mechanisms of *in vivo* modulation of granulomatous inflammation in murine schistosomiasis japonica. *Infection and Immunity* **52,** 513–518.

Olds, G. R., Olveda, R., Tracy, J. W. and Mahmoud, A. A. F. (1982). Adoptive transfer of modulation of granuloma formation and hepatosplenic disease in murine schistosomiasis japonica by serum from chronically infected animals. *Journal of Immunology* **128,** 1391–1393.

Olds, G. R., Finegan, C. and Kresina, T. F. (1986). Dynamics of hepatic glycosaminoglycan accumulation in murine *Schistosoma japonicum* infection. *Gastroenterology* **91,** 1335–1342.

Olivier, L. (1952). A comparison of infections in mice with three species of schistosomes, *Schistosoma mansoni*, *Schistosoma japonicum* and *Schistosomatium douthitti*. *American Journal of Hygiene* **55,** 22–35.

Ottesen, E. A., Hiatt, R. A., Cheever, A. W., Sotomayor, Z. R. and Neva, F. A. (1978). The acquisition and loss of antigen specific cellular immune responsiveness in acute and chronic schistosomiasis in man. *Clinical and Experimental Immunology* **33,** 38–56.

Parra, J. C., Lima, M. S., Gazzinelli, G. and Colley, D. G. (1988). Immune responses during human schistosomiasis mansoni. XV. Anti-idiotype T cells can recognize and respond to anti-SEA idiotypes directly. *Journal of Immunology* **140,** 2401–2405.

Pearce, E. J. and James, S. L. (1986). Post lung stage schistosomula of *Schistosoma mansoni* exhibit transient susceptibility to macrophage-mediated cytotoxicity *in vitro* that may relate to late phase killing *in vivo*. *Parasite Immunology* **8,** 513–527.

Pelley, R. P., Pelley, R. J., Hamburger, J., Peters, P. A. and Warren, R. S. (1976). *Schistosoma mansoni* soluble egg antigens. I. Identification and purification of three major antigens, and the employment of radioimmunoassay for their further characterization. *Journal of Immunology* **117,** 1553–1560.

Pesigan, T. P., Farooq, M., Hairston, N. G., Jauregui, J. J., Garcia, E. G., Santos, A. T., Santos, B. C. and Besa, A. A. (1958). Studies on *Schistosoma japonicum* in the Philippines. I. General consideration and epidemiology. *Bulletin of the World Health Organization* **18,** 345–455.

Phillips, S. M. and Lammie, P. J. (1986). Immunopathology of granuloma formation and fibrosis in schistosomiasis. *Parasitology Today* **2,** 296–312.

Pons, H. A., Morgan, J. S., Hutchinson, M. L., Rojkind, M., Groszmann, R. J. and Stadecker, M. J. (1989). Resistance to reinfection in experimental murine schistosomiasis: Role of porto-hepatic hemodynamics. *American Journal of Tropical Medicine and Hygiene* **41,** 189–197.

Reis, M. G. and Andrade, Z. A. (1987). Functional significance of periovular granuloma in schistosomiasis. *Brazilian Journal of Medical and Biological Research* **20,** 55–62.

Sadun, E. H. (1963). Immunization in schistosomiasis by previous exposure to homologous and heterologous cercariae by inoculation of preparations from schistosomes and by exposure to irradiated cercariae. *Annals of the New York Academy of Sciences* **113,** 418–439.

Sadun, E. H., Lin, S. S. and Williams, J. E. (1958). Studies on the host–parasite relationships in *Schistosoma japonicum*: The effects of single graded infections and the route of migration of schistosomula. *American Journal of Tropical Medicine and Hygiene* **7,** 434–449.

Saint, R. B., Beall, J. A., Grumont, R. J., Mitchell, G. F. and Garcia, E. G. (1986). Expression of *Schistosoma japonicum* antigens in *Escherichia coli*. *Molecular and Biochemical Parasitology* **18,** 333–342.

Sher, A., McIntyre, S. and Von Lichtenberg, F. (1977). *Schistosoma mansoni*: Kinetics and class specificity of hypergammaglobulinaemia induced during murine infection. *Experimental Parasitology* **41,** 415–422.

Sher, A., James, S., Correa-Oliveira, R., Hieny, S. and Pearce, E. (1989). Schistosome vaccines: Current progess and future prospects. *Parasitology* **98,** S61–S68.

Sidner, R. A., Carter, C. E. and Colley, D. G. (1987). Modulation of *Schistosoma japonicum* pulmonary egg granulomas with monoclonal antibodies. *American Journal of Tropical Medicine and Hygiene* **36,** 361–370.

Smith, D. B., Davern, K. M., Board, P. G., Tiu, W. U., Garcia, E. G. and Mitchell, G. F. (1986). Mr 26,000 antigen of *Schistosoma japonicum* recognized by resistant WEHI 129/J mice is a parasite glutathione *S*-transferase. *Proceedings of the National Academy of Sciences (USA)* **83,** 8703–8707.

Smith, M. A., Clegg, J. A., Kusel, J. R. and Webbe, G. (1975). Lung inflammation in immunity to *Schistosoma mansoni*. *Experientia* **31,** 26–28.

Smithers, S. R. and Doenhoff, M. (1982). Schistosomiasis. *In* "Immunology of Parasitic Infections" (S. Cohen and K. S. Warren, eds), 2nd edn, pp. 527–607. Blackwell, Oxford.

Smithers, S. R., Simpson, A. J., Yi, X., Omer-Ali, P., Kelly, C. and McLaren, D. J. (1987). The mouse model of schistosome immunity. *Acta Tropica* **12,** supplement, 21–30.

Stavitsky, A. B. (1987). Immune regulation of schistosomiasis japonica. *Immunology Today* **8,** 228–233.

Sueyasu, Y. (1920). On the migratory course of *Schistosoma japonicum* in the body of the final host. *Journal of the Kyoto Medical Association* **17,** 1–3 (cited by Sadun *et al.*, 1958).

Swanson, V. L. and Williams, J. E. (1963). Pathology of *Schistosoma japonicum* in the Taiwanese monkey (*Macaca cyclopsis*). I. Comparison of Formosan and Japanese strains. *American Journal of Tropical Medicine and Hygiene* **12,** 748–752.

Taylor, M. G. and Bickle, Q. D. (1986). Irradiated schistosome vaccines. *Parasitology Today* **2,** 132–134.

Tiu, W. U., Davern, K. M., Garcia, E. G., Moll, H. and Mitchell, G. F. (1989). Monoclonal antibodies reacting with *Schistosoma japonicum* eggs and their target epitopes. *Acta Tropica* **46,** 75–92.

Usawattanakul, W., Kamijo, T. and Kojima, S. (1982). Comparison of recovery of schistosomula of *Schistosoma japonicum* from lungs of mice and rats. *Journal of Parasitology* **68,** 783–790.

Van Snick, J. L. and Masson, P. L. (1980). Incidence and specificities of IgA and IgM anti-IgG autoantibodies in various mouse strains and colonies. *Journal of Experimental Medicine* **151,** 45–55.

Vignali, D. A., Klaus, S. N., Bickle, Q. D. and Taylor, M. G. (1989). Histological examination of the cellular reactions around schistosomula of *Schistosoma mansoni* in the lungs of sublethally irradiated and unirradiated immune and control rats. *Parasitology* **98,** 57–65.

Vogel, H. (1942). Uber Entwihling, Lebensdauer und Tod der Eier von Bilharzia japonica in Wirtsgewebe. *Deutsche Tropenmedizinische Zeitschrift* **46,** 57–69.

Vogel, H. and Minning, W. (1953). The acquired resistance of *Macacus rhesus* to *Schistosoma japonicum*. *Zeitschrift für Tropenmedezin und Parasitologie* **4,** 418–505.

Von Lichtenberg, F. (1987). Consequences of infection with schistosomes. *In* "The Biology of Schistosomes: From Genes to Latrines" (D. Rollinson and A. J. G. Simpson, eds), pp. 185–232. Academic Press, London and San Diego.

Von Lichtenberg, F. and Byram, J. E. (1980). Pulmonary cell reactions in natural and acquired host resistance to *Schistosoma mansoni*. *American Journal of Tropical Medicine and Hygiene* **29,** 1286–1300.

Von Lichtenberg, F., Sadun, E. H. and Bruce, J. I. (1963). Host response to eggs of *Schistosoma mansoni*. III. The role of egg in resistance. *Journal of Infectious Diseases* **133,** 113–122.

Von Lichtenberg, F., Erickson, D. G. and Sadun, E. H. (1973). Comparative histopathology of schistosoma granulomas in the hamster. *American Journal of Pathology* **72,** 149–178.

Von Lichtenberg, F., Sher, A. and McIntyre, S. (1977). A lung model of schistosome immunity in mice. *American Journal of Pathology* **87,** 105–120.

Ward, R. E. and McLaren, D. J. (1989). *Schistosoma mansoni*: Migration and attrition of challenge parasites in naive rats and rats protected with vaccine sera. *Parasite Immunology* **11,** 125–146.

Warren, K. S. (1982). The secret of immunopathogenesis of schistosomiasis: *In vivo* models. *Immunological Reviews* **61,** 189–213.

Warren, K. S. and Berry, E. G. (1972). Induction of hepatosplenic disease by single pairs of the Philippine, Formosan, Japanese and Chinese strains of *Schistosoma japonicum*. *Journal of Infectious Diseases* **126,** 482–491.

Warren, K. S., Boros, D. L., Hang, L. M. and Mahmoud, A. A. F. (1975). The *Schistosoma japonicum* egg granuloma. *American Journal of Pathology* **80,** 279–293.

Weiss, J. B., Aronstein, W. S. and Strand, M. (1987). *Schistosoma mansoni*: Stimulation of artificial granuloma formation *in vivo* by carbohydrate determinants. *Experimental Parasitology* **64,** 228–236.

Wheater, P. R. and Wilson, R. A. (1979). *Schistosoma japonicum*: A histological study of migration in the laboratory mouse. *Parasitology* **79,** 49–62.

Wilks, N. E. (1967). Lung to liver migration of schistosomes in the laboratory mouse. *American Journal of Tropical Medicine and Hygiene* **16,** 599–605.

Wilson, R. A. and Coulson, P. S. (1989). Where and how does immune elimination of *Schistosoma mansoni* occur? *Parasitology Today* **5,** 274–278.

Wilson, R. A., Coulson, P. S. and McHugh, S. M. (1983). A significant part of the "concomitant immunity" of mice to *Schistosoma mansoni* is a consequence of a leaky hepatic portal system, not immune killing. *Parasite Immunology* **5,** 595–601.

Woodruff, D. S., Merenlender, A. M., Upatham, E. S. and Viyanaut, V. (1987). Genetic variants and differentiation of three *Schistosoma* species from the Philippines, Laos and Peninsula Malaysia. *American Journal of Tropical Medicine and Hygiene* **36,** 345–354.

Wright, M. D., Tiu, W. U., Wood, S. M., Walker, J. C., Garcia, E. G. and Mitchell, G. F. (1988a). *Schistosoma mansoni* and *S. japonicum* worm numbers in 129/J mice of two types and dominance of susceptibility in F1 hybrids. *Journal of Parasitology* **74,** 618–622.

Wright, M. D., Rogers, M. V., Davern, K. M. and Mitchell, G. F. (1988b). *Schistosoma mansoni* antigens differentially recognized by resistant WEHI 129/J mice. *Infection and Immunity* **56,** 2948–2952.

Wu, S.-S., Peng, L.-X. and Zhao, W.-X. (1989). Cellular responses elicited by challenge schistosomula of *Schistosoma japonicum* in the skin of naive and chronically infected mice. *Chinese Journal of Parasitology and Parasitic Diseases* **7,** 259–262.

Wyler, D. J., Lammie, P. J., Michael, A. I., Rosenwasser, L. J. and Phillips, S. M. (1987). *In vitro* and *in vivo* evidence that autoimmune reactivity to collagen develops spontaneously in *Schistosoma mansoni* infected mice. *Clinical Immunology and Immunopathology* **44,** 140–148.

Xu, X.-J. and Ni, C.-H. (1987). Observation on the difference in susceptibility of *Oncomelania* snails to different isolates of *Schistosoma japonicum*. *Chinese Journal of Parasitology and Parasitic Diseases* **5,** 25–28.

Yolles, T. K., Moore, D. V., DeGuisti, D. L., Ripsom, C. A. and Meleney, H. E. (1947). A technique for the perfusion of laboratory animals for the recovery of schistosomes. *Journal of Parasitology* **33,** 419–426.

Zhang, J.-S. and Zhu, J.-H. (1989). Studies on antibody-dependent cell-mediated cytotoxicity to *Schistosoma japonicum* schistosomula in mice. *Chinese Journal of Parasitology and Parasitic Disease* **7,** 271–275.

Zhang, W.-C., Chen, J.-H., Zeng, Q.-R., Feng, D.-C., Cai, G.-D. and Wu, H.-N. (1983). Artificial induction of specific immune modulation of *Schistosoma japonicum* egg granulomas. *Journal of Parasitology and Parasitic Diseases* **1,** 154–157.

Influence of Pollution on Parasites of Aquatic Animals

R. A. KHAN

Department of Biology and Ocean Sciences Centre, Memorial University of Newfoundland, St John's, Newfoundland, Canada A1C 5S7

AND

J. THULIN

The National Environment Protection Board, Marine Section, Box 584, S-74071, Öregrund, Sweden

I. Introduction

Pollutants are being discharged continuously into the aquatic environment and there is increasing concern about their impact on the ecosystem. Some major long-term threats to the aquatic biota are chronic pollution in

ADVANCES IN PARASITOLOGY VOL. 30
ISBN 0-12-031730-3

estuarine and coastal areas originating from the land, and oil spills and seeps from tankers and drilling rigs offshore. A pollutant acting alone might not necessarily be harmful, but in combination with others could induce stress. During the last decade, it has been increasingly clear that several fish diseases and abnormalities occur with increased prevalences in wild fish populations in highly polluted areas. While it has been difficult to relate concentrations of specific pollutants in water and sediments directly to disease, there are indications of such a connection in some organisms (Hodgins *et al.*, 1977; Murchelano, 1982; Lindesjöö and Thulin, 1987; Overstreet, 1988). The literature on the possible relation between diseases and pollution in the aquatic environment has been comprehensively reviewed by Sindermann (1982), Mix (1986) and O'Connor and Huggett (1988). Most of the infectious diseases dealt with are caused by viruses and bacteria. However, fish generally harbour a wide range of ecto- and endoparasites. The latter as well as their hosts might be affected in a number of different ways by contaminants. Thus, pollutants might influence, directly or indirectly, the prevalence, intensity and pathogenicity of a parasite. Further, fish may, as a consequence of parasitization, show increased susceptibility to toxic effects. There is a developing body of literature, from experimental as well as from field studies, about the interrelation between pollution and fish parasites. The objective of this chapter is to review our present knowledge of this relationship. In attempting to do so, we examine how some different aquatic pollutants enter and affect the fish, which constitute the microhabitat of the parasites. We also deal with a series of published and unpublished studies concerned with the influence of pollutants on ecto- and endoparasites of aquatic organisms. Finally, we make suggestions for areas in need of further research.

II. Pollutants and their Entry into Fish

There are several categories of aquatic pollutants. Some of the more important are sewage, pesticides, polychlorinated biphenyls, heavy metals, pulp and paper effluents and petroleum aromatic hydrocarbons (PAHs). Some lubricating muds used in drilling operations contain heavy metals and, contaminated with petroleum, can also affect benthic fauna. More recently, acid rain has become a threat to the survival of biota living in freshwater habitats.

Pollutants enter aquatic organisms via three routes: the mouth, integument or gills. Those entering via the mouth pass through the gastrointestinal tract and are subsequently taken up into the blood or voided in the faeces, or both. Contaminants that enter through the integument and gills are carried by the blood to various tissues and organs. Lipophilic organic compounds, including many xenobiotics, are detoxified in the liver by oxidation catalysed by mixed function oxygenase and metabolites are either excreted in the urine or pass into the gall bladder before elimination in the faeces. Occasionally, toxification by the mixed function oxygenase system can occur, i.e. PAHs can be hydrolysed into a carcinogen, benzo(a)pyrene. Storage (bioaccumulation) might occur in tissues where lipid is stored (Connell and Miller, 1981). These authors cite a review which indicated greater storage and persistence of petroleum hydrocarbons in lipid-rich than in lipid-poor species of fish. However, some petroleum hydrocarbons (toluene, olefins, etc.) may accumulate in muscle and ultimately cause tainting in fish (Ogata *et al.*, 1987).

Metals are released as liquid wastes from urban and industrial developments, fossil fuels, mines and metal smelters, and leeched from soil by acid rain. Uptake by aquatic biota occurs as free ions across respiratory surfaces and in organic form through the food chain (Hodson, 1988). Subtle effects from long-term exposure include impairment of growth and reproduction and occasionally behaviour. Metallothionein (a low molecular weight protein rich in cysteine) is believed to detoxify heavy metals, especially copper and zinc, through competitive binding.

Bleached kraft mill effluents are also known to affect aquatic organisms. Andersson *et al.* (1988) reported profound effects of the effluents which, in perch (*Perca fluviatilis*), resulted in reduced gonadal growth, enlarged liver and induction of cytochrome P450-dependent activities. Metabolic disorders in the form of elevated levels of ascorbic acid and disturbed carbohydrate metabolism, a suppressed immune system and impaired gill function were also observed in fish collected in the vicinity of a discharge site.

III. Effects of Pollutants on Fish

Toxic effects of pollutants can be assessed at three different levels: that of the population (examining the fate of indicator species), the organism (biochemical and physiological changes, etc.) and the organ, tissue and cell (histopathological changes in structure and function). Subtle effects, especially following long-term (chronic) exposure to low (sub-lethal) levels, become apparent through disturbances in feeding, growth, behaviour and reproduction. PAHs have been reported to inhibit food consumption in winter flounder (*Pseudopleuronectes americanus*) and Atlantic cod (*Gadus morhua*) (Fletcher *et al.*, 1981, 1982; Kiceniuk and Khan, 1983, 1987). As a result, growth and body condition were significantly depressed in treated fish compared with controls. Affected cod swam on their sides in an erratic manner and tended to remain at the top of the water column. Flounder avoided burrowing into contaminated sediment during non-feeding periods and presumably in nature become easy prey to predators.

Many classes of pollutants are known to affect the organs and tissues of fish (Malins *et al.*, 1985). These include pesticides (Couch, 1975), heavy metals (Trump *et al.*, 1975; Tafanelli and Summerfelt, 1975; Bengtsson *et al.*, 1988), polychlorinated biphenyls (Freeman *et al.*, 1982), pulp mill effluent (Couillard *et al.*, 1988; Thulin *et al.*, 1989) and petroleum hydrocarbons (McCain *et al.*, 1978; Haensley *et al.*, 1982; Solangi and Overstreet, 1982; Khan and Kiceniuk, 1984). Low gonadal somatic indices and histological anomalies were reported in the ovaries of plaice (*Pleuronectes platessa*) after the *Amoco Cadiz* oil spill along the coast of Brittany, France (Lopez *et al.*, 1981). Similarly, atretic follicles were observed in the ovaries of fish collected near petroleum oil and gas platforms in the Gulf of Mexico (Scott *et al.*, 1981). PAHs are also known to produce lesions in male cod and affect testicular growth or delay spermatogenesis, or both (Fletcher *et al.*, 1982; Truscott *et al.*, 1983; Khan and Kiceniuk, 1984). In some instances, organs were enlarged (Fletcher *et al.*, 1981) or decreased in size (Kiceniuk and Khan, 1987). Excessive mucus secretion, capillary dilation, lamellar hyperplasia and fusion of gill filaments occur in fish exposed to PAHs (Solangi and Overstreet, 1982; Hawkes, 1977; Haensly *et al.*, 1982; Khan and Kiceniuk, 1984) and a variety of other pollutants (Walsh and Ribellin, 1975; Eller, 1975; Mallatt, 1985). Lesions also occur in the liver (Couch, 1975), kidney (Trump *et al.*, 1975) and spleen (Haensly *et al.*, 1982). A characteristic finding in the spleen is an increase of melanomacrophage centres following exposure to some pollutants (Haensly *et al.*, 1982; Solangi and Overstreet, 1982; Khan and Kiceniuk, 1984). Dokholyan *et al.* (1980) reported decreased levels of haemoglobin and numbers of erythrocytes in sturgeon exposed to PAHs. Some organic pollutants such as benzene can adversely affect haemopoietic tissue function (Taberski, 1983). PAHs also affect the skin of fish culminating in lesions associated with fin erosion, ulcers, papillomas and lymphocystosis.

The physiological mechanisms in fish are also influenced by pollutants. Sabo and Stegman (1977) reported glycogen and lipid depletion in *Fundulus heteroclitus* collected near an oil spill site. Triglyceride levels were also lower in oiled fish than in untreated specimens. Dey *et al.* (1983) noted that bile acids and other physiological components in flounder and cod were altered following exposure to petroleum. Fletcher *et al.* (1981, 1982) reported changes in electrolyte balance in flounder after exposure to oil-contaminated sediment. Krahn *et al.* (1986) showed a positive correlation between hepatic lesions in English sole (*Parophrys vetulus*) from Puget sound, Washington, USA and concentration of metabolites of aromatic hydrocarbons in bile. Polycyclic aromatic hydrocarbons are metabolized by the liver and excreted via bile in faeces, but some components can pass through intestinal cells and enter the enterohepatic circulation (Chipman, 1982). Because some metabolites such as benzo(a)pyrene are carcinogenic, it is not surprising that liver lesions were observed in sole taken from a heavily oil-polluted area, which contained high concentrations of aromatic (e.g. benzo(a)pyrene) and heterocyclic hydrocarbons (e.g. carbazoles) in both fish and sediment (Malins and Roubal, 1985).

Pollutants are also known to affect the immune response in animals (Wojdani and Alfred, 1984). A variety of pollutants in the aquatic environment can alter a fish's defence mechanisms (Zeeman and Brindley, 1981). Coagulation of mucus (Burton *et al.*, 1972), damage to tissue barriers (Di Michele and Taylor, 1978) and depression of cellular and humoral immunity may occur (Robohm and Nitkowski, 1974). Leucopenia has been reported in sturgeon (*Acipenser gueldenstaedti*) exposed to PAHs (Dokholyan *et al.*, 1980). Weeks and Warinner (1984) reported markedly reduced phagocytic efficiency by macrophages in two species of bottom feeding fish—the spot (*Leiostomus xanthurus*) and hogchoker (*Trinectes maculatus*)—taken from the polluted River Elizabeth in Virginia, USA. Low levels of four heavy, water-borne metals (nickel, zinc, copper and cadmium) suppressed the immune response of *Salmo trutta* and *Cyprinus carpio* during a 38-week exposure period (O'Neill, 1981). Heavy metals such as cadmium at a concentration of 50 parts/10^6 caused an increase in the proportion of eosinophils from 5 to 51% within 24 hours. Fries (1986) reported that three chemicals—benzo(a)pyrene, pentachlorophenol and hexachlorobenzene—suppressed the immune response in *Fundulus heteroclitus* 2 days after intraperitoneal injection. The rosette-forming cells (lymphocytes) in untreated immunized control fish were about twice as numerous as in the chemically treated fish. The chlorinated hydrocarbons suppressed the immune response more than the benzo(a)pyrene. It is likely that exposure to pollutants, which act as stressors, culminates in the release of cortisol (Wedemeyer, 1970; Pickering, 1981), a well-known immunosuppressant, the level of which fluctuates in stressed fish (Di Michele and Taylor, 1978). Fish exposed to pollutants also consume less food, which in turn can affect the level of the immune response. Additionally, it has been reported that the level of mixed function oxygenase, a detoxifying enzyme, decreases after immunosuppression (Beisel, 1982; Hansen *et al.*, 1982). Consequently, it is not surprising that long-term exposure to toxic substances that suppress the immune response is associated with a variety of diseases and abnormalities in aquatic animals (Snieszko, 1974; Sindermann, 1979).

Abnormalities and diseases associated with pollution have been reported from a number of regions and countries. Mix (1986), after critically reviewing published reports, concluded that cancerous diseases in fish from Puget

Sound (Washington) and Fox River (Illinois), both in the USA, and from Japan were associated with chemical contaminants in the environment. However, information linking pollution and tumours in fish from other areas in the USA was inadequate to draw any conclusions. Mix (1986) also cited surveys conducted in highly polluted areas in Yugoslavia (Sava River), Germany (Rhine and Elbe rivers) and Australia (Port Phillip Bay), which reported no connection between pollutants and neoplasms in fish. However, it was reported recently that there were high prevalences (20–50%) of diseased fish in the North Sea, German Bight and the River Elbe (Anon., 1987). On the other hand, only a small percentage (ca. 1%) of marine fish in the north-west Atlantic were diseased (Ziskowski *et al.*, 1987).

Recently, a workshop on the use of pathology in the study of the effects of contaminants was held by the International Council for the Exploration of the Sea (Thulin, 1986). Some Danish, German and Dutch workers provided evidence of a correlation between pollution and disease in fish, while others did not. Thus, Dethlefsen (1984, 1986) and Wolthaus (1984) noted a high prevalence of external disease (lymphocystosis, ulceration, epidermal papilloma, hyperplasia, etc.) on dab (*Limanda limanda*) in the centre of the German Bight where pollution was high. He cited the work of Vethaak, whose studies over a 2-year period showed the prevalences of external and internal abnormalities of 15 000 fish to be 18.7 and 5.8%, respectively. However, researchers from the British Isles could find no evidence of a correlation between pollution and disease (McVicar *et al.*, 1988). This was supported by the findings of Möller (1988), who was of the opinion that stress induced by environmental changes, selection pressures, inadequate nutrition, parasitism, fishing pressures and pollutants all contribute towards the diseased condition. While it was agreed that the problem of "linking pollution with disease is much more complex than hitherto considered" (Möller, 1987a, 1988), it was also considered that such a link was becoming more evident than previously. Finally, the workshop recommended that "the potential for the use of ectoparasites as indicators of environmental changes, including pollution effects, should be explored as a promising new approach" (Thulin, 1986).

Fish are hosts to many species of parasites (Möller and Anders, 1986) which under natural conditions cause little or no harm; however, the combined effect of parasites and pollutants could be synergistic and ultimately harmful.

IV. Influence of Pollutants on Ectoparasites

A. Ciliates

Overstreet and Howse (1977) noted infestations of a peritrich ciliate,

Epistylis sp., a facultative ectosymbiont, on a number of freshwater and marine fish taken from areas of low salinity but rich in organic waste. These authors also reported that secondary infections with systemic bacteria (*Aeromonas hydrophila*), producing haemorrhagic lesions, occurred only in fish from a polluted habitat. Esch *et al.* (1976) reported external lesions in centrarchid fish from a pond in South Carolina into which thermal water from a nuclear production facility flowed. Apparently, epizootic outbreaks of *Epistylis* sp. plus *A. hydrophila* had occurred some time previously. This study revealed external lesions in three species of fish, including small mouth bass (*Micropterus salmoides*), in which a direct correlation between thermal loading, body condition and prevalence of the *Epistylis* + *Aeromonas* complex occurred. It is believed that high water temperatures induce higher metabolic rates resulting in the reduction of immunity. Overstreet (1982), in a review of abiotic factors affecting marine parasitism, cited a number of studies which showed that high temperatures were conducive to increased prevalence and intensity of infestation of parasites.

Dabrowska (1974) studied the parasite fauna and histological changes in fish collected above and below outlets of waste effluents discharged into the Lyna and Walsza rivers in Poland. Trichodinid ciliate infections were observed in about 79% of eight species of fish collected from polluted areas in the Lyna River. Many (70%) of these fish had histological disorders in the gill epithelium and the gill circulatory system. No abnormality was observed in fish taken from apparently unpolluted areas.

Lehtinen *et al.* (1984) observed more heavy infections of trichodinid ciliates on the gills of flounder (*Platichthys flesus*) exposed for 2 months to low levels of bleached pulp kraft mill effluent than in fish exposed to effluents from mills using oxygen and chlorine in the bleaching process. Histopathological changes in the gill and liver correlated with chlorinic bleaching used in the process of pulp production. The chronic exposure of fish to bleached kraft pulp mill effluents can result in biochemical, physiological and pathological disturbances (Couillard *et al.*, 1988; Andersson *et al.*, 1988). Impaired gill function and suppression of the immune response might have contributed to the excessive infestation noted in the flounder.

The prevalence and intensity of trichodinid infections also increase on the gills of fish following exposure to water-soluble oil fractions (WSOF). The longhorn sculpin (*Myoxocephalus octodecemspinosus*) is parasitized by *Trichodina cottidarum* in the north-west Atlantic (Lom and Laird, 1969). Following exposure to oil-contaminated sediment for 12 weeks and a depuration period of 20 weeks, the prevalence and intensity of trichodinids were greater in these fish than in the controls (see Table 2 and Fig. 2). In two separate experiments, the mean number of parasites was greater in oil-treated sculpins than in untreated controls (Khan, unpublished data).

Additional evidence of impaired resistance in fish after exposure to WSOF was observed recently in Atlantic cod (Khan, 1990).Trichodinid infections are rare in adult cod 50 cm in length (prevalence, 2% of 107 fish). Following exposure to WSOF (ca. 100 parts/10^9) for 12 weeks and a depuration period of 14 weeks, both the prevalence and intensity of the parasite had increased substantially in the oil-treated fish compared with the controls. Eighty-eight percent of 24 oil-treated cod harboured a mean of 102.3 ± 3.4 (standard error) trichodinids per 10 gill filaments, whereas only 9% of 23 controls were infected with a mean of 0.9 ± 0.1 parasites. The oil-treated fish displayed gill lesions typical of those exposed to WSOF (Khan and Kiceniuk, 1984). In adult cod that harboured low-grade infections, it is probable that exposure to pollutants altered the hosts' resistance and, coupled with lesions in the gills which created an environment suitable for proliferation, ultimately led to an increased intensity. Because cod tend to "school", spread of the infection between hosts might occur readily.

An opportunity to test the relationship between parasitism and chronic exposure to crude petroleum in nature arose recently following a major oil spill (Khan, 1990). The supertanker *Exxon Valdez*, after running aground in Prince William Sound, Gulf of Alaska, spilled about 44 million litres of crude oil that was subsequently dispersed by strong winds and currents along several thousand square kilometres of coastline including the Kenai Peninsula. Samples of gill tissue from an intertidal sculpin, *Oligocottus* sp., were collected and examined from two areas, namely Seward, a site apparently not contaminated by the oil, and an oiled beach on the Pye Islands. Seventeen fish from the oil-free site showed no sign of gill abnormalities and only one specimen was parasitized by *Trichodina* sp. In contrast, 6 of 14 fish from the oil-contaminated area displayed severe hyperplasia of the branchial epithelium and harboured a mean number of 30.0 trichodinids per fish, while a mean of 14.3 parasites per fish was recorded for the entire group, which were all parasitized. Because these fish from the oil-contaminated area were exposed for a period of 3–4 months, observations on the degree of parasitism and gill hyperplasia are in agreement with those noted in laboratory studies. In the absence of any other contaminant in this pristine area, it appears that these ciliates are good indicators of petroleum hydrocarbons in the marine environment.

High mortality among six species of fish was reported in a highly polluted lake (Naini Tal) in the Himalayas each winter (from December to March) when up to 100 000 fish died (Das and Shrivastava, 1984). Municipal and domestic sewage, in addition to silt from erosion, was discharged from 23 rivers into the lake. Physicochemical and biological data indicated heavy pollution during the die-off period. This was associated with a low dissolved oxygen concentration, increases in alkalinity and NH_4-nitrogen concentrations, and toxins originating from poisonous plants which died in November and had been washed by rain and meltwater into the lake. Heavy infestation of fish with four species of parasites occurred between February and March. Two ciliates, *Chilodonella cyprini* and *Trichodina domerguei*, which "clogged the gill filaments", were associated with mortality, whereas three others, a myxozoon (*Chloromyxum esocinum*), a copepod (*Argulus japonicus*) and leeches (species unknown), were not connected with it. Infected fish underwent convulsions and displayed decreased fin movement before death, which occurred shortly thereafter. The authors noted heavy mucus secretion and a weak fishy odour from the greyish gills during the period of infestation. They concluded that mortality (80 000 fish) between December and January was caused by pollution and thereafter (between February and March) by intense infestation with parasites. One can conclude from this report that host defence mechanisms in fish which survived the initial die-off were compromised to an extent which permitted intense infestation by parasites that ultimately killed many fish.

Ichthyophthirius multifiliis is another ectoparasitic ciliate of many species of freshwater fish. Vladimirov and Flerov (1975) noted that two species of fish exposed to phenol and polychloropinene appeared more susceptible to infestation by the ciliate. The progeny of fish which survived also seemed to be more susceptible to the disease.

Ewing and Ewing (1982) studied the susceptibility of channel catfish (*Ictalurus punctatus*) to the ciliate *Ichthyophthirius multifiliis*, when exposed to sub-lethal concentrations of copper. A weak positive correlation between susceptibility to the parasite and concentration of dissolved copper was detected. The mean number of trophozoites per cm^2 of caudal fin was greater on copper-treated (0.45–1.6 mg $litre^{-1}$) fish (range 22–64) than on controls (range 22–42).

Mohan and Sommerville (1988) investigated the effect of sub-lethal levels of cadmium on the susceptibility and immune response of carp (*Cyprinus carpio*) to *I. multifiliis*. There appeared to be no significant difference in the intensity of the infection between cadmium-treated fish (7 days exposure to the metal before infection) and untreated controls. Fish immunized against the ciliate, exposed to cadmium for 10 days and challenged subsequently had a significantly greater intensity of the infection. Fish simultaneously exposed to *I. multifiliis* and cadmium took longer (14 days) to develop immunity than did the untreated controls. There appears to be little doubt that exposure to sub-lethal levels of cadmium alters the susceptibility and defence mechanisms of fish to parasites.

B. MONOGENEA

A study on the interrelation of water quality, gill parasites and gill pathology

TABLE 1 *Averages of some nutrients, trace metals and pesticides in water samples, and Monogenea and gill pathology of the three host species in south-east Biscayne Bay and south-west Biscayne Bay at Black Creek Canal from May 1975 to August 1976*[a]

	South-east Biscayne Bay				*South-west Biscayne Bay at Black Creek Canal*			
Component	*No. of samples*	*Minimum*	*Mean*	*Maximum*	*No. of samples*	*Minimum*	*Mean*	*Maximum*
Total ammonia nitrogen·($mg\,l^{-1}$)	6	0	0	0.012	11	0.03	0.45	1.0
Arsenic ($\mu g\,l^{-1}$)	6	0	0	0	3	2.0	2.0	2.0
Lead ($\mu g\,l^{-1}$)	6	0	0	0	4	4.0	12.5	20.0
Manganese ($\mu g\,l^{-1}$)	6	0	0	0	4	4.0	6.0	20.0
Mercury ($\mu g\,l^{-1}$)	6	0	0	0	4	0.2	0.28	0.5
Diazionon ($\mu g\,l^{-1}$)	6	0	0	0	3	0.02	0.026	0.06
2,4-D ($\mu g\,l^{-1}$)	6	0	0	0	3	0.00	0.09	0.27
Silvex ($\mu g\,l^{-1}$)	6	0	0	0	3	0.00	0.04	0.10
Parathion ($\mu g\,l^{-1}$)	6	0	0	0	2	0.00	0.01	0.02
Neodiplectanum wenningeri (no. per gill arch)	69	0	0.625	5	52	25	72.5	> 100
Ancyrocephalus sp. (no. per gill arch)	57	0.1	1.4	8	80	69	124.75	> 500
Ancyrocephalus parvus (no. per gill arch)	44	0.32	2.25	4.5	54	61	89.25	> 200
Pathological changes[b]	170	None	None	Slight	186	Moderate	Severe	Severe

[a] After Skinner (1982).
[b] Slight, mucus production above normal; moderate, heavy mucus production and epithelial hyperplasia; severe, fusion of lamellae, loss of structure.

in fish from Biscayne Bay, Florida, USA, revealed that three species of fish from the south-western region harboured heavy infestations of monogeneans on the gills (Skinner, 1982). Fish including the yellow fin mojarra *Gerres cinereus* (Gerreidae), grey snapper *Lutjanus griseus* (Lutjanidae) and timucu *Strongylura timucu* (Belonidae) from this polluted area, which was degraded by discharges of agricultural, industrial and urban wastes, showed excessive mucus secretion, epithelial hyperplasia, fused gill lamellae, clubbing (telangiectasis) and aneurisms. In contrast, the intensity of parasitism was considerably less in fish from the south-eastern part, which was less polluted (Table 1). Skinner (1982) suggested that exposure to the pollutants (ammonia, trace metals and pesticides) acted synergistically as a stressor and caused histopathological and physiological changes which altered host resistance to the parasites.

Among groups of Atlantic cod (*G. morhua*) exposed experimentally to WSOF, the prevalence and intensity of *Gyrodactylus* spp. were significantly

FIG. 1. Cross-section of the gill of an oil-treated cod (*Gadus morhua*) showing monogeneids (arrows) between secondary lamellae (× 70).

TABLE 2 *Influence of oil-contaminated sediments or water-soluble oil fractions on parasites*

Species	*Parasite's location*	*Host species*	*Fish group*	*No. of fish*	*Percentage parasitized*	*No. of parasites per fish*
Steringophorus furciger	Gastrointestinal tract	*P. americanus*[a] (flounder)	Control	19	94	15.4 ± 0.1
			Oil-treated	18	67	$5.9^b \pm 0.2$
Echinothynchus gadi	Gastrointestinal tract	*G. morhua*[a] (cod)	Control	21	100	4.9 ± 0.1
			Oil-treated (Hibernia)	21	71	$2.0^b \pm 0.2$
Gyrodactylus spp.	Gills	*G. morhua*[c] (cod)	Control	15	60	2.2 ± 0.1
			Oil-treated	14	100	$10.2^b \pm 0.2$
Trichodina cottidarum	Gills	*M. octodecem-spinosus*[d] (sculpin)	Control	21	48	1.1 ± 0.3
			Oil-treated	20	95	$19.0^b \pm 0.9$

[a] After Khan and Kiceniuk (1983).
[b] $P < 0.05$.
[c] After Khan and Kiceniuk (1988).
[d] Khan (unpublished data).

greater in the oil-treated groups than in controls 16–20 weeks after depuration (Fig. 1 and Table 2). Moreover, pathological changes in the gills were associated with exposure to WSOF and a greater intensity of parasites (Table 2). No difference was observed in fish that were not depurated (Khan and Kiceniuk, 1988). Species of *Gyrodactylus* are ovoviviparous. It is likely that the WSOF, which can affect the fish's defence mechanism as well as induce mucus cell hyperplasia, created a habitat conducive to parasitic infestation and reproduction. Thus, enhanced parasitism in branchial tissue, which appears to be more sensitive than others to cellular changes induced by pollutants, could be a useful indicator in a monitoring programme for biological effects.

C. COPEPODS

Lernaeocera branchialis occurs as an ectoparasite on the gills of gadoid fish but obtains its nutrients from their blood (Kabata, 1958; Mann, 1970; Sundnes, 1970). The parasite is known to induce blood and weight loss, affect reproduction and cause mortality (Hislop and Shanks, 1979; Khan, 1988). Groups of sub-adult and mature cod parasitized by *L. branchialis*, and uninfected controls, were exposed to WSOF for 12 weeks. In addition to weight loss, deaths occurred among both juvenile and adult cod parasitized and exposed to WSOF. In adults only, the liver somatic indices in both parasitized groups were significantly lower than in the controls (Khan, 1988). More recently, groups of second-year (19–25 cm) cod were exposed to WSOF (ca. 100 parts/10^9) about 6 weeks after infection (water temperature, 12°C) in a flow-through system. Within 2 weeks, seven of the oil-treated cod that harboured 2–4 *L. branchialis* died. This experiment was repeated with groups of cod of similar size approximately 12 weeks after infection. Again, 8 of 12 infected, oil-treated cod succumbed (Khan, unpublished data). These results corroborate other observations that gadoids infected with *L. branchialis* have a low tolerance to stressors such as low oxygen tension in water (Mann, 1970; Möller, 1987a), WSOF, etc.

D. GLOCHIDIA

Glochidia larvae of bivalve molluscs attach to fish as part of their cyclical development and sometimes cause damage to host tissue and mortality. Moles (1980) investigated the sensitivity of coho salmon fry (*Onchorhynchus kisutch*) parasitized by glochidia of a freshwater mussel (*Anodonta oregonensis*) to Prudhoe Bay crude oil fractions, toluene and naphthalene. Different levels of parasitism at various PAH concentrations were determined. Susceptibility to the pollutants increased linearly with increased parasitism.

It was also observed that fry infected with 20–35 glochidia were significantly more sensitive to each of the toxicants than were uninfected fish. Exposure to PAHs increases energy consumption, which is associated with detoxification and elimination of the metabolites (Rice *et al.*, 1977). Because the glochidia also utilize the energy resources of the host (Moles, 1980), the combined effect probably increased sensitivity by decreasing the fish's available energy.

V. INFLUENCE OF POLLUTANTS ON ENDOPARASITES

A. PROTOZOA

A histozoic myxozoon (*Myxobolus lintoni*) caused protruding growths in various tissues of *Cyprinodon variegatus* taken from polluted areas of Galveston Bay, Texas, USA. Overstreet and Howse (1977) subsequently reported infections in isolated stocks from stressed habitats and, more recently, Overstreet (1988) saw additional cases in polluted areas from Mississippi and Louisiana, USA. According to Paperna and Overstreet (1981), citing Sarig, a species of *Myxobolus* reached epizootic proportions in mullets (*Mugil cephalus* and *M. capito*) in contaminated ponds in Israel.

Studies were conducted to assess the influence of PAHs on a myxozoon (*Ceratomyxa acadiensis*) that parasitizes the gall bladder of winter flounder (*Pseudopleuronectes americanus*) (Khan *et al.*, 1986). Infections, although prevalent at times, are usually low-grade, especially during the feeding season (April to November). Adult flounder were exposed to oil-contaminated sediment (2600 $\mu g\ g^{-1}$) for 6 months and necropsied. The prevalence of the myxozoan infection was greater (80% of 20) in the oil-treated group than among the controls (28% of 21; see Table 3). Additionally, the intensity of the infection was greater (ca. $1 \times 10^2\ ml^{-1}$) among the oil-contaminated group than among the controls in which the parasites were too few to estimate their number accurately. One of us (R.A.K.) subsequently ascertained that season and starvation influenced the prevalence of myxozoan infections in flounder (Table 3). Fish collected from the field in January and March, when no feeding occurred, or those starved for 4 weeks during the feeding season (June), exhibited higher prevalences of infection than did flounder that were fed to satiation during June or November. Moreover, the prevalence of infection was greater (96%) in flounder exposed to oiled sediment for 3 months (June to August) than among the controls (45%). Exposure to oil-contaminated sediment inhibits feeding and the release of bile in flounder (Fletcher *et al.*, 1982), which also occurs during winter and conceivably could permit an accumulation of myxozoan spores. The possibility cannot be ruled out that the metabolites of PAHs, which are excreted via bile, and alterations of bile acids (Dey *et al.*, 1983), might enhance parasite nutrition and reproduction.

Some pollutants may cause a decrease in parasitism in a fish host. Narasimhamurti and Kalavati (1984) reported a maximum of 22.4% prevalence in 250 *Channa punctatus* examined at monthly intervals for a myxozoan gill parasite, *Henneguya waltairensis*. The infection was observed only in fish taken from an unpolluted tank and was absent from a tank polluted with sewage. The authors believed that high salinity, low oxygen levels and increased turbidity might have been responsible for the absence of the infection from the polluted tank.

TABLE 3 *The effects of season and starvation on the prevalence of myxozoan parasites* Ceratomyxa acadiensis *in the gall bladder of adult winter flounder,* Pseudopleuronectes americanus

Time of year	*State of fish*	*No. examined*	*No. infected*	*Percentage of fish infected*
January	Starved[a]	26	19	73
March	Starved[a]	17	13	76
June	Fed	27	6	22
June	Starved	23	17	74
June/August	Oiled/fed	24	23	96
June/August	Fed	28	12	45
November	Fed	22	10	45

[a] Winter flounder do not feed at these times of the year.

B. HAEMATOZOA

Studies were undertaken to determine the interaction of haematozoa in marine fish and exposure to PAHs. *Trypanosoma murmanensis* infects several species of marine fish in the north-west Atlantic and is transmitted by a piscicolid leech, *Johanssonia arctica* (Khan, 1976; Khan *et al.*, 1980a,b). The protozoon's effects vary from blood changes to mortality, especially in young fish (Khan, 1977, 1985). Juvenile and adult winter flounder (*Pseudopleuronectes americanus*) were exposed for 6 weeks to oil-contaminated sediment (2600–3200 $\mu g\, g^{-1}$) after infection with *T. murmanensis* (Khan, 1987a). A total of 73 of 82 (89%) juvenile flounder of the parasitized plus oil-treated group succumbed, in contrast to 39 (48%) in the trypanosome-infected group and 27 (33%) in the oil-treated group. Among adult fish, mortality was higher in the oil-treated group that was parasitized (46% of 63) than in the groups which were only oil-treated (17%) or trypanosome-infected (16%). Fish which survived the trypanosome infection and concurrent exposure to oil exhibited enlarged livers and spleens, decreased gut indices and, occasionally, low blood values. Similar results were observed among juvenile flounder exposed to oil-contaminated sediment for 6 weeks and subsequently infected with *T. murmanensis*. A total of 14 of 50 (28%) juveniles died after exposure to oil. All (18) of the survivors that were infected afterwards succumbed, whereas mortality was lower in the groups which were only infected (33%) or oil-treated (24%). Death among the oil-contaminated plus parasitized groups was associated with severe tail rot, excessive secretion of mucus on the body surface, depigmentation and haemorrhage from the caudal peduncle.

The relationship between various crude oil concentrations and concurrent trypanosome infection was studied in adult winter flounder. The death rate among adult fish after infection with *T. murmanensis* is generally low (Khan, 1985). However, exposure to oil-contaminated sediment with a concomitant trypanosome infection resulted in mortality which was greatest at the highest concentration (Table 4). The relationship is not quite linear but it does indicate that fewer deaths occurred at the lower concentrations. Moreover, mortality was consistently less in the uninfected oil-treated group than in the infected fish (Khan, unpublished data).

TABLE 4 *Mortality in adult winter flounder after exposure to various concentrations of oil-contaminated sediment with or without a concomitant trypanosome infection over a period of 8 weeks*

Approximate total hydrocarbon concentration ($\mu g\ g^{-1}$)	*No. died/no. exposed*[a]	
	Uninfected	*Infected*
0	0/60	3/60
75	6/60	9/60
300	9/60	23/60
600	9/60	24/60
1000	15/60	32/60
2200	34/56	44/58

[a] Each result represents three trials of about 20 fish per trial.

Sub-adult cod were more susceptible to trypanosome infection after chronic exposure to PAHs (50–100 $\mu g\ litre^{-1}$) for 12 weeks than adults (Khan, 1987a). About 68% of the juveniles succumbed, whereas no deaths occurred among the adult fish. Mortality was lower in the other groups of juvenile fish that were only oil-treated (47%) or only infected with *T. murmanensis* (11%). Additionally, condition (K) factor was significantly lower among both groups of oil-treated cod than in the control group. The prevalence and intensity of the trypanosome infection were always greater in the oil-treated groups than in the untreated fish. Blood factors were significantly lower in longhorn sculpin (*Myoxocephalus octodecemspinosus*) which were exposed to both PAHs and parasites simultaneously (Kiceniuk *et al.*, 1981). However, haematocrit and haemoglobin values in oil-treated plus parasitized cod did not decrease (Khan, 1987a). Because the haemoglobin and haematocrit values in oil-treated cod were significantly greater than those of controls, exposure to PAHs may have stimulated haemopoiesis, which compensated for the blood loss that accompanies an infection with *T. murmanensis* (Khan, 1985). Zbanyszek and Smith (1984) reported increased haematocrit, haemoglobin and erythrocyte values in rainbow trout (*Salmo gairdneri*) after acute exposure to PAHs.

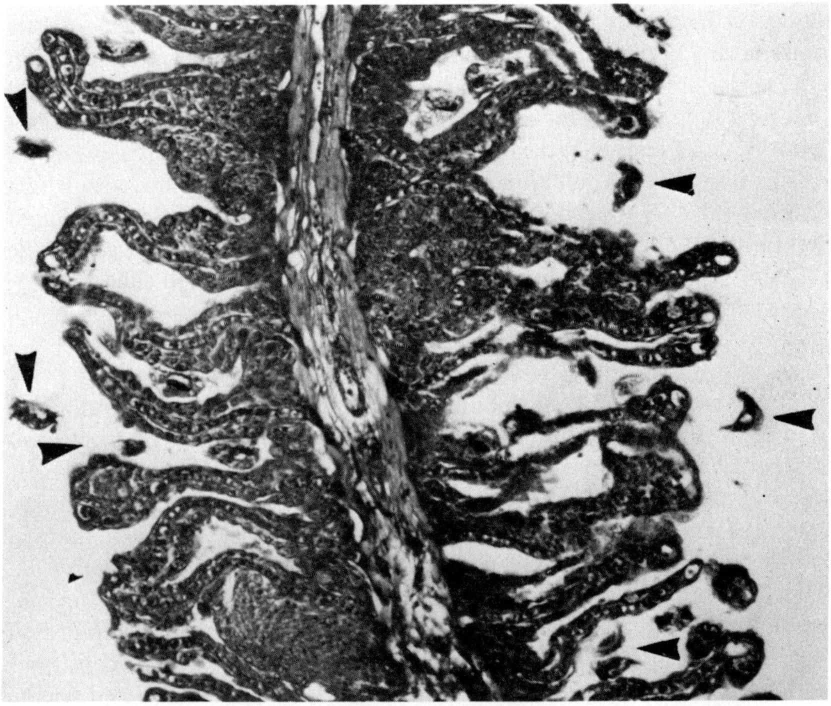

FIG. 2. Section of the gill of a longhorn sculpin (*Myoxocephalus octodecemspinosus*) showing trichodinids (arrows); note hyperplasia (×200).

Exposure of cod (*G. morhua*) to PAHs resulted in consistently lower gonadal somatic indices in both sexes of fish than in controls (Khan, 1987a). Although these values were lower on five of six occasions in the oiled plus parasitized groups than in those only oil-treated, the differences were not significant. Cod exposed to PAHs and held until spring (May) failed to spermiate or ovulate, whereas one or the other occurred in the untreated fish. The results of these pollutant plus parasite studies suggest that the combined effects can cause morbidity and mortality, and also affect reproduction (Khan, 1987a,b).

There was evidence of impaired resistance in both winter flounder and Atlantic cod after exposure to PAHs and infection with *T. murmanensis.* Parasites were readily demonstrable in the blood of oil-treated fish and parasitaemias were consistently greater than in untreated fish (Table 5). There were fewer leucocytes in smears of cardiac blood from oil-treated fish than in those from controls. Because exposure to PAHs is known to affect the immune system, probably the prolonged period of patency was associated with decreased production of immunocompetent cells.

One further study has provided evidence of a link between aquatic pollution and a fish blood parasite. Eiras (1987) suggested a relationship between the prevalence of *Haemogregarina bigemina* of *Blennius pholis* and pollution along the west coast of Portugal. Prevalence of the haemogregarine varied considerably in the four localities studied. However, it was greater at two sites where oil pollution was heavy (29–77%), less at one into which domestic effluents were discharged, and lowest (3%) at the least polluted site. Eiras (1987) attributed these differences to the abundance of the presumptive intermediate hosts (leeches) at the most polluted sites.

C. HELMINTHS

Pollutants tend to influence the prevalence, intensity or both of endoparasites in fish. Eure and Esch (1974) reported that the intensity of *Neoechinorhynchus cylindratus* in large-mouth bass, *Micropterus salmoides*, was greater in a river that received thermal effluents than in an unheated area during winter. In another study, a relationship was observed between thermal loading and parasitism in the mosquito fish *Gambusia affinis* (Aho *et al.*, 1976). The intensity of a strigeid trematode, *Ornithodiplostomum ptychocheilus*, which infects the brain and eyes, was greater in fish living in heated water than in others kept at ambient temperature. However, the density of another trematode (*Diplostomum scheuringi*), which was restricted to the body cavity, was lower in fish inhabiting warmer water. Hirshfield *et al.* (1983) observed an increase in prevalence (ca. four-fold) of the larval nematode, *Eustrongyloides* sp., in *Fundulus heteroclitus* taken from a discharge canal of a power plant located in Chesapeake Bay, Maryland, USA. Fish living near the intake area had fewer parasites. The authors attributed this increase in fish living in the discharge canal to an increased abundance of oligochaetes, suspected as the first intermediate host, in response to raised temperatures and organic enrichment (either natural or man-made) from the power plant.

Kussat (1969) noted that pollutants, discharged into the Bow River near

TABLE 5 *Effect of crude petroleum in sediment or as a water-soluble oil fraction on winter flounder* (Pseudopleuronectes americanus) *and Atlantic cod* (Gadus morhua) *concurrently infected with* Trypanosoma murmanensis[a]

Host species	*State of fish*	*No.*	*Mortality (%)*	*Prevalence (%)*	*Parasitaemia per ml*
Flounder (young)	Infected	82	48	60	—
	Infected-oiled	82	89	100	$4.7 \pm 1.0 \times 10^5$
Flounder (adult)	Infected	63	16	17	—
	Infected-oiled	63	46	91	$2.1 \pm 0.6 \times 10^4$
Cod (young)	Infected	18	11	44	—
	Infected-oiled	18	68	100	$1.3 \pm 0.5 \times 10^4$
Cod (adult)	Infected	16	0	38	$1.8 \pm 0.9 \times 10^4$
	Infected-oiled	19	0	79	$8.0 \pm 0.8 \times 10^4$

[a] From Khan (1987a).

the city of Calgary, Alberta, Canada, affected the parasite fauna of a catfish, *Catostomus commersoni*. A general biological survey conducted above (upstream) and below (downstream) the entry of industrial and domestic wastes into the river revealed a reduction in the number of invertebrate species in the bottom fauna and differences in dissolved oxygen, nitrogen and phosphate concentrations at the downstream locality. Catfish taken upstream were more heavily infected (prevalence and intensity) with the acanthocephalans *Octospinifer macilentus* and *Neoechinorhynchus cristatus* than fish originating downstream, and in some instances no parasites were observed in fish from the latter site. Intermediate hosts of the acanthocephalans are believed to be amphipods and ostracods, which were present only upstream. The author concluded that the polluted water represented a barrier to transmission of the parasites by excluding their intermediate hosts.

Overstreet and Howse (1977) reported that pollution influenced the prevalence and intensity of an acanthocephalan (*Dollfusentis chandleri*) in the Atlantic croaker (*Micropogonias undulatus*) in the Gulf of Mexico. In fish taken from a presumably uncontaminated bayou, about 42% of the fish harboured a mean of 28.1 parasites. Significantly fewer (about 14%) and lower (3.3 worms per fish) infestations were noted in the same species of fish collected over a 19-month period from the polluted River Pascagoula, near its mouth in Mississippi Sound. It appears that the intermediate host, a variety of amphipod, was not available as prey, which in turn affected the prevalence and intensity of the parasite in its host.

Möller-Buchner (1981) reported changes in parasite prevalence in a gobiid fish, *Pomatoschistus microps*, taken from the estuary of the River Elbe, which is heavily polluted. Three species of platyhelminths (*Diplostomum spathaceum*, *Tetracotyle* sp. and larvae of *Proteocephalus* sp.), two unidentified helminth cysts, and larval nematodes (*Raphidascaris* sp.) were observed. The prevalence and intensity of parasitism were high in this fish. None of the parasites, except *D. spathaceum*, was apparently present in the lower Elbe previously. The author attributed the poor nutritional condition of the fish collected at one site (Kollmar) not to the intensity of the parasites but to increased pollution which caused eutrophication and associated changes, e.g. oxygen depletion, and to a high rate of reproduction of potential intermediate hosts. It seems more likely, however, that pollution potentiated the effect of the parasites on their piscine hosts.

In some instances, specific pollutants have been shown to affect enteric parasites. Perevozchenko and Davydov (1974) observed that yearling carp infected with the cestode *Bothriocephalus gowkongensis* were more susceptible to DDT than uninfested fish. Boyce and Yamada (1977) reported that sockeye salmon (*Onchorhynchus nerka*) smolts infected with the intestinal cestode *Eubothrium salvelini* were more susceptible to zinc than non-infected smolts. Pascoe and Cram (1977) also noted that the three-spined stickleback (*Gasterosteus aculeatus*) was more susceptible to cadmium when parasitized by the cestode *Schistocephalus solidus*. The effects of parasitism by *S. solidus*, dietary restriction and exposure to cadmium were examined in sticklebacks. Fish exposed to all three stressors combined died earlier than those exposed to one or two of the stressors (Pascoe and Woodworth, 1980). Parasitism and a restricted diet also reduced survival compared with that of fish with dietary restriction only, but a combination of restricted diet and cadmium did not influence survival. The authors suggested that parasitism and dietary restriction, when combined, can decrease survival and reduce the fish's ability to detoxify and eliminate the pollutant. In contrast, McCahom *et al.* (1988) reported that the amphipod *Gammarus pulex*, parasitized by the acanthocephalan *Pomphorhynchus laevis*, was no more susceptible to exposure to lethal concentrations of cadmium than controls.

In oil-polluted waters, fish are exposed to different parts of the pollutant according to their habits. For example, pelagic or free-ranging fish such as Atlantic cod (*G. morhua*) tend to contact the water-soluble oily fraction (WSOF), whereas bottom-dwellers such as the winter flounder (*Pseudopleuronectes americanus*) are exposed to contaminated sediments which generally have a much higher hydrocarbon content.

Cod captured from coastal waters are often infected with the enteric acanthocephalan *Echinorhyncus gadi*, but after experimental exposure to WSOF (50–80 μg $litre^{-1}$) for 81–140 days, it was observed that the prevalence and intensity of the parasite infection were significantly lower compared to those of unexposed controls maintained in pollutant-free water (Table 3). Similarly, we found that flounders submerged in sediment contaminated with petroleum (2600 μg g^{-1}) in experimental tanks contained fewer specimens of the trematode *Steringophorus furciger*, compared to unexposed controls, after 34–160 days (Khan and Kiceniuk, 1983; see Table 2).

Marine fish normally drink sea water to osmoregulate and probably the experimental fish ingested PAHs during this process. PAHs are detoxified by the liver and the metabolites are excreted in the bile. Because bile acids and other physiological components are altered by exposure to oil pollution (Dey *et al.*, 1983), it may be that the parasites' environment in the fish gut was affected to such an extent that they were voided. However, the precise mechanism of worm-shedding has not yet been ascertained.

Increased susceptibility to a parasite was reported in blue gills (*Lepomis macrochirus*) exposed to an insecticide, heptachlor (Andrews *et al.*, 1966). The fish were held in earthen ponds and, within 28 days after exposure, large numbers of trematode metacercariae were observed in the liver and kidneys at low insecticide concentrations (0.0125–0.0375 parts/10^6). Samples taken later showed massive infestations in several organs including the kidney, spleen, liver, heart and reproductive organs. The lowest prevalence was noted in fish exposed to the highest concentration (0.05 parts/10^6), while no parasites were seen in the controls. It appeared that a reduction of the number of snails, which act as intermediate hosts, by the insecticide might have been responsible for the low prevalence of infection at the highest concentration. The authors suggested that heptachlor at the lowest concentrations weakened the fish to the extent that they became more susceptible to parasitic infestation than the controls. However, Van Valin *et al.* (1968) found no connection between the level of another insecticide (Mirex) and the degree of parasitism by diplostomid trematodes, but extensive damage was done to several of the internal organs of treated fish.

In a series of papers, Valtonen and co-workers have reported on the effects of effluents from a pulp and paper mill on fish parasites in a lake in central Finland (Valtonen and Koskivaara, 1987, 1989; Valtonen *et al.*, 1987a,b; Valtonen and Taskinen, 1988). They observed no adult trematodes in the intestine of roach, *Rutilus rutilus*, taken from Lake Vatia, Finland, which was polluted by pulp and paper effluents, whereas the parasites were present in fish obtained from three unpolluted (one oligotrophic and two eutrophic) lakes. The trematodes of this fish species and of perch (*Perca fluviatilis*) from these lakes were studied on a seasonal basis. *Sphaerostoma globiporum* was present in 30% of the fish from one of the eutrophic lakes, while it was absent from fish originating from the polluted lake. Similarly, the prevalence of two species of eye trematodes, *Tylodelphys clavata* and *Diplostomum* spp., in both fish species, and one species in the inner organs of perch, was lower in Lake Vatia than in the other lakes. The authors also noted massive infections of metacercariae on the fins and gill arches of roach from the eutrophic lakes only. These differences in prevalence were attributed to the sensitivity of the intermediate hosts to pollution, these hosts being absent or present in low numbers only in Lake Vatia.

On the other hand, the same authors reported that some parasite species of roach were found with increased prevalences in the polluted lake: the prevalences in the three unpolluted lakes and in the polluted lake were, respectively, 6 and 20% for the ciliate *Ichthiophthirius multifiliis*, 1 and 7% for the ciliate *Apisoma* sp., 2.5 and 13% for the ciliate *Trichodina* sp., 60 and 24% for the myxosporean *Myxidium rhodei*, 14 and 31% for the myxosporean *Myxobolus muelleri*, 1 and 7% for the monogenean *Dactylogyrus similis*, zero and 17% for *D. fallax*, zero and 13.2% for the monogeneans *Gyrodactylus gasterostei*, *G. carassii* and *G. vimbi*, 23 and 61% for the nematode larvae *Raphidascaris acus*, 12 and 23% for the acanthocephalan *Neoechinorhyncus rutili*, and 1 and 13% for the acanthocephalan *Acanthocephalus anquillae*.

The authors suggested that one reason for the increased prevalences of the protozoan and monogenean infections in the polluted lake may have been that the immunological response of the fish had been weakened due to the environmental stress caused by the resin acids or chlorinated organic substances present in the effluents from the pulp and paper mill. With the nematode and the acanthocephalans, there may have been an additional effect on the survival of their intermediate hosts in the stressed environment.

However, it is interesting that Valtonen and co-workers also found several parasite species to be unaffected by the differences in the four lakes. Among these were four additional species of *Dactylogyrus*, *Gyrodactylus prostae*, the roach cestodes *Proteocephalus torulosus* and *Caryophyllaeus laticeps*, the parasitic crustacean *Ergasilus sieboldi*, and *Argulus foliaceus*.

In a similarly polluted environment, in the effluent areas of a pulp mill situated in lake Vanern, Sweden, one of us found that the prevalence on perch (*P. fluviatilis*) of the copepods *Achtheres percarum* and *Caligus lacustris* increased with the distance from the point of effluent discharge (Thulin, 1983). A total of 243 perch, a relatively static fish, was examined from six equally distributed stations in a pollutant gradient from the pulp mill. None of the parasites was found on fish at the closest locality; *A. percarum* increased in prevalence up to 73.2% at the most distant locality, while the corresponding maximum proportion for *C. lacustris* was 10.7%. The latter species also exhibited the same trend on roach ($n = 212$), from zero up to 14.5%. On the other hand, *Argulus foliaceus* was found on 10% of the perch close to the pulp mill, on none of the fish from the intermediate localities, but on 25% of fish from the distant locality. On roach (*R. rutilus*), this parasite was found only at the distant locality (9.1%).

In a second investigation in effluent areas from two pulp mills, on the brackish Baltic coast of Sweden, one of which used chlorine for bleaching, Thulin *et al.* (1986, 1988) found a similar pattern with other ectoparasitic monogeneans. Both *Paradiplozoon homoion* and *Dactylogyrus* spp. on roach increased in both prevalence and intensity with the distance from both mills. This increase was most pronounced in the gradient from the mill which did not use bleach. Here the prevalence of *P. homoion* increased from 16 to 52%, and its mean intensity from 0.4 to 1.8 parasites per fish, from the least to the most distant site, 5000 m from the discharge point. The corresponding figures for *Dactylogyrus* sp. were 20 and 48% and 1.1 and 2.6%. However, the same trend was found in the gradient from the mill using chlorine for bleaching; the prevalence and intensity of *P. homoion* increased from 40% and 1.5 to 67% and 4.4 from localities 1 to 3, the last-mentioned situated 6000 m from the discharge point. The corresponding figures for *Dactylogyrus* spp. at this mill were 60% and 11.2, and 86% and 20.7.

The parasite fauna of perch (*P. fluviatilis*) were also examined from the same localities. The only significant difference found in the two gradients concerned the acanthocephalan *Neoechinorhynchus rutili*. This species was not found at all outside the unbleached mill, whereas it was found in 54% (mean intensity 2.6) of the 28 perch examined closest to the discharge point of the mill using chlorine, and in 18% (mean intensity 0.4) and 3% (mean intensity 0.1) with increasing distance from the mill.

Because the ectoparasites are in direct contact with the polluted water, there might well have been a direct negative effect on the parasite and its reproductive and survival capacity. The prevalence of *N. rutili* was obviously regulated through effects on its intermediate hosts, the gammarids. These results show many similarities with those of Valtonen and co-workers cited above, but discharges from pulp and paper mills contain so many components that experimental follow-up studies are needed to distinguish between the effects of the different pollutants.

Sakanari *et al.* (1984) studied the effect of sub-lethal concentrations of zinc and benzene on young (< 1-year-old) striped bass (*Morone saxatilis*) before and after infection with 25 larval nematodes (*Anisakis* sp.). The experimental protocol attempted to simulate three environmental situations: (i) fish enter a polluted area, acquire the infection and remain in the polluted area; (ii) infected fish enter and remain in the polluted area; and (iii) fish enter a polluted area, become parasitized and subsequently leave the polluted area. The results indicated that the combined effects of the pollutants and parasites caused a significant decrease in haematocrit values when fish were exposed for 4 weeks. The pollutants alone did not apparently affect antibody titres, although it is known that immunosuppression occurs in fish exposed to heavy metals and PAHs.

Several studies on the parasitic fauna of fish have recently been conducted in Poland in waters that were considered to be polluted. Reda (1989) observed no major difference between samples of bream (*Abramis brama*) taken from clean and polluted areas of the River Vistula near Warsaw. Some differences in the species composition, prevalence and density of parasites were attributed to the habitats rather than to pollution. Sulgostowska *et al.* (1987) noted that lower levels of infestation of flatfish by parasites occurred in Gdansk Bay, an area polluted by industrial sewage, than in two adjacent, "clean" areas of the south-east Baltic Sea. However, Rehulka (1983) reported that parasites taken from fish living in the middle course of the Ostravice River in Czechoslovakia, which is polluted by industrial wastes, survived the harmful effects of the toxicants (iron effluents, coal mud and petroleum wastes). Gardner and Yevich (1988) reported that winter flounder (*Pseudopleuronectes americanus*) exposed to chemically contaminated sediment in Black Rock Harbor, Connecticut, USA exhibited degenerative changes associated with trematode and microsporidian infections, but they provided no further details.

D. PARASITES OF MOLLUSCS

Few studies have examined the influence of pollutants on parasitized invertebrates. Such parasites, acting as a biological stressor, can sometimes potentiate the effect of the pollutant. Guth *et al.* (1977) studied the effect of *Schistosomatium douthitti* and *Trichobilharzia* sp. on the tolerance of snails, *Lymnaea stagnalis*, to lethal concentrations of zinc. Their results indicated that the parasites reduced the tolerance of the snail to lethal doses of zinc. This appeared to be more pronounced with *Trichobilharzia* sp., especially in the late stages of the infection and at high concentrations.

The oyster, *Crassostrea virginica*, is host to a number of protistan parasites including an apiocomplexan, *Perkinsus marinus*. Scott *et al.* (1985) reported that exposure of parasitized oysters to a combination of high salinity and chlorine-produced oxidants resulted in greater toxicity than in animals exposed to high salinity alone. The increase in mortality was correlated with parasitism.

IV. CONCLUSIONS

As pollutants continue to be released into the aquatic environment, the prevalence of diseases and abnormalities may eventually increase. Hepatic lesions in sole, *Parophrys vetulus*, have been reported from two areas—Puget Sound and Commencement Bay in Washington, USA—where toxic chemicals contaminate bottom sediment (Malins *et al.*, 1985; Krahn *et al.*, 1986, 1987; Becker and Ginn, 1987; Rhodes *et al.*, 1987). Overstreet (1988) reviewed aquatic pollution problems around the south-eastern coasts of the USA, calling attention to lesions in finned fish and in shellfish. Ziskowski *et al.* (1987) examined 85 000 fish from the north-west Atlantic and noted that the disease prevalence was 1.15%. This was a follow-up of an earlier study which showed a statistical relationship between pollution and disease conditions in winter flounder (*Pseudopleuronectes americanus*) from the New York Bight area (Ziskowski and Murchelano, 1975; Murchelano and Ziskowski, 1976; Murchelano, 1982).

Petroleum hydrocarbons also pose a major threat to commercial fisheries, although this is not generally accepted (McIntyre, 1982). However, evidence from natural oil spills indicates that affected fish are more prone to a variety of lesions and infections (Lopez *et al.*, 1981; Haensly *et al.*, 1982; Krahn *et al.*, 1986; West *et al.*, 1986; Brule, 1987; Stott *et al.*, 1981; Grizzle, 1986).

Little published information is currently available on the oil spill from the *Exxon Valdez* in Alaska in March 1989, but one of us (R.A.K.) observed, and also was shown photographic evidence of, the mortality of larval and adult herring (*Clupea pallasii*), large numbers of sea birds and otters. As a result of potential tainting of the flesh of several commercial fish species, the fishery in the Gulf of Alaska was closed for most of the year in 1989.

Recently, commercial quantities of petroleum have been discovered in a fertile fishing area (Grand Banks) where several commercial species harbour parasites. Chronic exposure in the form of small spills and seeps could not only impair the fish's natural defence mechanisms, resulting in increased parasite burdens (haematozoons, myxozoons and monogeneans), but also lead to susceptibility to opportunistic microorganisms that otherwise would be adequately controlled. There is a need, therefore, for constant monitoring of fish health in such areas, using lesions, histopathology and parasites as biological indicators.

Acid rain in recent years has had a major impact on the biota of lakes in several countries. Acidification is associated with the mobilization of aluminium and other trace metals which can be toxic to fish. Mucus production in both skin and gills is enhanced by acid stress (Daye and Garside, 1976; Zuchelkowski *et al.*, 1986). Chevalier *et al.* (1985) noted epithelial damage, which presumably decreased respiratory efficiency, in brook trout (*Salvelinus fontinalis*) in acidified lakes (pH 5.5) in Quebec, Canada. Exposure to pH values less than 5 is also associated with a decline of Na^+ and Cl^- ion concentration (Leivestad and Muniz, 1976). Turnpenny *et al.* (1987) conducted surveys in 60 upland streams in the UK, some of which were acidic, and concluded that water quality acting directly on fish was responsible for the decline of salmonids. The absence or paucity of salmonids was related to low pH and to high levels of aluminium or copper–zinc–lead toxicity. Chronic exposure to acidified water can therefore be classified as an environmental stress which can alter homeostasis through the release of catecholamines and corticosteroids (Wedemeyer, 1970; Pickering, 1981). Corticosteroids are known to have immunosuppressive properties which probably make the fish more susceptible to pathogenic organisms which eventually cause mortality.

Acidification has a profound effect on organisms living in lakes and rivers as well as on their parasites. A lake in Manitoba, Canada, with an initial pH of 6.8, was experimentally acidified from 1976 to 1981 to simulate the effects of acid precipitation. France and Graham (1985) observed changes in the prevalence of a microsporidian parasite (*Thelohania contejeani*) in crayfish (*Oreonectes virilis*), from 1.7% at pH 5.6 in 1979 to a maximum of 7.7% at pH 5.1 in 1981; in three reference lakes, not acidified experimentally, the prevalence was 0.3–0.6%. Mortality in crayfish occurred during winter. It is believed that enhanced microsporidiosis contributed to a reduction in the annual survival rates of 8% in 1979 and 18% in 1980. The rate of survival in 1980 was only one-half of that in the non-acidified lakes. The authors speculated that the increase in prevalence might have been associated with cannibalism, or that a low pH favoured the parasite's life-cycle, or that there was a decrease in host resistance caused by the sub-lethal acid stress, or that the latter two effects operated together. Because the infection can become latent in some animals without evidence of disease, acid stress might activate the parasite which ultimately kills its host.

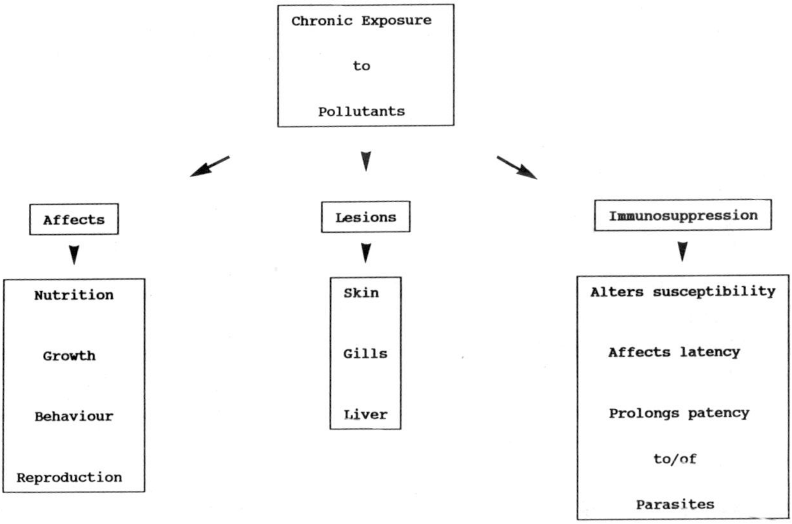

FIG. 3 The effects of sub-lethal concentrations of pollutants on aquatic organisms following chronic exposure.

The influence of chronic exposure to one or more pollutants, such as crude petroleum, heavy metals, etc., on aquatic animals might cause a number of changes (as shown in Fig. 3). There is indirect evidence from a number of studies that increased susceptibility occurs after long-term exposure to pollutants. This is especially evident in fish infected with trichodinids and monogeneans, as their intensity increases substantially and is associated with lesions caused by the contaminants. Additionally, patency is prolonged in haematozoan infections in fish in which organ and tissue changes have been demonstrated. Although many enteric parasites increase in prevalence and intensity, one or more factors might influence these changes. Thus, immunosuppression might represent one of the major underlying causes affecting susceptibility, latency and patency of some parasitic infections when animals are exposed to pollutants. The extent to which this occurs will hinge on the concentration of the pollutant(s), time of exposure and the age and species of fish.

There are several aspects of pollution–parasite research which require further study. There is little information which relates the intensity of the parasitic infection(s) to synergistic or antagonistic effects of the contaminant and associated histopathological changes which, in combination, might affect survival (see Skinner, 1982). A knowledge of the parasite fauna of fish of the area under investigation is therefore a major prerequisite. More information is also required on the following topics: (i) relationship between the concentration of the xenobiotic(s) in the tissues of fish and the parasite burden; (ii) the interaction of PAHs and heavy metals (see Sakanari *et al.*, 1984); (iii) the transfer of pollutants by the food web to fish via bioaccumulation in bivalves, crustaceans [one study has shown a reduction of feeding rate and growth in pink salmon (*Onchorhynchus gorbuscha*) during and after exposure to oil-contaminated prey, the brine shrimp (*Artemia salina*): see Schwartz, 1985], annelids, etc., and their ultimate effect on endoparasites; (iv) the effect of pollutants on the parasite fauna of sea birds, porpoises, whales and seals following chronic exposure or via the food chain; and (v) the effect of acid rain on the parasite fauna of fish living in lakes with low pH values (see Chevalier *et al.*, 1985).

VII. Summary

We have tried to draw attention to an increasing body of evidence (from several publications) that parasites of fish might be useful indicators of pollution. Several types of pollutants, including domestic sewage, pesticides, polychlorinated biphenyls, heavy metals, pulp and paper effluents, petroleum aromatic hydrocarbons, acid rain, and others, are known to affect aquatic animals. Many of the latter are parasitized and, under natural environmental conditions, most fish parasites are believed to cause little or no harm. However, chronic exposure to pollutants over a period of time causes biochemical, physiological and behavioural host changes that ultimately can influence the prevalence and intensity of parasitism. Some of these changes include host nutrition, growth and reproduction. Macroscopic lesions might not always be apparent, but subtle disorders in several specific tissues and organs might occur. Pollutants might promote increased parasitism in aquatic animals, especially fish, by impairing the host's immune response or favouring the survival and reproduction of the intermediate hosts. Alternatively, decreased parasitism might ensue through toxicity of the pollutant to free-living stages and intermediate hosts or by alteration of the host's physiology. Experimental studies indicate that the numbers of ectoparasites such as trichodinid ciliates and monogeneans increase significantly on the gills following exposure to a pollutant, and this is supported by field data on other ciliates and monogeneans where evidence of pollution has been clearly demonstrated. There is also evidence that endoparasitic protozoons, such as myxozoons, microsporans and haematozoons, all of which are capable of proliferating in their hosts, increase substantially in prevalence and intensity when interacting with pollutants. The period of patency might also be prolonged in haematozoan infections. Most reports of pollution effects on endoparasites suggest increased parasitism in fish hosts. This also applies to fish living in areas which receive thermal effluents. Parasites might in turn enhance their hosts' susceptibility to pollutants, and information in support of this view is accumulating. Finally, immunosuppression represents one of the underlying mechanisms influencing increased parasitism. Thus, while published information suggests more than a casual connection between fish parasites and pollution, further research is needed to establish the cause-and-effect relationship and at the same time take cognizance of histopathological effects of the toxic agents and their concentrations in water. Areas for future research are recommended.

Acknowledgements

We are grateful to Ms Mary Boland for typing the manuscript. Support was provided by the National Science and Engineering Council of Canada to R.A.K. and by Orvar and Gertrud Nybelin's foundation for fish biology research to J.T. This is Ocean Sciences Centre contribution No. 31.

References

Aho, J. M., Gibbons, J. W. and Esch, G. W. (1976). Relationship between thermal loading and parasitism in the mosquito fish. *In* "Thermal Ecology II" (G. W. Esch and R. W. McFarlane, eds), pp. 213–218. Technical Information Center, Energy Research and Development Agency, Washington, D.C. (Conf. 750425).

Andersson, T., Förlin, L., Hardig, J. and Larsson, A. (1988). Biochemical and physiological disturbances in fish inhabiting coastal waters polluted with bleached kraft mill effluents. *Marine Environmental Research* **24,** 233–236.

Andrews, A. K., Van Valin, C. C. and Stebbings, B. E. (1966). Some effects of heptachlor on bluegills, *Lepomis macrochirus*. *Transactions of the American Fisheries Society* **95,** 297–303.

Anonymous (1987). Fish disease in the Thames estuary. *Marine Pollution Bulletin* **15,** 198.

Becker, D. S. and Ginn, T. C. (1987). Hepatic lesions in English sole (*Paraphrys vetulus*) from Commencement Bay, Washington (USA). *Marine Environmental Research* **23,** 153–173.

Beisel, W. R. (1982). Single nutrients and immunity. *American Journal of Clinical Nutrition* **35,** 417–468.

Bengtsson, Å., Bengtsson, B. E. and Lithner, G. (1988). Vertebral defects in fourhorn sculpin, *Myoxocephalus quadricornis* L., exposed to heavy metal pollution in the Gulf of Bothnia. *Journal of Fish Biology* **33,** 517–529.

Boyce, N. P. and Yamada, S. B. (1977). Effects of a parasite, *Eubothrium salvelini* (Cestoda: Pseudophyllidea), on the resistance of juvenile sockeye salmon, *Oncorhynchus nerka*, to zinc. *Journal of the Fisheries Research Board of Canada* **34,** 706–709.

Brule, T. (1987). The reproductive biology and pathological changes of the plaice *Pleuronectes platessa* (L.) after the "Amoco Cadiz" oil spill along the north-west coast of Brittany. *Journal of the Marine Biological Association of the United Kingdom* **67,** 237–247.

Burton, D. T., Jones, A. M. and Cairns, J. (1972). Acute zinc toxicity to rainbow trout (*Salmo gairdneri*): Confirmation of the hypothesis that death is related to tissue hypoxia. *Journal of the Fisheries Research Board of Canada* **29,** 1463–1466.

Chevalier, G., Cauthier, L. and Moreau, G. (1985). Histopathological and electron microscopic studies of gills of brook trout, *Salvelinus fontinalis*, from acidified lakes. *Canadian Journal of Zoology* **63,** 2062–2070.

Chipman, J. K. (1982). Bile as a source of potential reactive metabolites. *Toxicology* **25,** 99–111.

Connell, D. W. and Miller, G. J. (1981). Petroleum hydrocarbons in aquatic ecosystems—behaviour and effects of sublethal concentrations: Part 2. *CRC Critical Reviews in Environmental Control* **12,** 105–162.

Couch, J. A. (1975). Histopathological effects of pesticides and related chemicals on the livers of fishes. *In* "The Pathology of Fishes" (W. E. Ribelin and G. Migaki, eds), pp. 559–584. University of Wisconsin Press, Madison, Wisconsin.

Couillard, C. M., Berman, R. A. and Panisset, J. C. (1988). Histopathology of rainbow trout exposed to bleached kraft pulp mill effluent. *Archives of Environmental Contamination and Toxicology* **17,** 319–323.

Dabrowska, H. (1974). Proba oceny stanu zdrowotnego ryb w rzece L'ynie i Walszy na tle ich zanieczyszcaenia (An attempt to evaluate the state of health of fish from the Lyna and Walza Rivers in connection to their pollution). *Pregl ad Zoologiczyn* **18,** 390–395 (in Polish).

Das, M. C. and Shrivastava, A. K. (1984). Fish mortality in Naini Tal Lake (India) due to pollution and parasitism. *Hydrobiological Journal* **20,** 60–64.

Daye, P. G., and Garside, E. T. (1976). Histopathologic changes in superficial tissue of brook trout, *Salvelinus fontinalis* (Mitchill), exposed to acute and chronic levels of pH. *Canadian Journal of Zoology* **54,** 2140–2155.

Dethlefsen, V. (1984). Diseases in North Sea fishes. *Helgolander Meeresuntersuchungen* **37,** 353–374.

Dethlefsen, V. (1986). Studies of pathology in relation to pollution. *In* "Report of the ICES Workshop on the Use of Pathology in Studies of the Effects of Contaminants" (J. Thulin, ed.), pp. 22–35. International Council for the Exploration of the Sea, Öregrund (C.M. 1986/E.40 Ref. F).

Dey, A. C., Kiceniuk, J. W., Williams, U. P., Khan, R. A. and Payne, J. F. (1983). Long term exposure of marine fish to crude petroleum. 1. Studies on liver lipids and fatty acids in cod (*Gadus morhua*) and winter flounder (*Pseudopleuronectes americanus*). *Comparative Biochemistry and Physiology* **75,** 93–101.

Di Michele, L. and Taylor, M. H. (1978). Histopathological and physiological responses of *Fundulus heteroclitus* to naphthalene exposure. *Journal of the Fisheries Research Board of Canada* **35,** 1060–1066.

Dokholyan, V. K., Schleyfer, G. S., Akhmedova, T. P. and Magomedov, A. K. (1980). The influence of dissolved petroleum products on the activity of some species of fish of the Caspian Sea. *Journal of Ichthyology* **20,** 128–133.

Eiras, J. C. (1987). Occurrence of *Haemogregarina bigemina* (Protozoa: Apicomplexa) in *Blennius pholis* (Pisces: Blenniidae) along the Portuguese west coast. *Journal of Fish Biology* **30,** 597–603.

Eller, L. L. (1975). Gill lesions in freshwater teleosts. *In* "The Pathology of Fishes" (W. E. Ribelin and G. Migaki, eds), pp. 305–330. University of Wisconsin Press, Madison, Wisconsin.

Esch, G. W., Hazen, T. C., Dimock, R. V., Jr and Gibbons, J. W. (1976). Thermal effluent and the epizootiology of the ciliate *Epistylis* and the bacterium *Aeromonas* in association with centrarchid fish. *Transactions of the American Microscopical Society* **95,** 687–693.

Eure, H. E. and Esch, G. W. (1974). Effects of thermal effluent on the population dynamics of helminth parasites in largemouth bass. *In* "Thermal Ecology" (J. W. Gibbons and R. R. Shavitz, eds), pp. 207–215. US Atomic Energy Commission Symposium Series, Augusta, Georgia (Conf. 730505).

Ewing, M. S. and Ewing, S. A. (1982). Susceptibility to *Ichthyophthirius multifiliis* of channel catfish exposed to sublethal concentrations of copper. *In* "Parasites—Their World and Ours", *Molecular and Biochemical Parasitology*, supplement, p. 401.

Fletcher, G. L., Kiceniuk, J. W. and Williams, U. P. (1981). Effects of oiled sediments on mortality, feeding and growth of winter flounder (*Pseudopleuronectes americanus*). *Marine Ecology Progress Series* **4,** 91–96.

Fletcher, G. L., King, M. J., Kiceniuk, J. W. and Addison, R. F. (1982). Liver hypertrophy in winter flounder following exposure to experimentally oiled sediments. *Comparative Biochemistry and Physiology C* **73,** 457–462.

France, R. L. and Graham, L. (1985). Increased microsporidian parasitism of the crayfish *Orconectes virilis* in an experimentally acidified lake. *Water, Air and Soil Pollution* **26,** 129–136.

Freeman, H. C., Sangland, G. and Flemming, B. (1982). The sublethal effects of a polychlorinated biphenyl (aroclor 1254) diet on Atlantic cod (*Gadus morhua*). *Science of the Environment* **24,** 1–11.

Fries, C. R. (1986). Effects of environmental stressors and immunosuppressants on immunity in *Fundulus heteroclitus*. *American Zoologist* **26,** 271–282.

Gardner, G. R. and Yevich, P. P. (1988). Comparative histopathological effects of chemically contaminated sediment on marine organisms. *Marine Environmental Research* **24,** 311–316.

Grizzle, J. M. (1986). Lesions in fishes captured near drilling platforms in the Gulf of Mexico. *Marine Environmental Research* **18,** 267–276.

Guth, D. J., Blankespoor, H. D. and Cairns, J., Jr (1977). Potentiation of zinc stress caused by parasitic infection of snails. *Hydrobiologia* **55,** 225–229.

Haensly, W. E., Neff, J. M., Sharp, J. R., Morris, A. C., Bedgood, M. F. and Beom, P. D. (1982). Histopathology of *Pleuronectes platessa* L. from Aber Wrac'h and Aber Benoit, Brittany, France: Long-term effects of the *Amoco Cadiz* crude oil spill. *Journal of Fish Diseases* **5,** 365–391.

Hansen, M. A., Fernandes, G. and Good, R. A. (1982). Nutrition and immunity. *Annual Review of Nutrition* **2,** 151–177.

Hawkes, J. W. (1977). The effects of petroleum hydrocarbon exposure on the structure of fish tissues. *In* "Fate and Effects of Petroleum Hydrocarbons in Marine Organisms and Ecosystems" (D. A. Wolfe, ed.), pp. 115–128. Pergamon Press, New York.

Hirshfield, M. F., Morin, R. P. and Hepner, D. J. (1983). Increased prevalence of larval *Eustrongyloides* (Nematoda) in the mummichog, *Fundulus heteroclitus* (L.) from the discharge canal of a power plant in the Chesapeake Bay. *Journal of Fish Biology* **23,** 135–142.

Hislop, J. R. G. and Shanks, A. M. (1979). Recent investigations on reproductive biology of the haddock, *Melanogrammus aeglefinus* of the northern North Sea and the effects on fecundity of infection with the copepod parasite *Lernaeocera branchialis. Journal du Conseil pour l'Exploration de la Mer* **39,** 244–251.

Hodgins, H. O., McCain, B. B. and Hawkes, J. W. (1977). Marine fish and invertebrate diseases, host disease resistance, and pathological effect of petroleum. *In* "Effects of Petroleum on Arctic and Subarctic Marine Environments and Organisms" (D. C. Malins, ed.). Vol. 2, pp. 95–173. Academic Press, London and San Diego.

Hodson, P. V. (1988). The effect of metal metabolism on uptake, disposition and toxicity in fish. *Aquatic Toxicology* **11,** 3–18.

Kabata, Z. (1958). *Lernaeocera obtusa* n. sp.: Its biology and its effects on the haddock. *Department of Agriculture and Fisheries, Scotland* **3,** 1–26.

Khan, R. A. (1976). The life cycle of *Trypanosoma murmanensis* Nikitin. *Canadian Journal of Zoology* **54,** 1840–1849.

Khan, R. A. (1977). Blood changes in Atlantic cod (*Gadus morhua*) infected with *Trypanosoma murmanensis. Journal of the Fisheries Research Board of Canada* **34,** 2193–2196.

Khan, R. A. (1985). Pathogenesis of *Trypanosoma murmanensis* in marine fish of the northwestern Atlantic following experimental transmission. *Canadian Journal of Zoology* **63,** 2141–2144.

Khan, R. A. (1987a). Effects of chronic exposure to petroleum hydrocarbons on two species of marine fish infected with a hemoprotozoan, *Trypanosoma murmanensis. Canadian Journal of Zoology* **65,** 2703–2707.

Khan, R. A. (1987b). Crude oil and parasites of fish. *Parasitology Today* **3,** 99–100.

Khan, R. A. (1988). Experimental transmission, development and effects of a parasitic copepod, *Lernaeocera branchialis* on Atlantic cod, *Gadus morhua. Journal of Parasitology* **74,** 586–599.

Khan, R. A. (1990). Parasitism in marine fish after chronic exposure to petroleum hydrocarbons in the laboratory and to the Exxon Valdez oil spill. *Bulletin of Environmental Contamination and Toxicology* **44,** 759–763.

Khan, R. A. and Kiceniuk, J. (1983). Effects of crude oil on the gastrointestinal parasites of two species of marine fish. *Journal of Wildlife Diseases* **19,** 253–258.

Khan, R. A. and Kiceniuk, J. W. (1984). Histopathological effects of crude oil on Atlantic cod following chronic exposure. *Canadian Journal of Zoology* **62,** 2038–2043.

Khan, R. A. and Kiceniuk, J. W. (1988). Effect of petroleum aromatic hydrocarbons on monogeneids parasitizing Atlantic cod, *Gadus morhua* L. *Bulletin of Environmental Contamination and Toxicology* **41,** 94–100.

Khan, R. A., Barrett, M. and Murphy, J. (1980a). Blood parasites of fish from the northwestern Atlantic Ocean. *Canadian Journal of Zoology* **58,** 770–781.

Khan, R. A., Murphy, J. and Taylor, D. (1980b). Prevalence of a trypanosome in Atlantic cod (*Gadus morhua*) especially in relation to stocks in the Newfoundland area. *Canadian Journal of Fisheries and Aquatic Sciences* **37,** 1467–1475.

Khan, R. A., Bowring, W. R., Burgeois, C., Lear, H. and Pippy, J. H. (1986). Myxosporean parasites of marine fish from the continental shelf off Newfoundland and Labrador. *Canadian Journal of Zoology* **64,** 2218–2226.

Kiceniuk, J. W. and Khan, R. A. (1983). Toxicology of chronic crude oil exposure: Sublethal effect on aquatic organisms. *In* "Aquatic Toxicology" (J. O. Nriagu, ed.), pp. 425–436. John Wiley, New York.

Kiceniuk, J. W. and Khan, R. A. (1987). Effect of petroleum hydrocarbons on Atlantic cod, *Gadus morhua*, following chronic exposure. *Canadian Journal of Zoology* **65,** 490–494.

Kiceniuk, J. W., Khan, R. A., Dawe, M. and Williams, U. (1981). Examination of interaction of trypanosome infection and crude oil exposure on hematology of the longhorn sculpin (*Myoxocephalus octodecemspinosus*). *Bulletin of Environmental Contamination and Toxicology* **28,** 435–438.

Krahn, M. M., Rhodes, L. D., Meyers, M. S., Moore, L. K., MacLeod, W. D., Jr and Malins, D. C. (1986). Association between metabolites of aromatic compounds in bile and the occurrence of hepatic lesions in English sole (*Paraphrys vetulus*) from Puget Sound, Washington. *Archives of Enviromental Contamination and Toxicology* **15,** 61–67.

Krahn, M. M., Burrows, D. G., MacLeod, W. D., Jr and Malins, D. C. (1987). Determination of individual metabolites of aromatic compounds in hydrolysed bile of English sole (*Parophrys vetulus*) from polluted sites in Puget Sound, Washington. *Archives of Environmental Contamination and Toxicology* **6,** 511–522.

Kussat, R. H. (1969). A comparison of aquatic communities in the Bow River above and below sources of domestic and industrial wastes from the city of Calgary. *Canadian Fish Culture* **40,** 3–31.

Lehtinen, K. J., Notini, M. and Landler, L. (1984). Tissue damage and parasite frequency in flounders *Platichthys flesus* chronically exposed to bleached kraft pulp mill effluents. *Annales Zoologici Fennici* **21,** 23–28.

Leivestad, H. and Muniz, I. P. (1976). Fish kill at low pH in a Norwegian river. *Nature* **259,** 391–392.

Lindesjöö, E. and Thulin, J. (1987). Fin erosion of perch (*Perca fluviatilis*) in a pulp mill effluent. *Bulletin of the European Association of Fish Pathology* **7,** 11–13.

Lom, J. and Laird, M. (1969). Parasitic protozoa from marine and euryhaline fish of Newfoundland and New Brunswick. I. Peritrichous ciliates. *Canadian Journal of Zoology* **47,** 1367–1380.

Lopez, E., Peignoux-Deville, J., Lallier, F., Martelly, E. and Fontaine, Y. (1981). Anguilles contaminées par les hydrocarbures après l'échouage de l'Amoco Cadiz. Modifications histopathologiques des ovaries, des branchies et de glandes endocrines. *Comptes Rendus des Séances de l'Academie des Sciences, Série 3,* **292,** 407–411.

Malins, D. C. and Roubal, W. T. (1985). Free radicals derived from nitrogen-containing xenobiotics in sediments and liver and bile of English sole from Puget Sound, Washington. *Marine Environmental Research* **17,** 205–210.

Malins, D. C., Krahn, M. M., Brown, D. W., Rhodes, L. D., Myers, M. S., McCain, B. B. and Chan, S.-L. (1985). Toxic chemicals in marine sediment and biota from Mukilteo, Washington: Relationships with hepatic neoplasms and other hepatic lesions in English sole (*Parophrys vetulus*). *Journal of the National Cancer Institute* **74,** 487–494.

Mallatt, J. (1985). Fish gill structural changes induced by toxicants and other irritants: A statistical review. *Canadian Journal of Aquatic Science and Fisheries* **42,** 630–648.

Mann, H. (1970). Copepoda and Isopoda as parasites of marine fishes. *In* "A Symposium on Diseases of Fishes and Shell Fishes" (S. F. Snieszko, ed.), pp. 177–188. American Fisheries Society, Washington, D.C., Special Publication no. 5.

McCahom, C. P., Brown, A. F. and Pascoe, D. (1988). The effect of the acanthocephalan *Pomphorhynchus laevis* (Müller 1976) on the acute toxicity of cadmium to its intermediate host, the amphipod, *Gammarus pulex* (L.). *Archives of Environmental Contamination and Toxicology* **17,** 239–243.

McCain, B. B., Hodgins, H. O., Gronlund, W. D., Hawkes, J. W., Brown, D. W., Meyers, M. S. and Vandermeulen, J. H. (1978). Bioavailability of crude oil from experimentally oiled sediments to English sole (*Parophrys vetulus*) and pathological consequences. *Journal of the Fisheries Research Board of Canada* **35,** 657–664.

McIntyre, A. D. (1982). Oil pollution and fisheries. *Philosophical Transactions of the Royal Society of London B* **297,** 401–411.

McVicar, A. H., Bruno, D. W. and Fraser, C. O. (1988). Fish diseases in the North Sea in relation to sewage sludge dumping. *Marine Pollution Bulletin* **19,** 169–173.

Mix, M. C. (1986). Cancerous diseases in aquatic animals and their association with environmental pollutants: A critical literature review. *Marine Environmental Research* **20,** 1–141.

Mohan, C. V. and Sommerville, C. (1988). Effect of cadmium on susceptibility and immune response of common carp to the protozoan *Ichthyophthirius multifilis. Fifth European Multicolloquium of Parasitology*, p. 107.

Moles, A. (1980). Sensitivity of parasitized coho salmon fry to crude oil, toluene, and naphthalene. *Transactions of the American Fisheries Society* **109,** 293–297.

Möller, H. (1987a). Pollution and parasitism in the aquatic environment. *International Journal for Parasitology* **17,** 353–361.

Möller, H. (1987b). The marine ecologist—scientist or advocate of nature? *Marine Pollution Bulletin* **18,** 267–270.

Möller, H. (1988). The problem of quantifying long-term changes in the prevalence of tumours and non-specific growths of fish. *Journal du Conseil International pour l'Exploration de la Mer* **45,** 33–38.

Möller, H. and Anders, K. (1986). "Diseases and Parasites of Marine Fishes". Möller, Kiel.

Möller-Buchner, V. J. (1981). Untersuchungen zum Parasitenbefall von *Pomatoschistus microps* (Gobiidae, Pisces) in der Unterelbe. *Archiv für Hydrobiologie* **61,** supplement, 59–83.

Murchelano, R. A. (1982). Some pollution-associated diseases and abnormalities of marine fishes and shellfishes: A perspective for the New York Bight. *In* "Ecological Stress and the New York Bight: Science and Management" (G. F. Meyer, ed.), pp. 327–346. Estuarine Research Foundation, Columbia, South Carolina.

Murchelano, R. A. and Ziskowski, J. (1976). Fin rot disease studies in the New York Bight. *American Society of Limnology and Oceanography: Special Symposium* **2,** 329–336.

Narasimhamurti, C. C. and Kalavati, C. (1984). Seasonal variation of the myxosporidian, *Henneguya waltairensis* parasitic in the gills of the fresh water fish, *Channa punctatus* Bl. *Archiv für Protistenkunde* **128,** 351–356.

O'Connor, J. M. and Huggett, R. J. (1988). Aquatic pollution problems, North Atlantic coast, including Chesapeake Bay. *Aquatic Toxicology* **11,** 163–190.

Ogata, M., Miyake, Y. and Fujisawa, K. (1987). Oily smell and oil components in fish flesh reared in seawater containing heavy oil. *Oil and Chemical Pollution* **3,** 329–341.

O'Neill, J. G. (1981). The humoral immune response of *Salmo trutta* L. and *Cyprinus carpio* L. exposed to heavy metals. *Journal of Fish Biology* **19,** 297–306.

Overstreet, R. M. (1982). Abiotic factors affecting marine parasitism. *In* "Parasites—Their World and Ours" (D. F. Mettrick and S. S. Desser, eds), pp. 36–39. Fifth

International Congress of Parasitology. Elsevier Biomedical Press, Amsterdam.

Overstreet, R. M. (1988). Aquatic pollution problems, southeastern U.S. coasts: Histopathological indicators. *Aquatic Toxicology* **11,** 213–239.

Overstreet, R. M. and Howse, H. D. (1977). Some parasites and diseases of estuarine fishes in polluted habitats of Mississippi. *Annals of the New York Academy of Science* **298,** 427–462.

Paperna, I. and Overstreet, R. M. (1981). Parasites and diseases of mullets (Mugilidae). *In* "Aquaculture of Grey Mullets" (O. H. Oren, ed.). International Biological Program publication no. 26, pp. 411–493. Cambridge University Press, Cambridge.

Pascoe, D. and Cram, P. (1977). The effect of parasitism on the toxicity of cadmium to the three-spined stickleback, *Gasterosteus aculeatus* L. *Journal of Fish Biology* **10,** 467–472.

Pascoe, D. and Woodworth, I. (1980). The effects of joint stress on sticklebacks. *Zeitschrift für Parasitenkunde* **62,** 159–163.

Perevozchenko, I. I. and Davydov, O. N. (1974). DDT i ego metabolity u nekotorykh tsestod ryb (DDT and its metabolites in some cestodes in fishes). *Gidrobiologicheskii Zhurnal* **10,** 86–90 (in Russian).

Pickering, A. (ed.) (1981). "Stress and Fish". Academic Press, London and San Diego.

Reda, E. S. A. (1989). An analysis of parasite fauna of bream *Abramis brama* (L.), in Vistula near Warszawa in relation to the character of fish habitat. 1. Review of parasite species. *Acta Parasitologica Polonica* **32,** 309–326.

Rehulka, J. (1983). The parasites of some fish species of the polluted coarse zone of the Ostravice River. *Prace Vuhr Vodnany* **12,** 63–68.

Rhodes, L. D., Myers, M. S., Gronlund, W. D. and McCain, B. B. (1987). Epizootic characteristics of hepatic and renal lesions in English sole, *Parophrys vetulus*, from Puget Sound. *Journal of Fish Biology* **31,** 395–407.

Rice, S. D., Thomas, R. E. and Short, J. W. (1977). Effect of petroleum hydrocarbons on breathing and coughing rates and hydrocarbon uptake–depuration in pink salmon fry. *In* "Physiological Responses of Marine Biota to Pollutants" (F. J. Vernberg, A. Calabrese, F. P. Thurberg and W. B. Vernberg, eds), pp. 259–277. Academic Press, London and San Diego.

Robohm, R. A. and Nitkowski, M. F. (1974). Physiological response of the cunner, *Tautogolabrus adspersus* to cadmium. IV. Effects of the immune system. *In* "U.S. Department of Commerce, NOAA Technical Report NMFS SSRF-681", pp. 15–20.

Sabo, D. and Stegman, J. (1977). Some metabolic effects of petroleum hydrocarbons in marine fish. *In* "Physiological Responses of Marine Biota to Pollutants" (F. J. Vernberg, A. Calabrese, F. P. Thurberg, and W. B. Vernberg, eds), pp. 279–288. Academic Press, London and San Diego.

Sakanari, J. A., Moser, M., Reilly, C. A. and Yoshino, T. P. (1984). Effects of sublethal concentrations of zinc and benzene on striped bass, *Morone saxafilis* (Walbaum), infected with larval *Anisakis* nematodes. *Journal of Fish Biology* **24,** 553–563.

Schwartz, J. P. (1985). Effect of oil-contaminated prey on the feeding and growth rate of pink salmon fry (*Oncorhynchus gorbuscha*). *In* "Marine Pollution and Physiology: Recent Advances" (F. J. Vernberg, F. P. Thurberg, A. Calabrese and W. B. Vernberg, eds), pp. 459–476. University of South Carolina Press, Columbia, South Carolina.

Scott, G. I., Middaugh, D. P. and Sammons, T. I. (1985). Interactions of chlorine-produced oxidants (CPO) and salinity in affecting lethal and sublethal effects in the eastern or American oyster, *Crassostrea virginica* (Gmelin) infected with the protistan parasite, *Perkinsus marinus*. *In* "Marine Pollution and Physiology: Recent Advances (F. J. Vernberg, F. P. Thurberg, A. Calabrese and W. B. Vernberg, eds), pp. 351–376. University of South Carolina Press, Columbia, South Carolina.

Sindermann, C. J. (1979). Pollution-associated diseases and abnormalities of fish and shellfish: A review. *Fishery Bulletin* **76,** 717–749.

Sindermann, C. J. (1982). Implications of oil pollution in production disease in marine organisms. *Philosophical Transactions of the Royal Society of London B* **297,** 385–399.

Skinner, R. H. (1982). The interrelation of water quality, gill parasites, and gill pathology of some fishes from South Biscayne Bay, Florida. *Fishery Bulletin* **80,** 269–280.

Snieszko, S. F. (1974). The effects of environmental stress on outbreaks of infectious diseases of fishes. *Journal of Fish Biology* **6,** 197–208.

Solangi, M. A. and Overstreet, R. M. (1982). Histopathological changes in two estuarine fishes, *Menidia beryllina* (Cope) and *Trinectes maculatus* (Bloch and Schneider), exposed to crude oil and its water-soluble fractions. *Journal of Fish Diseases* **5,** 13–35.

Stott, G. G., McArthur, N. H., Tarpley, R., Jacobs, V. and Sis, R. F. (1981). Histopathologic surveys of ovaries of fish from petroleum production and control sites in the Gulf of Mexico. *Journal of Fish Biology* **18,** 261–269.

Sulgostowska, T., Banaczyk, G. and Grabda-Kazubska, B. (1987). Helminth fauna of flatfish (Pleuronectiformes) from Gdansk Bay and adjacent areas (south-east Baltic). *Acta Parasitologica Polonica* **31,** 231–240.

Sundnes, G. (1970). "*Lernaeocera branchialis* (L.) on Cod (*Gadus morhua* L.) in Norwegian Waters". Institute of Marine Research, Bergen, Norway.

Taberski, K. (1983). "Histological effects of benzene on juvenile striped bass (*Morone saxatilis*)". M.Sc. Thesis, California State University, San Francisco, California.

Tafanelli, R. and Summerfelt, R. C. (1975). Cadmium-induced histopathological changes in goldfish. *In* "The Pathology of Fishes" (W. E. Ribelin and G. Migaki, eds), pp. 613–645. University of Wisconsin Press, Madison, Wisconsin.

Thulin, J. (1983). "Sjukdomar och parasiter hos fisk utanför Skoghallsverken". Statens Naturvårdsverk, Öregrund.

Thulin, J. (ed.) (1986). "Report of the ICES Workshop on the Use of Pathology in Studies of the Effects of Contaminants". International Council for the Exploration of the Sea, Öregrund (C.M. 1986/E.40, Ref. F).

Thulin, J., Höglund, J. and Lindesjöö, E. (1986). "Sjukdomar och parasiter hos abborre, mört och gers i Norrsundet—och Sandarnerecipienten." Statens Naturvårdsverk, Öregrund.

Thulin, J., Höglund, J. and Lindesjöö, E. (1988). Diseases and parasites of fish in a bleached kraft mill effluent. *Water Science Technology* **20,** 179–180.

Thulin, J., Höglund, J. and Lindesjöö, E. (1989). "Fish Diseases in Coastal Waters of Sweden. (in Swedish: English summary and conclusion). Statens Naturvårdsverk, Öregrund.

Trump, B. F., Jones, R. T. and Sahaphong, S. (1975). Cellular effects of mercury on fish kidney tubules. *In* "The Pathology of Fishes" (W. Ribelin and G. Migaki, eds), pp. 585–612. University of Wisconsin Press, Madison, Wisconsin.

Truscott, B., Walsh, J. M., Burton, M. P., Payne, J. F. and Idler, D. R. (1983). Effect of acute exposure to crude petroleum on some reproductive hormones in salmon and flounder. *Comparative Biochemistry and Physiology* **75C,** 121–130.

Turnpenny, A. W. H., Sadler, K., Aston, R. J., Milner, A. G. P. and Lynam, S. (1987). The fish populations of some streams in Wales and northern England in relation to acidity and associated factors. *Journal of Fish Biology* **31,** 415–434.

Valtonen, E. T. and Koskivaara, M. (1987). The effect of environmental stress on trematodes of perch and roach in central Finland. *In* "Second International Symposium of Ichthyoparasitology", p. 102. Tihany, Hungary.

Valtonen, E. T. and Koskivaara, M. (1989). Effects of effluent from a paper and pulp mill on parasites of the roach in central Finland. *Soviet–Finnish Symposium on Fish Parasites of North-Western Europe, Petrozavodsk, USSR, 10–14/1, 1988*, pp. 163–168.

Valtonen, E. T. and Taskinen, J. (1988). *Rhipidocotyle campanula* in its first and second intermediate hosts in central Finland; associated with pollution? *Abstract of the Vth European Multicolloquium of Parasitology*, p. 110, Budapest, Hungary.

Valtonen, E. T., Brummer-Korvenkontio, H. and Koskivaara, M. (1987a). Parasites of roaches in four lakes in central Finland in relation to environmental stress. *Proceedings of the 13th Scandinavian Symposium of Parasitology. Information* **19,** 52. Institute of Parasitology. Åbo Akademi, Finland.

Valtonen, E. T., Koskivaara, M. and Brummer-Korvenkontio, H. (1987b). Parasites of fishes in central Finland in relation to environmental stress. *Biological Research Report of the University of Jyvåskylå* **10,** 129–130.

Van Valin, C. C., Andrews, A. K. and Eller, L. L. (1968). Some effects of mirex on two warm-water fishes. *Transactions of the American Fisheries Society* **97,** 185–192.

Vladimirov, V. L. and Flerov, B. A. (1975). Susceptibility to ichthyophthiriasis in fish following exposure to phenol and polychloropinene poisoning. *Biologiya Vnutrennikh Vod Informatsionnyi Byulleten'* **25,** 35–37 (in Russian).

Walsh, A. H. and Ribelin, W. E. (1975). The pathology of pesticide poisoning. *In* "The Pathology of Fishes" (W. E. Ribelin and G. Migaki, eds), pp. 559–584. University of Wisconsin Press, Madison, Wisconsin.

Wedemeyer, G. (1970). The role of stress in the disease resistance of fishes. *In* "A Symposium on Diseases of Fishes and Shellfishes" (S. F. Snieszko, ed.), pp. 30–35. American Fisheries Society, Washington, D.C.

Weeks, B. A. and Warriner, J. E. (1984). Effects of toxic chemicals on macrophage phagocytosis in two estuarine fishes. *Marine Environmental Research* **14,** 327–335.

West, W. R., Smith, P. A., Booth, G. M., Ulise, S. A. and Lee, M. L. (1986). Determination of genotoxic polycyclic aromatic hydrocarbons in a sediment from the Black River (Ohio). *Archives of Environmental Contamination and Toxicology* **15,** 241–249.

Wojdani, A. and Alfred, L. J. (1984). Alterations in cell-mediated immune functions induced in mouse splenic lymphocytes by polycyclic aromatic hydrocarbons. *Cancer Research* **44,** 942–945.

Wolthaus, B.-G. (1984). Seasonal changes in frequency of diseases in dab, *Limanda limanda*, from the southern North Sea. *Helgolander Meeresuntersuchungen* **37,** 375–387.

Zbanyszek, R. and Smith, L. S. (1984). The effect of water-soluble aromatic hydrocarbons on some haematological parameters in rainbow trout, *Salmo gairdneri* Richardson, during acute exposure. *Journal of Fish Biology* **24,** 545–552.

Zeeman, M. G. and Brindley, W. A. (1981). Effects of toxic agents upon fish immune systems: A review. *In* "Immunologic Considerations in Toxicology" (R. P. Sharma, ed.), Vol. 2, pp. 1–60. CRC Press, Boca Raton, Florida.

Ziskowski, J. and Murchelano, R. (1975). Fin erosion in winter flounder (*Pseudopleuronectes americanus*) from the New York Bight. *Marine Pollution Bulletin* **6,** 26–28.

Ziskowski, J. J., Despres-Patanjo, L., Murchelano, R. A., Howe, A. B., Ralph, D. and Atran, S. (1987). Diseases in commercially valuable fish stocks in the northwest Atlantic. *Marine Pollution Bulletin* **18,** 496–504.

Zuchelkowski, E. M., Lantz, R. C. and Hinton, D. E. (1986). Skin mucous cell response to acid stress in male and female brown bullhead catfish, *Ictalurus nebulosus* (Lesueur). *Aquatic Toxicology* **8,** 139–148.

Index

Figures in *italic* type; Tables in **bold** type.

Cumulative Index of Titles

Cumulative Index of Authors